ROBOTICS FOR CELL MANIPULATION AND CHARACTERIZATION

ROBOTICS FOR CELL MANIPULATION AND CHARACTERIZATION

Edited by

CHANGSHENG DAI
Professor, Dalian University of Technology, China

GUANQIAO SHAN
Postdoctoral Fellow, University of Toronto, Canada

YU SUN
Professor, University of Toronto, Canada

ACADEMIC PRESS
An imprint of Elsevier

ELSEVIER

Academic Press is an imprint of Elsevier
125 London Wall, London EC2Y 5AS, United Kingdom
525 B Street, Suite 1650, San Diego, CA 92101, United States
50 Hampshire Street, 5th Floor, Cambridge, MA 02139, United States
The Boulevard, Langford Lane, Kidlington, Oxford OX5 1GB, United Kingdom

Notices
Knowledge and best practice in this field are constantly changing. As new research and experience
broaden our understanding, changes in research methods, professional practices, or medical
treatment may become necessary.

Practitioners and researchers must always rely on their own experience and knowledge in evaluating
and using any information, methods, compounds, or experiments described herein. In using such
information or methods they should be mindful of their own safety and the safety of others, including
parties for whom they have a professional responsibility.

To the fullest extent of the law, neither the Publisher nor the authors, contributors, or editors, assume
any liability for any injury and/or damage to persons or property as a matter of products liability,
negligence or otherwise, or from any use or operation of any methods, products, instructions, or ideas
contained in the material herein.

ISBN 978-0-323-95213-2

For information on all Academic Press publications
visit our website at https://www.elsevier.com/books-and-journals

Publisher: Mara E. Conner
Acquisitions Editor: Sonnini R. Yura
Editorial Project Manager: Tim Eslava
Production Project Manager: Kamesh R
Cover Designer: Christian Bilbow

Typeset by STRAIVE, India

Contents

PART I   Robotic cell manipulation

1. Introduction of robotics for cell manipulation and characterization   **3**

Guanqiao Shan, Changsheng Dai, Zhuoran Zhang, Xian Wang, and Yu Sun

2. Robotic cell injection with force sensing and control   **19**

Yuzhang Wei and Qingsong Xu

3. Robotic orientation control and enucleation of cells   **47**

Lin Feng, Wei Zhang, Chunyuan Gan, Chutian Wang, Hongyan Sun, Yiming Ji, and Luyao Wang

Contributors

Prajwal Agrawal
Acoustic Robotics Systems Laboratory, Institute of Mechanical Systems, Department of Mechanical and Process Engineering, ETH Zurich, Rushlikon, Switzerland

Daniel Ahmed
Acoustic Robotics Systems Laboratory, Institute of Mechanical Systems, Department of Mechanical and Process Engineering, ETH Zurich, Rushlikon, Switzerland

Changsheng Dai
School of Mechanical Engineering, Dalian University of Technology, Ganjingzi District, Dalian, Liaoning, China

Xingzhou Du
School of Science and Engineering, The Chinese University of Hong Kong; Shenzhen Institute of Artificial Intelligence and Robotics for Society (AIRS), Shenzhen, China

Yue Du
Institute of Robotics and Automatic Information System, Tianjin Key Laboratory of Intelligent Robotics, Nankai University, Tianjin; Institute of Intelligence Technology and Robotic Systems, Shenzhen Research Institute of Nankai University, Shenzhen, China

Amir M. Esfahani
Department of Bioengineering, School of Medicine, Johns Hopkins University, Baltimore, MD, United States

Lin Feng
School of Mechanical Engineering & Automation; Beijing Advanced Innovation Center for Biomedical Engineering, Beihang University, Beijing, China

Yaqi Feng
State Key Laboratory of Robotics, Shenyang Institute of Automation; Institutes for Robotics and Intelligent Manufacturing, Chinese Academy of Sciences, Shenyang; University of Chinese Academy of Sciences, Beijing, China

Craig R. Forest
George W Woodruff School of Mechanical Engineering, Georgia Institute of Technology, Atlanta, GA, United States

Chunyuan Gan
School of Mechanical Engineering & Automation, Beihang University, Beijing, China

Huijun Gao
Research Institute of Intelligent Control and Systems, Harbin Institute of Technology, Harbin, China

Edison Gerena
Sorbonne Université, CNRS, Institut des Systèmes Intelligents et de Robotique, Paris, France

Zahra Ghorbanikharaji
Acoustic Robotics Systems Laboratory, Institute of Mechanical Systems, Department of Mechanical and Process Engineering, ETH Zurich, Rushlikon, Switzerland

Huiying Gong
Institute of Robotics and Automatic Information System, Tianjin Key Laboratory of Intelligent Robotics, Nankai University, Tianjin; Institute of Intelligence Technology and Robotic Systems, Shenzhen Research Institute of Nankai University, Shenzhen, China

Sinan Haliyo
Sorbonne Université, CNRS, Institut des Systèmes Intelligents et de Robotique, Paris, France

Andrew Jenkins
Department of Pharmaceutical Sciences, University of Saint Joseph, West Hartford, CT, United States

Yiming Ji
School of Mechanical Engineering & Automation, Beihang University, Beijing, China

Xiaowei Jin
Department of Mechanical and Materials Engineering, University of Nebraska-Lincoln, Lincoln, NE, United States

Mi Li
State Key Laboratory of Robotics, Shenyang Institute of Automation; Institutes for Robotics and Intelligent Manufacturing, Chinese Academy of Sciences, Shenyang; University of Chinese Academy of Sciences, Beijing, China

Jiaxin Liu
Key Laboratory of Biomimetic Robots and Systems, Beijing Institute of Technology, Ministry of Education, Beijing, China

Lianqing Liu
State Key Laboratory of Robotics, Shenyang Institute of Automation; Institutes for Robotics and Intelligent Manufacturing, Chinese Academy of Sciences, Shenyang; University of Chinese Academy of Sciences, Beijing, China

Yaowei Liu
Institute of Robotics and Automatic Information System, Tianjin Key Laboratory of Intelligent Robotics, Nankai University, Tianjin; Institute of Intelligence Technology and Robotic Systems, Shenzhen Research Institute of Nankai University, Shenzhen, China

Yuezhen Liu
School of Science and Engineering, The Chinese University of Hong Kong, Shenzhen, China

Yingqi Meng
Jiading District Central Hospital Affiliated Shanghai University of Medicine and Health Sciences, Shanghai, PR China

Grayson Minnick
Department of Mechanical and Materials Engineering, University of Nebraska-Lincoln, Lincoln, NE, United States

Riley E. Perszyk
George W Woodruff School of Mechanical Engineering, Georgia Institute of Technology; Department of Pharmacology and Chemical Biology, Emory University School of Medicine, Atlanta, GA, United States

Jordan Rosenbohm
Department of Mechanical and Materials Engineering, University of Nebraska-Lincoln, Lincoln, NE, United States

Bahareh Tajvidi Safa
Department of Mechanical and Materials Engineering, University of Nebraska-Lincoln, Lincoln, NE, United States

Adnan Shakoor
Control and Instrumentation Engineering Department, King Fahd University of Petroleum and Minerals, Dhahran, Saudi Arabia

Guanqiao Shan
Department of Mechanical and Industrial Engineering, University of Toronto, Toronto, ON, Canada

Zhan Shi
Acoustic Robotics Systems Laboratory, Institute of Mechanical Systems, Department of Mechanical and Process Engineering, ETH Zurich, Rushlikon, Switzerland

Hongyan Sun
School of Mechanical Engineering & Automation, Beihang University, Beijing, China

Mingzhu Sun
Institute of Robotics and Automatic Information System, Tianjin Key Laboratory of Intelligent Robotics, Nankai University, Tianjin; Institute of Intelligence Technology and Robotic Systems, Shenzhen Research Institute of Nankai University, Shenzhen, China

Yu Sun
Department of Mechanical and Industrial Engineering, University of Toronto, Toronto, ON, Canada

Stephen F. Traynelis
Department of Pharmacology and Chemical Biology, Emory University School of Medicine, Atlanta, GA, United States

Chutian Wang
School of Mechanical Engineering & Automation, Beihang University, Beijing, China

Huaping Wang
Intelligent Robotics Institute, School of Mechatronical Engineering, Beijing Institute of Technology, Beijing, China

Jintian Wang
Department of Mechanical and Industrial Engineering, University of Toronto, Toronto, ON, Canada

Luyao Wang
School of Mechanical Engineering & Automation, Beihang University, Beijing, China

Xian Wang
Hospital for Sick Children; Department of Mechanical and Industrial Engineering, University of Toronto, Toronto, ON, Canada

Yuzhang Wei
Department of Electromechanical Engineering, University of Macau, Taipa, Macau, China

Yupan Wu
School of Microelectronics, Northwestern Polytechnical University, Xi'an; Research & Development Institute of Northwestern Polytechnical University in Shenzhen; Yangtze River Delta Research Institute of NPU, Taicang, PR China

Mingyang Xie
College of Automation Engineering, Nanjing University of Aeronautics and Astronautics, Nanjing, China

Qingsong Xu
Department of Electromechanical Engineering, University of Macau, Taipa, Macau, China

Ruiguo Yang
Department of Mechanical and Materials Engineering; Nebraska Center for Integrated Biomolecular Communication, University of Nebraska-Lincoln, Lincoln, NE, United States

Mighten C. Yip
George W Woodruff School of Mechanical Engineering, Georgia Institute of Technology, Atlanta, GA, United States

Jiangfan Yu
School of Science and Engineering, The Chinese University of Hong Kong; Shenzhen Institute of Artificial Intelligence and Robotics for Society (AIRS), Shenzhen, China

Xinghu Yu
Ningbo Institute of Intelligent Equipment Technology Co. Ltd., Ningbo, China

Wei Zhang
School of Mechanical Engineering & Automation, Beihang University, Beijing, China

Yujie Zhang
Institute of Robotics and Automatic Information System, Tianjin Key Laboratory of Intelligent Robotics, Nankai University, Tianjin; Institute of Intelligence Technology and Robotic Systems, Shenzhen Research Institute of Nankai University, Shenzhen, China

Zhiyuan Zhang
Acoustic Robotics Systems Laboratory, Institute of Mechanical Systems, Department of Mechanical and Process Engineering, ETH Zurich, Rushlikon, Switzerland

Zhuoran Zhang
School of Science and Engineering, The Chinese University of Hong Kong, Shenzhen, Longgang, Shenzhen, China

Qili Zhao
Institute of Robotics and Automatic Information System, Tianjin Key Laboratory of Intelligent Robotics, Nankai University, Tianjin; Institute of Intelligence Technology and Robotic Systems, Shenzhen Research Institute of Nankai University, Shenzhen, China

Xin Zhao
Institute of Robotics and Automatic Information System, Tianjin Key Laboratory of
Intelligent Robotics, Nankai University, Tianjin; Institute of Intelligence Technology and
Robotic Systems, Shenzhen Research Institute of Nankai University, Shenzhen, China

Songlin Zhuang
Yongjiang Laboratory, Ningbo, China; Department of Mechanical Engineering, University
of Victoria, Victoria, BC, Canada

Preface

The manipulation and characterization of cells has wide applications in both clinical treatment and biological studies. The treatment of infertility requires the manipulation of sperm and egg, and the characterization of cardiomyocytes helps in drug screening, to name just two applications. Advances in robotics offer higher accuracy and spatiotemporal resolution for cell manipulation and characterization than manual operation, and promise improved treatment for patients, as well as new discoveries in cell biology.

This book introduces state-of-the-art robotic techniques to achieve cell manipulation and characterization, and presents their applications in areas such as in vitro fertilization, tissue engineering, cell mechanics, and health monitoring. A variety of enabling techniques are described, including computer vision, sensing and control, and field modeling. The topics in this book include:

- Chapter 1 reviews the development of robotic cell manipulation and characterization, and outlooks its future direction.
- Chapter 2 discusses the application of force sensing in precise cell injection with an aim to achieve minimal cell damage.
- Chapter 3 describes robotic orientation control and enucleation of cells by magnetically driven microrobots, providing a compact and high-throughput cell manipulation platform.
- Chapter 4 introduces the use of robotic cell manipulation for in vitro fertilization, with emerging techniques to reduce cell deformation and increase success rate.
- Chapter 5 discusses robotic cell transport in tissue engineering, with the aim of advancing regenerative medicine.
- Chapter 6 describes robotic intracellular biopsy, which expands the manipulation scale to subcellular level.
- Chapter 7 introduces robotic cell manipulation with optical tweezers, providing a noncontact manipulation strategy.
- Chapter 8 discusses magnetically driven robots used inside the human body for clinical treatment, allowing access to regions not accessible by traditional surgical tools.
- Chapter 9 describes robotic cell injection to characterize cell response to drug molecules for pharmaceutical applications.

- Chapter 10 introduces automated cell aspiration to study cell mechanics and genetics in a quantitative manner.
- Chapter 11 discusses microfabricated platforms to study cell mechanical properties with nanonewton scale sensing.
- Chapter 12 describes intracellular mechanics characterization by magnetically driven robots with a size of nanometers.
- Chapter 13 introduces robotic methods to utilize atomic force microscopy to investigate cell mechanics.
- Chapter 14 discusses robotic manipulation of zebrafish larva for orientation control and injection to determine appropriate disease therapy.
- Chapter 15 describes the high-throughput method of using acoustic fields to separate cells for health monitoring.
- Chapter 16 introduces the recent development of applying electrical fields for cell separation and characterization.

We thank all the chapter contributors for their great contributions. The past decades have witnessed significant development of robotic techniques in cell manipulation and characterization. It is our hope that this book can provide an updated and comprehensive reference for this fast-growing area with widespread applications in medicine and biology.

Changsheng Dai
Dalian, China

Guanqiao Shan
Toronto, Canada

Yu Sun
Toronto, Canada

PART I

Robotic cell manipulation

CHAPTER 1

Introduction of robotics for cell manipulation and characterization

Guanqiao Shan[a], Changsheng Dai[b], Zhuoran Zhang[c], Xian Wang[d], and Yu Sun[a]

[a]Department of Mechanical and Industrial Engineering, University of Toronto, Toronto, ON, Canada
[b]School of Mechanical Engineering, Dalian University of Technology, Ganjingzi District, Dalian, Liaoning, China
[c]School of Science and Engineering, The Chinese University of Hong Kong, Shenzhen, Longgang, Shenzhen, China
[d]Hospital for Sick Children, Toronto, ON, Canada

1 Introduction

Robotic cell manipulation has attracted wide attention since the 2000s and been intensively investigated in the past two decades [1]. The rapid development of robotic cell manipulation benefits from the mature robotic techniques at macroscales, the booming of microtechnologies, and the growing demands for precision manipulation in many disciplines, such as biotechnology and medicine. After two decades of development and evolution, robotic cell manipulation has brought unprecedented capabilities to manipulate and characterize cells with enhanced accuracy, efficiency, and consistency.

In 2002, a microrobot with three degrees of freedom (DOFs) was first developed to automate single cell injection using a micropipette [2]. In the mid-2000s, force sensors were integrated on the end effectors to either measure or control the force during cell manipulation [3–5]. In the 2010s, a large number of field driven microrobotic systems, such as optical tweezers, magnetic microrobots, acoustic tweezers, and PatcherBot, were developed for different tasks of cell manipulation [6–9]. After 2010, the development of robotic cell manipulation systems became more targeted toward specific biological or clinical uses, such as precision cell surgery [10], single-cell genetic testing [11], drug delivery [12], and cancer cell treatment [13]. Table 1.1 summarizes representative milestones in robotic cell manipulation and characterization.

This chapter introduces the fundamentals of robotic cell manipulation, reviews representative major developments over the past two decades, and discusses potential future trends of this field. In Section 2, how to utilize

Robotics for Cell Manipulation and Characterization
https://doi.org/10.1016/B978-0-323-95213-2.00008-9
3

Table 1.1 Summary of representative milestones in robotic cell manipulation and characterization.

Year	Milestone	Reference(s)
2002	Robotic single cell injection by a three-DOF microrobot	[2]
2003	Multiaxis force sensor for mechanical property characterization of biomembranes	[3]
2005	High-throughput robotic cell injection	[14]
2008	Two-axis microgripper for nanonewton force-controlled grasping of biological cells	[4]
2009	Acoustic tweezers for 2D cell patterning	[15]
	Robotic injection of embryonic stem cells into blastocyst hosts	[16]
	FluidFM for single cell injection	[5]
2010	Robotic selection and placement of suspended cells	[17]
2011	Robotic cell surgery for clinical in vitro fertilization	[10]
2012	Optical tweezers for 2D cell patterning	[6, 18]
	Magnetic nanowire control for cell manipulation	[7]
	3D rotational control of single cells	[19]
2013	Magnetic microrobots for in vitro transport of individual cells	[20]
2014	Robotic isolation and placement of adherent cells	[21]
	Robotic single-cell nanobiopsy for subcellular genomic analysis	[11]
	Magnetic microrobots for in situ force sensing	[22]
	Robotic cell injection for drug testing	[12]
2016	Magnetically motorized sperm cells for assisted fertilization	[23]
	Acoustic tweezers for 3D cell manipulation	[8]
2017	3D magnetic tweezers for intraembryonic measurement	[24]
	Robotic whole-cell patch clamp for cell electrophysiological characterization	[25]
2018	Magnetic microrobots for in vivo transport of individual cells	[26]
	Magnetic swarm control for hyperthermia therapy of cancer cells	[13]
	3D magnetic tweezers for robotic intracellular manipulation	[27]
	Ultrasound elastography for ultrafast imaging of cell elasticity	[28]
2019	PatcherBot for high-throughput intracellular patch clamp	[9]
2020	Magnetically controlled microrobots for characterization of neural connectivity	[29]
	Robotic somatic cell nuclear transfer	[30]
2021	Magnetically controlled microrobots for targeted drug delivery in vivo	[31, 32]
2022	Magnetic swarm control for intracellular measurement with enhanced signal-to-noise ratio	[33]

end effectors and remote physical fields for cell manipulation is introduced. In Section 3, different robotic systems for cell characterization are discussed. In Section 4, the future trends of robotic cell manipulation are explored.

2 Robotic cell manipulation

2.1 Robotic end effectors

End effectors are tools mounted on a manipulator (e.g., positioning arm/stage) to directly or indirectly interact with the target objects. For robotic cell manipulation, micropipettes, AFM probes, and microgrippers are commonly used for cell injection, translocation, nanobiopsy, and mechanical property characterization.

2.1.1 Micropipettes

Glass micropipettes are commonly used for cell aspiration and injection. The diameter of a micropipette ranges from submicrometers to a few hundred micrometers. To perform cell manipulation, micropipettes are connected to a pneumatic or hydraulic pump for pressure control at the micropipette tip. The polished glass micropipettes have a large and smooth contact area, which reduces the friction and pressure applied to the target cells, causing minimal cell damage. Therefore, they are standard tools used in clinics for cell holding and aspiration [34]. Fig. 1.1A shows a micropipette used to aspirate human sperm for clinical in vitro fertilization (IVF) treatment.

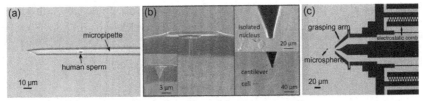

Fig. 1.1 Examples of end effectors commonly used for robotic cell manipulation and characterization. (A) A micropipette was used to aspirate human sperm for clinical IVF treatment. (B) An AFM probe with a long needle tip was used to mechanically characterize cell nuclei in situ. (C) A two-axis microgripper was used for nanonewton force-controlled grasping of biological cells. *((B) Reproduced from H. Liu, J. Wen, Y. Xiao, J. Liu, S. Hopyan, M. Radisic, C.A. Simmons, Y. Sun, In situ mechanical characterization of the cell nucleus by atomic force microscopy, ACS Nano 8 (4) (2014) 3821–3828, with permission from ACS Publications.)*

Anis et al. [17] developed a robotic manipulation system for automated selection and transfer of individual living cells. The system used the visual feedback for micropipette positioning and an open-loop control of fluid flow. To reduce the aspiration time and avoid cell loss, Lu et al. [35] and Liu et al. [36] used a PID controller for fluid control. To accurately control the cell position inside the micropipette, Zhang et al. [37, 38] developed an optimal and robust control for cell aspiration using micropipettes of different sizes. To characterize cell mechanical properties, Shojaei-Baghini et al. [39] designed a robotic system for automated pressure control and cell length measurement inside the micropipette. Robotic systems were also introduced for cell injection [2, 16, 40]. Force sensors were integrated with micropipette for injection force measurement and control [41–43]. Due to the fabrication limitations, it is difficult to fabricate a micropipette with a diameter less than 500 nm, and these relatively large diameters can cause considerable cell damage during cell injection [44]. An alternative is the nanopipette, with 100 nm diameter, which is made of a quartz capillary [11, 45].

2.1.2 AFM probes
AFM probes are powerful tools used for cell characterization. The force applied to these probes can be accurately sensed and controlled, and the force resolution is at piconewton level [47]. Using these probes, a robotic system is able to simultaneously measure the structures and the mechanical properties of individual living cells with nanometer spatial resolution and millisecond temporal resolution [48]. Fig. 1.1B shows an AFM probe used to mechanically characterize cell nuclei in situ [46]. Modifications have been made by etching hollow channels inside the AFM probes to provide pressure control. Sztilkovics et al. [49] reported successfully measuring single-cell adhesion force by applying vacuum pressure at the modified AFM tip and performing single cell detachment. The fluid volume can also be accurately controlled to conduct single-cell nanobiopsy or fluid delivery [47, 50].

2.1.3 Microgrippers
Unlike micropipettes and AFM probes, which fix or hold cells with vacuum pressure, microgrippers use two or more fingers to grasp cells for transloca-tion and microsurgery. All grasping fingers can achieve 1-DOF or 2-DOF movement simultaneously or independently. Fig. 1.1C shows a two-axis microgripper for nanonewton force-controlled grasping of biological cells [4]. To avoid cell damage, force sensors with resolution at nanonewton level

have been integrated, allowing closed-loop force control during cell grasping. Since biological cells can stick to the manipulation tools, without a positive pressure, it is challenging for microgrippers to release the cells compared to micropipettes and AFM probes. To solve this problem, additional actuators [51] or liquid bridges [52] have been introduced to provide mechanical thrust to the cell or physical isolation of the fingers for successful cell release.

2.2 Field-driven manipulation

In field-driven manipulation, actuation fields, such as magnetic, optical, acoustic, electric, or fluidic fields (see Fig. 1.2), are used to directly interact with target cells or control mobile microrobots for cell manipulation. Unlike the end effectors mentioned in Section 2.1, the microrobots used in field-driven manipulation are not mechanically tethered to the manipulators.

2.2.1 Magnetic manipulation

Magnetic fields show high potential in cell manipulation due to their fast response, wide control bandwidth, and biological compatibility [53]. Two types of magnetic fields, namely magnetic gradient and magnetic

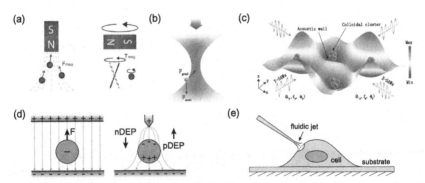

Fig. 1.2 Principles of actuation fields for robotic micromanipulation. (A) Gradient-based and torque-based magnetic fields. (B) Optical field. (C) Acoustic field. (D) Electric field. (E) Fluidic field. *((A) Reproduced from X. Wang, J. Law, M. Luo, Z. Gong, J. Yu, W. Tang, Z. Zhang, X. Mei, Z. Huang, L. You, et al., Magnetic measurement and stimulation of cellular and intracellular structures, ACS Nano 14 (4) (2020) 3805–3821, with permission from ACS Publications; (B) Reproduced from C.J. Bustamante, Y.R. Chemla, S. Liu, M.D. Wang, Optical tweezers in single-molecule biophysics, Nat. Rev. Methods Primers 1 (1) (2021) 1–29, with permission from Spring Nature; (C) Reproduced from S. Yang, Z. Tian, Z. Wang, J. Rufo, P. Li, J. Mai, J. Xia, H. Bachman, P.-H. Huang, M. Wu, et al., Harmonic acoustics for dynamic and selective particle manipulation, Nat. Mater. 21 (5) (2022) 540–546, with permission from Spring Nature.)*

torque, are mainly used to control microrobots for cell manipulation and characterization, as shown in Fig. 1.2A.

In gradient-based manipulation, the magnetic force F used to control a microrobot can be described by $F = m\nabla B$, where m is the magnetic moment of the microrobot and B is the magnetic flux density at the position of the microrobot. The microrobot with the magnetic pole can move along the field gradient continuously in a static or time-dependent magnetic field. The magnetic field can be generated by electromagnetic coils or permanent magnets. The magnetic gradient can be controlled using electromagnetic coils. However, the thermal effect of the coils causes magnetic field drift and thus degrades the control accuracy. Commonly used electromagnetic systems include multipole magnetic tweezers [27], magnetic resonance imaging-based systems [54], and MiniMag [55]. Magnetic fields generated by permanent magnets are less feasible in real-time magnetic gradient control. However, permanent systems generate less heat as well as stronger gradients. The main disadvantage of gradient-based manipulation is the small force applied to the microrobots (a few piconewtons). According to the scaling law, the magnetic force is proportional to the third power of the microrobots' dimensions. A strong magnetic gradient is required for gradient-based manipulation in complex environments, for example, translocation in fluid flow.

In torque-based manipulation, the magnetic torque T used to control a microrobot can be described by $T = m \times B$. The microrobot with the magnetic pole rotates in the magnetic field until the dipole direction is aligned with the magnetic field direction. The rotation of microrobots with different shapes enables their locomotion on a 2D surface or in a 3D space. Wire-shaped or dumbbell-shaped microrobots are able to walk on the substrate surface in rotating magnetic fields. Their locomotion relies on the friction from the substrate or the induced flow differences due to the wall effect in a liquid environment. Compared to spherical beads, larger torques can be generated on wire-shaped or dumbbell-shaped microrobots due to their high-aspect-ratio shapes [56]. To increase the friction between the substrate and the microrobots, additional acoustic fields can be used to press the microrobots against the substrate [57]. The large friction enhances the upstream propulsion against background fluid flows. Helical microrobots have also been used in rotating magnetic fields. The magnetic torque applied to a helical structure rotates it around its helical axis for 3D propulsion [23]. Since the torque is determined by m and B (in contrast to ∇B), it is easier to generate a large torque than a large force on a microrobot, leading to wider applications of torque-based manipulation. The magnetic gradient and torque can also be used together for micromanipulation, where the torque is used to adjust the orientation of the microrobot and the gradient is used to move it to the target location [20].

2.2.2 Optical manipulation
Optical manipulation utilizes focused laser beams to trap and transport cells with a nanometer resolution. The optical system has high compatibility with microscopic systems for visual feedback. In optical manipulation, a laser beam generates a gradient force and a scattering force on the target cell, and traps the cell near the narrowest focal point (i.e., beam waist), as shown in Fig 1.2B. By controlling the position of the beam waist based on the visual feedback, the optical system works as nanotweezers to accurately manipulate the trapped cell. The force generated by optical tweezers is typically at piconewton scale with a subpiconewton resolution. The high resolution enables accurate force control and position control for robotic cell manipulation. However, the small force limits the application of optical systems to measurement of cell mechanical properties, where forces at nanonewton level are typically required. In addition to optical tweezers, laser beams with relatively high power can be used in an automated visual tracking system to immobilize motile cells for single cell selection and translocation [58].

2.2.3 Acoustic manipulation
Acoustic manipulation utilizes acoustic waves to noninvasively manipulate and characterize cells and tissues. The acoustic system is able to penetrate deep into the tissue for in vivo cell manipulation. In this technique, piezoelectric or interdigital transducers are deposited on a piezoelectric substrate to generate acoustic waves. Cells are pushed to either acoustic nodes or antinodes in the field, as shown in Fig. 1.2C. Different wave types are used for different manipulation tasks. Acoustic travelling waves are applied for upstream propulsion [59] and high-resolution imaging [60]. Acoustic streams are widely used to rotate cells three-dimensionally [61, 62]. Acoustic standing waves have been proposed to trap cells and form different patterns, such as sheets, lines, and islands [63]. Compared with magnetic tweezers and optical tweezers, acoustic tweezers generate larger forces (e.g., 150 pN vs. <50 pN on a 5 μm particle/cell) based on their working principle [64]. The periodic acoustic waves enable the swarm control of cells while optical tweezers are usually used to manipulate single cells [65].

2.2.4 Electric manipulation

Electric manipulation utilizes DC or AC electric fields to sort and characterize cellular and subcellular structures. In a uniform field, charged structures (e.g., DNA) move toward oppositely charged electrodes by electrophoresis, and in a nonuniform field, dielectrophoresis (DEP) force is generated on neutral structures based on the polarizability differences between the structures and the surrounding medium (see Fig. 1.2D). The electric force applied to the target cells or cellular structures depends on the electrical properties of both medium and targets, and also the size and shape of the target structures. Using DEP, different kinds of cells, such as bacteria, yeast cells, red blood cells, and cancer cells, can be sorted with high throughput [66, 67]. DEP also enables extraction of subcellular structures, such as DNA, RNA, and glucosidase, for cell metabolism analysis [68, 69] and measures the cell electrical properties such as membrane conductance, membrane capacitance, and cytoplasmic conductivity [70, 71].

2.2.5 Fluidic manipulation

In fluidic manipulation, a fluid field can be generated by external oscillation, motion of the particles inside the medium, or pressure control inside microchannels (e.g., micropipettes and AFM probes). Meyer et al. [72] used a surface transducer to generate vibration at the micropipette tip, which enabled oocyte rotation for manipulation in IVF treatment. Zhang et al. [7] demonstrated that rotating magnetic nanowires were capable of propulsion and steering near a solid surface and create a fluid flow for cell transfer and rotation. In addition, fluid flow controlled by microchannels was also proposed for rotating and transporting single or multiples cells [19, 38]. Qualifying and controlling the fluidic force applied to the target object during manipulation can be challenging. Shan et al. [73] modeled the fluid dynamics inside different-sized micropipettes, which took oil compressibility and tube elasticity into consideration. Based on the dynamics model, an adaptive controller was developed to accurately position different-sized cells inside micropipettes. Due to fabrication limitations, it is difficult to manufacture a micropipette with a diameter less than 500 nm, and these relatively large diameters can cause large cell damage during cell injection [44].

3 Robotic cell characterization

3.1 Mechanical characterization

Cell mechanical properties are key parameters, which reflect cell growth, differentiation, and fate. To characterize cell mechanics, glass micropipettes

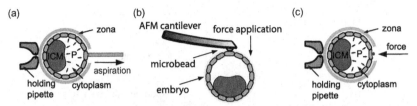

Fig. 1.3 Examples of mechanical characterization of a blastocyst. (A) Micropipette aspiration. (B) AFM probing. (C) Indentation by a magnetically controlled bead. *(Reproduced from X. Wang, Z. Zhang, H. Tao, J. Liu, S. Hopyan, Y. Sun, Characterizing inner pressure and stiffness of trophoblast and inner cell mass of blastocysts, Biophys. J. 115 (12) (2018) 2443–2450, with permission from Elsevier.)*

are widely used due to their easy operation and high throughput. In automated micropipette aspiration, a suction pressure is applied to the micropipette tip and the motion of the cells inside the micropipette is automatically measured [39], as shown in Fig. 1.3A. Varieties of models can be used to analyze the cell mechanics, such as Young's modulus, viscosity, and tension, according to the pressure and the cell motion [74]. The accuracy of the micropipette aspiration depends on the model selection, pressure control, and micropipette tip shape and thickness. Force sensors have been integrated with the micropipettes for force measurement [75]. Cell mechanical properties can be calculated based on the indentation depth and the applied force. AFM system is the standard tool for cell mechanics characterization with piconewton-level force resolution. It is able to measure local and global mechanical properties using different-shaped tips and indentation models [76], as shown in Fig. 1.3B. Micropipettes and AFM probes are tethered to the micromanipulators. It is challenging for these methods to measure the intracellular structures for a long period without disturbing the cell functions. To solve this problem, magnetic beads have been used for measurement by indenting the intracellular structures in a precalibrated magnetic field [77], as shown in Fig. 1.3C. The accuracy of this method depends on the calibration accuracy and stability of the magnetic field.

3.2 Intracellular structure characterization

Characterizing the intracellular structures is of vital importance to study the natural cellular process, disease progression, and drug effects. A majority of robotic systems have been developed to either extract the contents of living cells or directly conduct measurement in situ with minimal invasiveness and high efficiency. Fig. 1.4 shows representative intracellular structure characterization and their measurement techniques [44]. Glass micropipettes are

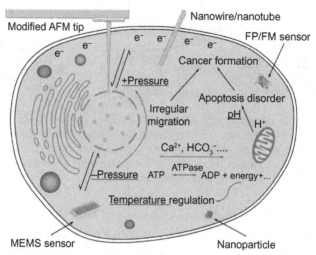

Fig. 1.4 Schematic showing representative intracellular structure characterization and their measurement techniques. *(Reproduced from J. Liu, J. Wen, Z. Zhang, H. Liu, Y. Sun, Voyage inside the cell: microsystems and nanoengineering for intracellular measurement and manipulation, Microsyst. Nanoeng. 1 (1) (2015) 1–15, with permission from Spring Nature.)*

commonly used to extract intracellular structures of large cells. Liu et al. [30] and Zhao et al. [40] used micropipettes of 15–20 μm in diameter to extract the nucleus from a porcine oocyte with high success rate. Shakoor et al. [78] used 2–3 μm micropipettes in combination with a microfluidic device for automated extraction of mitochondria and nucleus from human acute pro-myelocytic leukemia cells and human fibroblast cells. The relatively large diameters can cause significant cell damage during cell injection [44]. An alternative is the nanopipette with 100 nm diameter with an electrode integrated inside the nanopipette tip. Using DEP, femtoliter quantities of intracellular material can be extracted [11]. Similarly, AFM probes with hollow channels [5, 47] and carbon nanotubes [79] have been used to withdraw the intracellular contents for minimally invasive cell monitoring. Compared to nanopipettes and nanotubes, AFMs are able to provide force feedback when the tip penetrates cellular structures, such as cell membrane, which can be used to further reduce the cell damage. To improve the throughput, nanostraws have been used with microfluidic devices. Cells are patterned in the microfluidic device and multiple nanostraws are used for extraction from a high number of cells at the same time. The extraction process can last for several days for an extended period of real-time cell monitoring. Many other approaches, including the injection of nanoparticles [33], fluorescent markers [80], or MEMS sensors [81] into cells, can be used for intracellular measurement. These approaches provide high signal-to-noise ratio, but are usually limited to prespecified targets [82].

4 Summary and outlook

Since the early 2000s, robotic techniques have been developed to automate single cell manipulation, enabling cell translocation, rotation, and mechanics characterization. During the 2010s, robotic intracytoplasmic sperm injection was invented for clinical IVF, initializing robotic surgery at cellular level. Following this more complex manipulation techniques, such as microinjection, nanobiopsy, and in situ force sensing, were developed for manipulation of intracellular structures. More recently, robotic intracellular manipulation and characterization have been rapidly progressing and moving from in vitro demonstration to in vivo verification. Although robotic cell manipulation is still a relatively young field, the fundamental physics and varieties of robotic techniques are now in place after 20 years of evolution. It can be foreseen that more tangible impacts of techniques will be developed for biological study and clinical treatment.

References

[1] Z. Zhang, X. Wang, J. Liu, C. Dai, Y. Sun, Robotic micromanipulation: fundamentals and applications, Annu. Rev. Control Robot. Auton. Syst. 2 (1) (2019) 181–203.
[2] Y. Sun, B.J. Nelson, Biological cell injection using an autonomous microrobotic system, Int. J. Robot. Res. 21 (10–11) (2002) 861–868.
[3] Y. Sun, K.-T. Wan, K.P. Roberts, J.C. Bischof, B.J. Nelson, Mechanical property characterization of mouse zona pellucida, IEEE Trans. Nanobiosci. 2 (4) (2003) 279–286.
[4] K. Kim, X. Liu, Y. Zhang, Y. Sun, Nanonewton force-controlled manipulation of biological cells using a monolithic MEMS microgripper with two-axis force feedback, J. Micromech. Microeng. 18 (5) (2008) 055013.
[5] A. Meister, M. Gabi, P. Behr, P. Studer, J. Vörös, P. Niedermann, J. Bitterli, J. Polesel-Maris, M. Liley, H. Heinzelmann, FluidFM: combining atomic force microscopy and nanofluidics in a universal liquid delivery system for single cell applications and beyond, Nano Lett. 9 (6) (2009) 2501–2507.
[6] H. Chen, D. Sun, Moving groups of microparticles into array with a robot-tweezers manipulation system, IEEE Trans. Robot. 28 (5) (2012) 1069–1080.
[7] L. Zhang, T. Petit, K.E. Peyer, B.J. Nelson, Targeted cargo delivery using a rotating nickel nanowire, Nanomed. Nanotechnol. Biol. Med. 8 (7) (2012) 1074–1080.
[8] F. Guo, Z. Mao, Y. Chen, Z. Xie, J.P. Lata, P. Li, L. Ren, J. Liu, J. Yang, M. Dao, Three-dimensional manipulation of single cells using surface acoustic waves, Proc. Natl. Acad. Sci. USA 113 (6) (2016) 1522–1527.

[9] I. Kolb, C.R. Landry, M.C. Yip, C.F. Lewallen, W.A. Stoy, J. Lee, A. Felouzis, B. Yang, E.S. Boyden, C.J. Rozell, PatcherBot: a single-cell electrophysiology robot for adherent cells and brain slices, J. Neural Eng. 16 (4) (2019) 046003.

[10] Z. Lu, X. Zhang, C. Leung, N. Esfandiari, R.F. Casper, Y. Sun, Robotic ICSI (intra-cytoplasmic sperm injection), IEEE Trans. Biomed. Eng. 58 (7) (2011) 2102–2108.

[11] P. Actis, M.M. Maalouf, H.J. Kim, A. Lohith, B. Vilozny, R.A. Seger, N. Pourmand, Compartmental genomics in living cells revealed by single-cell nanobiopsy, ACS Nano 8 (1) (2014) 546–553.

[12] J. Liu, V. Siragam, Z. Gong, J. Chen, M.D. Fridman, C. Leung, Z. Lu, C. Ru, S. Xie, J. Luo, Robotic adherent cell injection for characterizing cell-cell communication, IEEE Trans. Biomed. Eng. 62 (1) (2014) 119–125.

[13] B. Wang, K.F. Chan, J. Yu, Q. Wang, L. Yang, P.W.Y. Chiu, L. Zhang, Reconfigurable swarms of ferromagnetic colloids for enhanced local hyperthermia, Adv. Funct. Mater. 28 (25) (2018) 1705701.

[14] H. Matsuoka, T. Komazaki, Y. Mukai, M. Shibusawa, H. Akane, A. Chaki, N. Uetake, M. Saito, High throughput easy microinjection with a single-cell manipulation supporting robot, J. Biotechnol. 116 (2) (2005) 185–194.

[15] J. Shi, D. Ahmed, X. Mao, S.-C.S. Lin, A. Lawit, T.J. Huang, Acoustic tweezers: patterning cells and microparticles using standing surface acoustic waves (SSAW), Lab Chip 9 (20) (2009) 2890–2895.

[16] L.S. Mattos, E. Grant, R. Thresher, K. Kluckman, Blastocyst microinjection automation, IEEE Trans. Inform. Technol. Biomed. 13 (5) (2009) 822–831.

[17] Y.H. Anis, M.R. Holl, D.R. Meldrum, Automated selection and placement of single cells using vision-based feedback control, IEEE Trans. Autom. Sci. Eng. 7 (3) (2010) 598–606.

[18] A.G. Banerjee, S. Chowdhury, W. Losert, S.K. Gupta, Real-time path planning for coordinated transport of multiple particles using optical tweezers, IEEE Trans. Autom. Sci. Eng. 9 (4) (2012) 669–678.

[19] C. Leung, Z. Lu, X.P. Zhang, Y. Sun, Three-dimensional rotation of mouse embryos, IEEE Trans. Biomed. Eng. 59 (4) (2012) 1049–1056.

[20] E.B. Steager, M. Selman Sakar, C. Magee, M. Kennedy, A. Cowley, V. Kumar, Automated biomanipulation of single cells using magnetic microrobots, Int. J. Robot. Res. 32 (3) (2013) 346–359.

[21] O. Guillaume-Gentil, T. Zambelli, J.A. Vorholt, Isolation of single mammalian cells from adherent cultures by fluidic force microscopy, Lab Chip 14 (2) (2014) 402–414.

[22] W. Jing, D. Cappelleri, A magnetic microrobot with in situ force sensing capabilities, Robotics 3 (2) (2014) 106–119.

[23] M. Medina-Sánchez, L. Schwarz, A.K. Meyer, F. Hebenstreit, O.G. Schmidt, Cellular cargo delivery: toward assisted fertilization by sperm-carrying micromotors, Nano Lett. 16 (1) (2016) 555–561.

[24] X. Wang, M. Luo, H. Wu, Z. Zhang, J. Liu, Z. Xu, W. Johnson, Y. Sun, A three-dimensional magnetic tweezer system for intraembryonic navigation and measurement, IEEE Trans. Robot. 34 (1) (2017) 240–247.

[25] L.A. Annecchino, A.R. Morris, C.S. Copeland, O.E. Agabi, P. Chadderton, S.R. Schultz, Robotic automation of in vivo two-photon targeted whole-cell patch-clamp electrophysiology, Neuron 95 (5) (2017) 1048–1055.

[26] J. Li, X. Li, T. Luo, R. Wang, C. Liu, S. Chen, D. Li, J. Yue, S.-H. Cheng, D. Sun, Development of a magnetic microrobot for carrying and delivering targeted cells, Sci. Robot. 3 (19) (2018) eaat8829.

[27] X. Wang, M. Luo, C. Ho, Z. Zhang, Q. Zhao, C. Dai, Y. Sun, Robotic intracellular manipulation: 3D navigation and measurement inside a single cell, in: 2018 IEEE International Conference on Robotics and Automation (ICRA), IEEE, 2018, pp. 2716–2721.

[28] P. Grasland-Mongrain, A. Zorgani, S. Nakagawa, S. Bernard, L.G. Paim, G. Fitzharris, S. Catheline, G. Cloutier, Ultrafast imaging of cell elasticity with optical microelastography, Proc. Natl. Acad. Sci. USA 115 (5) (2018) 861–866.

[29] E. Kim, S. Jeon, H.-K. An, M. Kianpour, S.-W. Yu, J.-Y. Kim, J.-C. Rah, H. Choi, A magnetically actuated microrobot for targeted neural cell delivery and selective connection of neural networks, Sci. Adv. 6 (39) (2020) eabb5696.

[30] Y. Liu, X. Wang, Q. Zhao, X. Zhao, M. Sun, Robotic batch somatic cell nuclear transfer based on microfluidic groove, IEEE Trans. Autom. Sci. Eng. 17 (4) (2020) 2097–2106.

[31] H. Zhang, Z. Li, C. Gao, X. Fan, Y. Pang, T. Li, Z. Wu, H. Xie, Q. He, Dual-responsive biohybrid neutrobots for active target delivery, Sci. Robot. 6 (52) (2021) eaaz9519.

[32] Y. Dai, X. Bai, L. Jia, H. Sun, Y. Feng, L. Wang, C. Zhang, Y. Chen, Y. Ji, D. Zhang, Precise control of customized macrophage cell robot for targeted therapy of solid tumors with minimal invasion, Small 17 (41) (2021) 2103986.

[33] X. Wang, T. Wang, X. Chen, J. Law, G. Shan, W. Tang, Z. Gong, P. Pan, X. Liu, J. Yu, Microrobotic swarms for intracellular measurement with enhanced signal-to-noise ratio, ACS Nano 16 (7) (2022) 10824–10839.

[34] G. Shan, Z. Zhang, C. Dai, H. Liu, X. Wang, W. Dou, Y. Sun, Robotic cell manipulation for blastocyst biopsy, in: 2022 International Conference on Robotics and Automation (ICRA), IEEE, 2022, pp. 7923–7929.

[35] Z. Lu, Single cell deposition and patterning with a robotic system, PLoS One 5 (10) (2010) e13542.

[36] J. Liu, C. Shi, J. Wen, D. Pyne, H. Liu, C. Ru, J. Luo, Automated vitrification of embryos: a robotics approach, IEEE Robot. Autom. Mag. 22 (2) (2015) 33–40.

[37] X.P. Zhang, C. Leung, Z. Lu, N. Esfandiari, R.F. Casper, Y. Sun, Controlled aspiration and positioning of biological cells in a micropipette, IEEE Trans. Biomed. Eng. 59 (4) (2012) 1032–1040.

[38] Z. Zhang, J. Liu, X. Wang, Robotic pick-and-place of multiple embryos for vitrification, IEEE Robot. Autom. Lett. 2 (2) (2016) 570–576.

[39] E. Shojaei-Baghini, Y. Zheng, Y. Sun, Automated micropipette aspiration of single cells, Ann. Biomed. Eng. 41 (6) (2013) 1208–1216.

[40] Q. Zhao, J. Qiu, Z. Feng, Y. Du, Y. Liu, Z. Zhao, M. Sun, M. Cui, X. Zhao, Robotic label-free precise oocyte enucleation for improving developmental competence of cloned embryos, IEEE Trans. Biomed. Eng. 68 (8) (2020) 2348–2359.

[41] H. Xie, H. Zhang, J. Song, X. Meng, J. Geng, L. Sun, Living cell manipulation and in situ nanoinjection based on frequency shift feedback using cantilevered micropipette probes, IEEE Trans. Autom. Sci. Eng. 17 (1) (2019) 142–150.

[42] Y. Xie, D. Sun, C. Liu, H.Y. Tse, S.H. Cheng, A force control approach to a robot-assisted cell microinjection system, Int. J. Robot. Res. 29 (9) (2010) 1222–1232.

[43] Y. Wei, Q. Xu, Design and testing of a new force-sensing cell microinjector based on small-stiffness compliant mechanism, IEEE/ASME Trans. Mechatron. 26 (2) (2020) 818–829.

[44] J. Liu, J. Wen, Z. Zhang, H. Liu, Y. Sun, Voyage inside the cell: microsystems and nanoengineering for intracellular measurement and manipulation, Microsyst. Nanoeng. 1 (1) (2015) 1–15.

[45] Y. Nashimoto, Y. Takahashi, Y. Zhou, H. Ito, H. Ida, K. Ino, T. Matsue, H. Shiku, Evaluation of mRNA localization using double barrel scanning ion conductance microscopy, ACS Nano 10 (7) (2016) 6915–6922.

[46] H. Liu, J. Wen, Y. Xiao, J. Liu, S. Hopyan, M. Radisic, C.A. Simmons, Y. Sun, In situ mechanical characterization of the cell nucleus by atomic force microscopy, ACS Nano 8 (4) (2014) 3821–3828.

[47] O. Guillaume-Gentil, R.V. Grindberg, R. Kooger, L. Dorwling-Carter, V. Martinez, D. Ossola, M. Pilhofer, T. Zambelli, J.A. Vorholt, Tunable single-cell extraction for molecular analyses, Cell 166 (2) (2016) 506–516.

[48] M. Li, N. Xi, Y. Wang, L. Liu, Advances in atomic force microscopy for single-cell analysis, Nano Res. 12 (4) (2019) 703–718.

[49] M. Sztilkovics, T. Gerecsei, B. Peter, A. Saftics, S. Kurunczi, I. Szekacs, B. Szabo, R. Horvath, Single-cell adhesion force kinetics of cell populations from combined label-free optical biosensor and robotic fluidic force microscopy, Sci. Rep. 10 (1) (2020) 1–13.

[50] O. Guillaume-Gentil, E. Potthoff, D. Ossola, C.M. Franz, T. Zambelli, J.A. Vorholt, Force-controlled manipulation of single cells: from AFM to FluidFM, Trends Biotechnol. 32 (7) (2014) 381–388.

[51] B.K. Chen, Y. Zhang, Y. Sun, Active release of microobjects using a MEMS microgripper to overcome adhesion forces, J. Microelectron. Syst. 18 (3) (2009) 652–659.

[52] A. Vasudev, A. Jagtiani, L. Du, J. Zhe, A low-voltage droplet microgripper for microobject manipulation, J. Micromech. Microeng. 19 (7) (2009) 075005.

[53] B.J. Nelson, I.K. Kaliakatsos, J.J. Abbott, Microrobots for minimally invasive medicine, Annu. Rev. Biomed. Eng. 12 (2010) 55–85.

[54] M. Vonthron, V. Lalande, G. Bringout, C. Tremblay, S. Martel, A MRI-based integrated platform for the navigation of micro-devices and microrobots, in: 2011 IEEE/RSJ International Conference on Intelligent Robots and Systems, IEEE, 2011, pp. 1285–1290.

[55] S. Schuerle, S. Erni, M. Flink, B.E. Kratochvil, B.J. Nelson, Three-dimensional magnetic manipulation of micro-and nanostructures for applications in life sciences, IEEE Trans. Magn. 49 (1) (2012) 321–330.

[56] Q. Zhou, T. Petit, H. Choi, B.J. Nelson, L. Zhang, Dumbbell fluidic tweezers for dynamical trapping and selective transport of microobjects, Adv. Funct. Mater. 27 (1) (2017) 1604571.

[57] D. Ahmed, A. Sukhov, D. Hauri, D. Rodrigue, G. Maranta, J. Harting, B.J. Nelson, Bioinspired acousto-magnetic microswarm robots with upstream motility, Nat. Mach. Intell. 3 (2) (2021) 116–124.

[58] Z. Zhang, C. Dai, X. Wang, C. Ru, K. Abdalla, S. Jahangiri, C. Librach, K. Jarvi, Y. Sun, Automated laser ablation of motile sperm for immobilization, IEEE Robot. Autom. Lett. 4 (2) (2019) 323–329.

[59] A.D.C. Fonseca, T. Kohler, D. Ahmed, Ultrasound-controlled swarmbots under physiological flow conditions, Adv. Mater. Interfaces 9 (26) (2022) 2200877.

[60] G. Jin, H. Bachman, T.D. Naquin, J. Rufo, S. Hou, Z. Tian, C. Zhao, T.J. Huang, Acoustofluidic scanning nanoscope with high resolution and large field of view, ACS Nano 14 (7) (2020) 8624–8633.

[61] D. Ahmed, T. Baasch, B. Jang, S. Pane, J. Dual, B.J. Nelson, Artificial swimmers propelled by acoustically activated flagella, Nano Lett. 16 (8) (2016) 4968–4974.

[62] N.F. Läubli, J.T. Burri, J. Marquard, H. Vogler, G. Mosca, N. Vertti-Quintero, N. Shamsudhin, A. DeMello, U. Grossniklaus, D. Ahmed, 3D mechanical characterization of single cells and small organisms using acoustic manipulation and force microscopy, Nat. Commun. 12 (1) (2021) 1–11.

[63] Y. Yang, T. Ma, S. Li, Q. Zhang, J. Huang, Y. Liu, J. Zhuang, Y. Li, X. Du, L. Niu, Self-navigated 3D acoustic tweezers in complex media based on time reversal, Research 2021 (2021) 9781394.

[64] P. Li, Z. Mao, Z. Peng, L. Zhou, Y. Chen, P.-H. Huang, C.I. Truica, J.J. Drabick, W.S. El-Deiry, M. Dao, Acoustic separation of circulating tumor cells, Proc. Natl. Acad. Sci. USA 112 (16) (2015) 4970–4975.

[65] L. Yang, J. Yu, S. Yang, B. Wang, B.J. Nelson, L. Zhang, A survey on swarm micro-robotics, IEEE Trans. Robot. 38 (3) (2022) 1531–1551.

[66] H. Moncada-Hernández, B.H. Lapizco-Encinas, Simultaneous concentration and separation of microorganisms: insulator-based dielectrophoretic approach, Anal. Bioanal. Chem. 396 (5) (2010) 1805–1816.

[67] Y. Kang, D. Li, S.A. Kalams, J.E. Eid, DC-dielectrophoretic separation of biological cells by size, Biomed. Microdevices 10 (2) (2008) 243–249.

[68] R. Pan, M. Xu, J.D. Burgess, D. Jiang, H.-Y. Chen, Direct electrochemical observation of glucosidase activity in isolated single lysosomes from a living cell, Proc. Natl. Acad. Sci. USA 115 (16) (2018) 4087–4092.

[69] B.P. Nadappuram, P. Cadinu, A. Barik, A.J. Ainscough, M.J. Devine, M. Kang, J. Gonzalez-Garcia, J.T. Kittler, K.R. Willison, R. Vilar, Nanoscale tweezers for single-cell biopsies, Nat. Nanotechnol. 14 (1) (2019) 80–88.

[70] B.G. Hawkins, C. Huang, S. Arasanipalai, B.J. Kirby, Automated dielectrophoretic characterization of Mycobacterium smegmatis, Anal. Chem. 83 (9) (2011) 3507–3515.

[71] H.O. Fatoyinbo, N.A. Kadri, D.H. Gould, K.F. Hoettges, F.H. Labeed, Real-time cell electrophysiology using a multi-channel dielectrophoretic-dot microelectrode array, Electrophoresis 32 (18) (2011) 2541–2549.

[72] D. Meyer, M.L.P. Colon, H.V. Alizadeh, L. Su, B. Behr, D.B. Camarillo, Orienting oocytes using vibrations for in-vitro fertilization procedures, in: 2019 International Conference on Robotics and Automation (ICRA), IEEE, 2019, pp. 4837–4843.

[73] G. Shan, Z. Zhang, C. Dai, X. Wang, L. Chu, Y. Sun, Model-based robotic cell aspiration: tackling nonlinear dynamics and varying cell sizes, IEEE Robot. Autom. Lett. 5 (1) (2020) 173–178.

[74] B. González-Bermúdez, G.V. Guinea, G.R. Plaza, Advances in micropipette aspiration: applications in cell biomechanics, models, and extended studies, Biophys. J. 116 (4) (2019) 587–594.

[75] A. Shulev, I. Roussev, K. Kostadinov, Force sensor for cell injection and characterization, in: 2012 International Conference on Manipulation, Manufacturing and Measurement on the Nanoscale (3M-NANO), IEEE, 2012, pp. 335–338.

[76] J. Chen, Nanobiomechanics of living cells: a review, Interface Focus 4 (2) (2014) 20130055.

[77] X. Wang, Z. Zhang, H. Tao, J. Liu, S. Hopyan, Y. Sun, Characterizing inner pressure and stiffness of trophoblast and inner cell mass of blastocysts, Biophys. J. 115 (12) (2018) 2443–2450.

[78] A. Shakoor, M. Xie, T. Luo, J. Hou, Y. Shen, J.K. Mills, D. Sun, Achieving automated organelle biopsy on small single cells using a cell surgery robotic system, IEEE Trans. Biomed. Eng. 66 (8) (2018) 2210–2222.

[79] R. Singhal, Z. Orynbayeva, R.V. Kalyana Sundaram, J.J. Niu, S. Bhattacharyya, E.A. Vitol, M.G. Schrlau, E.S. Papazoglou, G. Friedman, Y. Gogotsi, Multifunctional carbon-nanotube cellular endoscopes, Nat. Nanotechnol. 6 (1) (2011) 57–64.

[80] Y. Wang, J.Y.-J. Shyy, S. Chien, Fluorescence proteins, live-cell imaging, and mechanobiology: seeing is believing, Annu. Rev. Biomed. Eng. 10 (2008) 1–38.

[81] R. Gómez-Martínez, A.M. Hernández-Pinto, M. Duch, P. Vázquez, K. Zinoviev, E.J. de La Rosa, J. Esteve, T. Suárez, J.A. Plaza, Silicon chips detect intracellular pressure changes in living cells, Nat. Nanotechnol. 8 (7) (2013) 517–521.

[82] S.G. Higgins, M.M. Stevens, Extracting the contents of living cells, Science 356 (6336) (2017) 379–380.

CHAPTER 2

Robotic cell injection with force sensing and control

Yuzhang Wei and Qingsong Xu
Department of Electromechanical Engineering, University of Macau, Taipa, Macau, China

1 Introduction

Cell microinjection normally uses a micropipette to transport foreign materials into living cells [1, 2]. Microinjection was introduced in the first half of the last century and it can deliver diverse materials (e.g., DNA, RNAi, sperm, proteins, toxins, and drug compounds) with certain volume into cells. Therefore, microinjection has broad applications in genetics, transgenics, molecular biology, drug discovery, reproductive studies, and other biomedical areas [3–6].

The significance of cell injection for human beings has been well demonstrated [7]. In fact, there are many methods to introduce foreign materials into cells, such as chemical method, vehicular method (including erythrocyte fusion and vesicle fusion [8]), electrical method (including electroporation [9]), and mechanical method (including microinjection, hyposmotic shock [10], sonication [11], and microprojectiles [12]). Especially, microinjection shows some distinct advantages, such as quantitative delivery, high reproducibility, high viability, high reliability and high efficiency, making it a standard method for delivering materials into embryos and cells [13–16]. Generally, microinjection adopts a fine micropipette to penetrate through a cell membrane, and then introduce foreign materials into the cell by a pulse of pressure [17].

1.1 Conventional manual microinjection

Currently, microinjection is mainly performed manually. However, it is an intensive work in biological research [3]. As for microinjection of zebrafish embryos, there are mainly four steps to complete the microinjection. First, a human operator would identify the embryos under the field of view of a microscope, and then carefully place the tip of the micropipette close to the embryo. Meanwhile, the embryo is usually fixed with specially designed

Robotics for Cell Manipulation and Characterization
https://doi.org/10.1016/B978-0-323-95213-2.00006-5

19

petri dish or another micropipette. Second, the operator should manually penetrate the embryo fast by hand, or control a manipulator to produce a thrust movement, when the micropipette initially contacts with the embryo. Third, after penetrating through the membrane, the operator should continue moving the micropipette until it penetrates into the yolk of the embryo. Fourth, after arriving at the desired location, foreign materials can be released through a pulse of hydraulic pressure or compressed air pressure through the microsyringe [18]. Obviously, a successful microinjection needs high skill, enough experience, and high concentration.

Generally, training a proficient operator needs several months or up to 1 year, and hence, the process is very time-consuming [19–22]. Nevertheless, a skilled operator can only achieve a success rate of approximately 80% [23, 24]. In addition to long training time and concentration, five key requirements influence the success rate, that is, an accurate injection point, proper speed variation, appropriate penetration trajectory, proper penetration force, and no contamination [25]. However, human operators cannot meet these tough requirements over long periods of concentration, resulting in relatively low success rate and poor repeatability. Therefore, manual microinjection is only feasible for injecting a few cells at a time [13].

1.2 Robotic cell microinjection

Compared with manual microinjection, robotic microinjection is more productive, more reliable, more repeatable, and free of fatigue, which produces a higher success rate and survival rate [26]. Robotic microinjection adopts automation techniques to deliver foreign materials into living cells with a fine needle [4]. Robot microinjection systems mainly include a piercing mechanism (including precise positioning manipulator and microinjector), cell holder (e.g., petri dish or micropipette), injection control loop, machine vision and other sensors, user interface, and an environment control system for adjusting cell cultivation conditions (e.g., temperature, pH value, and humidity) [27]. The robotic system can operate consistently and efficiently, and therefore guarantee the reproducibility and high throughput of cell injection. Moreover, it is possible to handle special cases that cannot be completed by traditional techniques [28]. With the increasing of automation level, human involvement can be reduced and the speed of microinjection increased, allowing scientists to concentrate on analyzing the results. Furthermore, automated robotic injection can offer more reliable results with high accuracy [17].

1.3 Force-assisted robotic cell microinjection

Currently, the robotic microinjection system consists of two categories, that is, pure vision-based microinjection system and force-based microinjection system. The pure vision-based microinjection is not reliable, because it has three inherent disadvantages. First, as this method cannot directly detect the injection force, excessive force could cause irreversible damage to cells. Second, it is hard for this method to distinguish the phenomenon of surface sliding from successful penetration. Third, it takes a long time for capturing and processing images, and hence, cannot achieve real-time feedback.

By contrast, force sensor-based microinjection can offer real-time force feedback, which can be used to overcome the aforementioned disadvantages and achieve reliable control process. Thus, force-assisted robotic microinjection is vital for scenarios where large amounts of cells need to be injected in a limited time with accurate operation.

In the remaining sections, the recent developments in force-assisted robotic cell microinjection are surveyed, and current challenges and future developments are discussed. In Sections 2 and 3, microinjection of adherent cells and suspended cells are introduced, respectively. Microforce sensors for cell microinjection are reviewed in Section 4. Section 5 summarizes the current challenges and future development of cell micromanipulation. Finally, Section 6 concludes the chapter.

2 Microinjection of adherent cells

Adherent cells normally stick to a petri dish and present a cell population. Alternatively, suspended cells grow loosely, as shown in Fig. 2.1. Apart from blood cells and germ cells, a majority of the remaining cells of the human body are adherent cells. Thus, the research on microinjection of adherent cells is meaningful for drug screening and disease study.

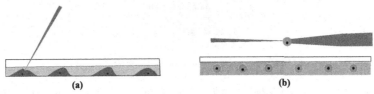

(a)   (b)

Fig. 2.1 Sketch diagram of microinjection of (A) adherent cells and (B) suspended cells [17].

The size of adherent cells (10–20 μm in diameter) is 5–10 times smaller than suspended cells (e.g., oocytes). Consequently, microinjection of adherent cells requires a more precise manipulator and more accurate feedback controller. Additionally, it is difficult to recognize adherent cells due to their small size and accumulation into population, as this requires high performance of machine vision and force sensing systems [17]. Moreover, a fine injection capillary (<1 μm) is necessary to inject the small adherent cells, whereas, it is not necessary to prepare holding devices for adherent cells, because they grow at the bottom of petri dish.

Concerning the microinjection of adherent cells, there are some practicable commercial devices, for example, microinjection robots provided by Eppendorf, Newport, Kleindiek, Narishige, Cellbiology Trading, etc. Among them, Eppendorf offers a prevalent adherent microinjection robot NI2 (see Fig. 2.2), which realizes semiautomated microinjection. It is mainly composed of an inverted microscope, an axial injector actuated by a pizeoelectric actuator, a microinjector driven by pneumatic, two stepping-motor-driven stages with three axes, a video system with CCD camera and monitor, and user interface for controlling the microinjection process. During the microinjection process, the microscope can automatically focus on cells and the injecting pipette. The positioning control (consisting of

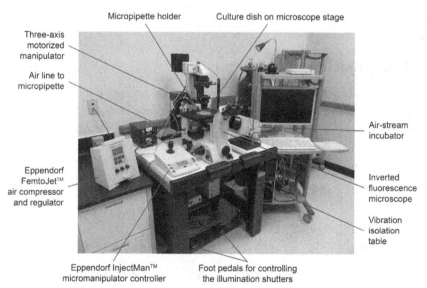

Fig. 2.2 Semiautomated adherent cell injection system (NI2), which is composed of microscope, motorized stage and micromanipulator, pressure microinjector, and camera system [16].

visual servo control based on image processing and height control for calibrating microinjection height) can achieve a relatively high success rate. The built-in software can control the microinjection process well, and the maximum microinjection speed reaches up to 1500 cells per hour. Additionally, this system can realize retrieval of injected cells and multiple injections for the same type of cell [29].

As these commercial microinjection devices are semiautomatic, extensive manual work is still needed. Due to the involvement of human operators, the microinjection speed is limited, which is not feasible when large numbers of cells need to be injected in a limited time and stable transfected cell lines need to be produced. Furthermore, the success rate would be influenced by the performance of the operators. Importantly, these devices cannot monitor the condition of the micropipette, for example, clogging and small breakages. Consequently, these individual devices may give rise to low throughput, low success rate, and decreased reliability of the results [7].

In order to improve the automation level and success rate, force sensors are equipped in similar devices. For example, an impedance sensor developed by Kallio et al. [17] was used to detect the touch force between cells and the injecting micropipette and monitor the condition of the micropipette, such as breakage or clogging. Zhang et al. [29] proposed an impedance force sensor for adherent cell injection (see Fig. 2.3). Moreover, an atomic

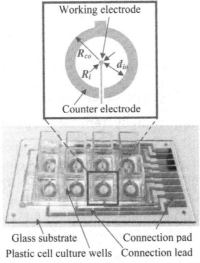

Fig. 2.3 Impedance force sensors for adherent cell injection [29].

force microscope (AFM) was used to realize microinjection of adherent cells with force control by Meister et al. [30].

In summary, it is challenging to develop an automated microinjection robot for adherent cells. In practice, robustness should be improved to deal with the uncertainties caused by biology. Additionally, flexibility should be enhanced to cope with different types of cells and constantly changing states of cell population.

3 Microinjection of suspended cells

As for suspended cells, there are three popular types, that is, Drosophila embryo, zebrafish embryo, and mouse oocytes. In this section, the technical details of microinjection for suspended cells are introduced.

3.1 Requirement of suspended cell injection

Suspended cells are unconstrained in a nutrient solution. During microinjection, they need to be fixed with extra equipment (e.g., specially designed petri dish), resulting in a more complicated and time-consuming microinjection process. During the process, injection failure occurs when there is a big collision between cell and micropipette or cell movement occurs. For example, the fine microinjection pipette can easily be broken off during the spermatozoon microinjection into mouse embryos (approximately 98 μm in diameter) [31].

Currently, the microinjection of suspended cells is mainly performed manually [29], despite the limitations of manual microinjection as mentioned earlier. In contrast, robotic microinjection can complete the intensive work with higher success rate and survival rate. Thus, many researchers are concentrating on the development of robotic microinjection systems.

3.2 Robotic microinjection system for suspended cells

Regarding commercial microinjection robots for suspended cells, there are some semiautomated or teleoperated robotic microinjection systems, for example, the systems provided by Eppendorf, Narishige, Cellbiology Trading, etc. Nevertheless, skilled human operators are still needed to manipulate the robots. In the literature, Li et al. [32] proposed an automated microinjection robot system for suspended cells, as shown in Fig. 2.4. In this system, the microinjection operation can be performed by four steps as follows:

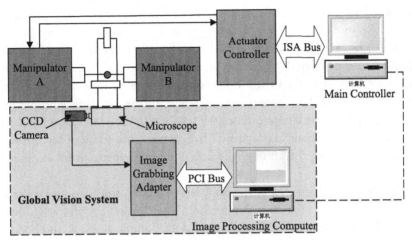

Fig. 2.4 Scheme of an automated suspended-cell microinjection system.

- First, the microinjection pipette, suspended cell, and holding pipette are recognized by machine vision.
- Second, the suspended cell is moved to the desired microinjection area along with the holding device.
- Third, the microinjection pipette is driven to penetrate the cell membrane, arrive at the desired position, and then release foreign materials.
- Fourth, the injected cell is transported back to the petri dish.

The microinjection process will be repeated to inject the remaining cells.

The microinjection process solely depends on visual-servo positioning control. Nevertheless, this process is not reliable due to three inherent disadvantages. First, it is hard to protect fragile cells and the micropipette from excessive force. Second, it is difficult to distinguish the sliding phenomenon from successful microinjection via planar image. Third, it is challenging to realize real-time control, because it takes a certain amount of time to process multiple images [21].

3.3 Force-assisted robotic microinjection of suspended cells

To implement more reliable control, force feedback control can provide real-time control in a simpler way. As compared with sole visual-servo control, microinjection with force sensing has three distinct advantages. First, the force signals during microinjection can help judge whether a penetration is successful or not, because there is a dramatic drop in the force signals after gradual increase during the microinjection process. The judgment is important to save precious foreign materials and to provide more reliable results. Second, a force sensor can provide real-time force feedback for diverse cells in a simpler way, as long as the microinjection force lies in the detection range of the force sensor, which would improve the flexibility and microinjection speed with shorter time of image processing. Third, the value of microinjection force is vital to understand the mechanical properties of cells and distinguish between healthy cells and injured ones, because they have different mechanical properties. Importantly, the value of force feedback is essential in protecting fragile cells and micropipettes from excessive force.

In summary, compared with visual-servo control, force-assisted robotic microinjection can provide real-time force feedback in a simpler way, leading to the improvement of flexibility, reliability, and robustness, and hence, the evaluation of success rate and survival rate.

4 Microforce sensors for cell microinjection

Many researchers focus on the development of microforce sensors for cell microinjection. There are mainly five prevalent types of microforce sensors. It is notable that the microinjection force of suspended cell generally lies in the μN–mN range [18].

4.1 Vision-based force sensors

As microinjection is usually performed under the field of view of a microscope, visual feedback is the most widely adopted sensing method in current microinjection systems [33]. Vision-based force sensors employ image processing and an accurate cell model to detect the microinjection force [18]. Image processing is mainly adopted to detect the deformation of soft objects, such as cell membrane, cell hold devices, and microinjector. The measured deformation is applied as the input of a force estimation algorithm, and then, the microinjection force can be obtained by an appropriate cell model [34]. In the following, several popular image processing methods and cell models are presented.

4.1.1 Image processing for vision-based force sensor
For detecting the microinjection force, image processing aims to measure the deformation of the target object. This procedure mainly consists of a localization process and a recognition process.

For the recognition of cells and micropipettes, matching method is a popular algorithm, in which, pattern matching and feature matching are the two most prevalent matching methods. The pattern matching (also called template matching) performs direct localization by measuring the similarity degree between a template and a collected image [34, 35]. Regarding the feature matching method, the target features (e.g., active contour [36], brightness [22], and amplitude spectra) are extracted first, and then the extracted features are matched in the feature space. The active contour recognition method, which always obtains closed contours, has been widely applied. It is mainly composed of the processes of edge detection, shape modeling, segmentation, pattern recognition, and object tracking [37]. There are also some other contour searching methods, such as the second-derivative zero-crossing detector and computational approach based on the Canny criteria [38]. These methods normally require preprocessing and postprocessing to obtain the connected and closed contours [39].

4.1.2 Cell model for vision-based force sensor

Cell model is normally based on a cell's initial geometric information obtained from image processing, cell mechanical property generated from a priori knowledge, and a predefined coordinate system [34]. Then, the microinjection force can be derived from the updated cell boundary information. In the literature, the current cell models can be mainly divided into three types, that is, microscale continuum model, energetic model, and nanoscale structural model [23].

The microscale continuum model assumes that the biological cell is described by a one- or two-phase continuum model [40]. A typical model is shown in Fig. 2.5A. This model can obtain the mechanical properties of the cells, and gain the distribution of stresses and strains of the cells [41].

The energetic model assumes that the cell consists of multiple cytoskeleton structures with energy budget [42]. A typical model is illustrated in Fig. 2.5B. The model works based on percolation theory and polymer physics models for large deformations. It is independent of coordinate system selection and cytoskeleton architecture details, as energy is a scalar quantity. However, the physical property of the model is not sufficiently consistent with experimental results.

The nanoscale structural model adopts tensegrity structures to represent the cell. It has two types, that is, spectrin-network model and cytoskeletal model. Spectrin-network model contains a specific microstructural network for large deformations [43], and a typical model is shown in Fig. 2.5C.

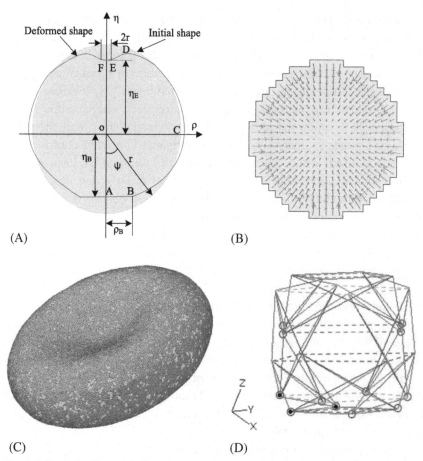

Fig. 2.5 Typical cell models for vision-based force sensing. (A) Microscale continuum cell model [41], (B) energetic cell model [42], (C) spectrin-network model [43], and (D) cytoskeletal model [23].

Cytoskeletal model considers the cytoskeleton and cytoskeleton contractile forces as the primary structural component and central role, respectively. A typical such model is depicted in Fig. 2.5D. The nanoscale structural model can characterize many nonlinear deformations of biological cells. It treats the cell as a system of microfilament, microtubule, and actin, along with force distributed inside cells [44].

In addition, to measure the deformation of cells, Liu et al. [33, 45] computed the microinjection force by making use of the deflection of a cell holder. Karimirad et al. [46] calculated the force applied on a spherical biological cell by using a well-trained artificial neural network. Nevertheless, this method ignores the inner structure of the cell, and it is also sensitive to parameter variation.

4.1.3 Advantages and limitations

The vision-based force sensors have been widely used owing to three distinct advantages. First, they can offer global force (rather than local force) feedback generated by contact force sensors [23]. Second, for some specific conditions, vision-based force sensors are irreplaceable when the contact force sensors find it challenging or impossible to detect the microinjection force [46]. Third, by making use of the microscope in the microinjection system, they can provide the force information without an additional equipment [24].

However, vision-based force sensors are sensitive to inevitable parameter variation. Moreover, the accuracy of the deduced force is heavily dependent on the adopted cell model. Meanwhile, a priori knowledge of the mechanical properties is necessary to establish and identify the cell model [47]. The mechanical properties of a cell are nonlinear and time-varying, which makes it impossible to accurately provide the microinjection force by a priori knowledge-based model. Additionally, the sampling frequency is relatively low, which is limited by the image processing and force estimation algorithm. This results in a low microinjection speed. Furthermore, the estimation performance is restricted by the small depth of field of view for the microscope [29].

In summary, the vision-based force sensor is only appropriate for cells with simple structure and stable mechanical properties. Moreover, such sensors are only suitable for scenarios that do not require high accuracy.

4.2 Capacitive force sensors

Currently, the capacitive force sensors for measuring microinjection force are generally based on microelectromechanical systems (MEMS). MEMS techniques can integrate the custom-built microinjectors with capacitive force sensors [22]. The MEMS capacitive force sensors usually work based on comb drives (i.e., the parallel-plate capacitors). There are two sensing modes, that is, transverse mode and lateral mode. In the transverse mode, the capacitance is changed due to the variation of gap size between the two parallel plates (see Fig. 2.6A). In the lateral mode, the capacitance is changed due to the variation of the overlapping area. Generally, the transverse mode can realize a higher resolution than the lateral mode, which is

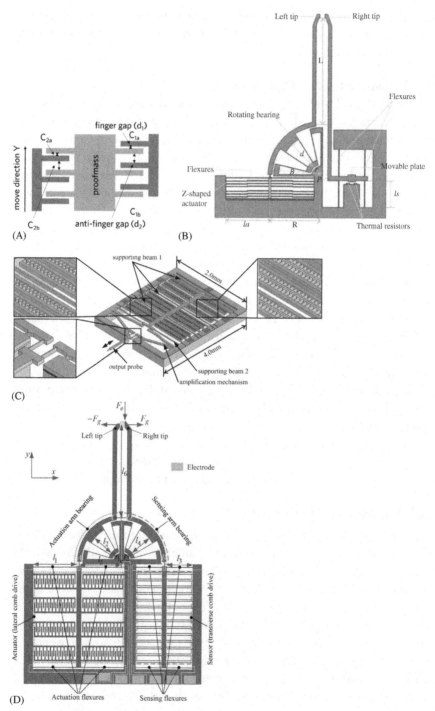

Fig. 2.6 Typical designs of capacitive force sensors. (A) Details of comb drives [48], (B) solid model of a capacitive nanomanipulator [49], (C) capacitive MEMS griper [50], and (D) single capacitive sensor for two-axis force sensing [51].

achieved at the cost of worse linearity. In practice, the change of capacitances can be converted to the variation of DC voltage by using an electronic circuit.

The force measurement range of a capacitive force sensor is from organism level (mN) to subcellular level (pN) [52]. Moreover, MEMS capacitive force sensors exhibit wide bandwidth, high sensitivity, and small footprint size. In particular, they can achieve multiple degree-of-freedom (DOF) measurements [51]. A typical solid model is given in Fig. 2.6B.

In the literature, Sun and Nelson [52] adopted the transverse mode with a large overlapping area to construct a 2-DOF capacitive force sensor, which obtained a very high sensitivity, that is, 0.01 and 0.24 μN. The proposed sensor was used to measure the mechanical properties of mouse oocyte and embryo zona pellucida [53]. Nevertheless, a relatively small stroke constricts its application in cell manipulation. Muntwyler et al. [54] designed a three-axis capacitive force sensor with two-mode readout electronics, generating an adjustable force measurement range from ±20 to ±200 μN. Additionally, the manufacturing process has been modified on a double silicon-on-insulator (SOI) substrate, which reduces the fabrication complexity of multiaxis sensors. Yang and Xu [50] proposed a MEMS capacitive gripper (see Fig. 2.6C). Moreover, Xu [51] used one set of transverse comb drives to sense the forces on two perpendicular directions, which provided a high resolution of 0.61 μN (see Fig. 2.6D).

In comparison with conventional piezoelectric force sensors and strain-gauge sensors, capacitive force sensors are more stable and more sensitive, and exhibit no hysteresis. As compared with AFM, they are also more stable and more compact. In particular, MEMS capacitive force sensors offer several outstanding properties, such as wide force range, multi-DOF measurement, compact size, and high sensitivity. Nevertheless, the fabrication process is relatively complicated, which imposes high requirements in terms of equipment, resulting in relatively high cost per unit. Furthermore, MEMS capacitive force sensors are very fragile and would be broken in case of any mistake or uncontrolled accident. In addition, the MEMS capacitive force sensors can only be used to manipulate the cells rather than injecting foreign materials into the cells, because they cannot integrate the required injection equipment due to their low load-bearing ability.

4.3 Optical-based force sensors

Optical-based force sensors generally use a light source (e.g., light emitting diode [LED], laser, or halogen lamp) to illuminate a load-sensitive medium (e.g., microcantilever or grating). A photodetector (e.g., photodiode or CCD camera) is adopted to measure the ranks of illumination, refractive index, or spectrum of the reflected light from the load-sensitive medium. The force can be computed from the change of measured results along with some known properties [55].

AFM is a typical optical-based force sensor, which adopts a cantilever as the load-sensitive medium (see Fig. 2.7A and B). During measurement, the photodiode would amplify and sense the small displacement of the cantilever, and the force can be calculated by the measured displacement and known spring constant of the cantilever. However, the results are affected

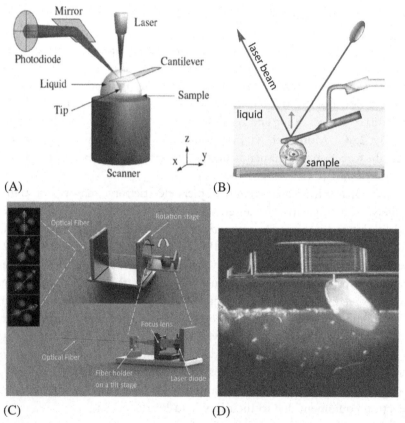

Fig. 2.7 Typical designs of optical force sensors. (A) Scheme of optical-based cantilever force sensor by using AFM [53], (B) cell injection and force sensing by using an AFM [55], (C) optical force sensor using laser trap [56], and (D) micrograting-based optical force sensor [57].

by mechanical vibration, instability of laser point, and shot noise [58]. The system can be adapted to achieve ultrahigh sensitive and higher resolution (pN–nN [33]) by applying cantilevers with lower stiffness.

Laser trap is another traditional optical force sensor [56] (see Fig. 2.7C). It can achieve the resolution of nN or sub-nN level. During operation, the light energy should be high enough (close to UV spectrum) to realize high dissipation of visible light in aqueous solutions. However, this method may cause damage to the cell and affect the genetic materials under high energy and heat. This concern may be solved by adopting a new wavelength close to infrared (IR) spectrum [52].

By using a micrograting as the load-sensitive medium, Zhang et al. [57] reported a 1D optical force sensor for measuring the dynamic microinjection force of Drosophila embryos (see Fig. 2.7D). A miniaturized optical encoder with high resolution and linear range was used to measure the displacement of micrograting. The microinjection force can be deduced from the measured displacement and known spring constant. The resolution of the realized force sensor can be smaller than 1 µN within a range of 10 µN. Additionally, Loh et al. [59] presented an automated AFM-based microinjection and obtained the microinjection force of 36 nN.

Optical-based force sensors have extremely high resolution (nN to sub-nN) and exhibit immunity to electromagnetic effects. Concerning a non-contact optical force sensor, it can solve the contradiction between sensitivity and linearity. However, the limitations of sensing the microinjection force by adopting an AFM are also obvious. First, a set of complex transmission and receiving equipment costs a lot, and accurate optical alignment and adjustment are also necessary [46]. Second, the measurement range is small, limited by the deflection of the photodiode. Third, the measured results would be affected by the reflection and refraction of cultural medium in a petri dish. Fourth, the AFM cannot provide real-time imaging and can only manipulate a cell rather than injecting it [33].

4.4 Piezoresistive force sensors

The working principle of piezoresistive force sensors lies in the fact that the resistance changes with a force applied. There are two types of piezoresistive sensors suitable for cell microinjection, that is, piezoresistive strain-gauge and etched silicon membrane. A Wheatstone bridge circuit is normally used to measure the change of resistance of a piezoresistive strain-gauge. The resolution is easily affected by the noise, thermal drift, power consumption, and the bridge configuration of the sensor [21]. Piezoresistive sensor with an etched silicon membrane can achieve a resolution of μN level [60].

In the literature, a commercial piezoresistive force sensor (model: AE801, from Kronex Technologies Corp.) was used to measure the micro-injection force of zebrafish embryo [21] (see Fig. 2.8A). During the micro-injection, the dramatic drop of force signal indicated a successful penetration. Shulev et al. [61] presented a vertical microinjector equipped with piezoresistive microforce sensor, which obtained a resolution of sub-μN (see Fig. 2.8B). Beutel et al. [7] proposed a piezoresistive microforce sensor, which was composed of etched silicon membrane and gained a resolution of 120 μN (see Fig. 2.8C). The presented sensor can autocalibrate and monitor the status of the microinjection pipette (i.e., pipette breakage and cell sticking), which improves the robustness of the microinjection system. However, the force value was generated only by considering the displacement of the needle, rather than the deformation of the membrane. Stavrov et al. [62] presented an axial piezoresistive microforce sensor composed of silicon (see Fig. 2.8D). Interestingly, the measurement range can be adjusted by changing the amplification gain factor of on-board electronics. It varied from several tens of μN to several hundred mN with the resolution in nN and μN level, respectively. The main disadvantages of the previous work are that the described systems can only puncture the cell membrane without the ability to inject materials into cells. This is because they cannot bear the necessary equipment for injection, such as a connector for an injection pulse. Additionally, misalignments may occur in the process of manual assembly for some piezoresistive force sensors, which causes less reliable results [63].

4.5 Piezoelectric force sensors

Concerning the piezoelectric force sensors for cell microinjection, polyvinylidene fluoride (PVDF) film is the most widely used material in sensing the microinjection force. The PVDF film works based on the forward piezoelectric effect to generate electric charge under the applied force. Currently, there are mainly three types of PVDF force sensors, that is, cantilever-PVDF sensor, simply supported PVDF sensor, and fixed-guided PVDF sensor.

As for a cantilever-PVDF force sensor, the PVDF film serves as a cantilever beam, with one end fixed on a manipulator and a micropipette attached to the other free end. As the PVDF film is relatively soft, the free end will be deformed a lot when a small force is applied, leading to high sensitivity. Pillarisetti et al. [64] performed zebrafish embryo microinjection by

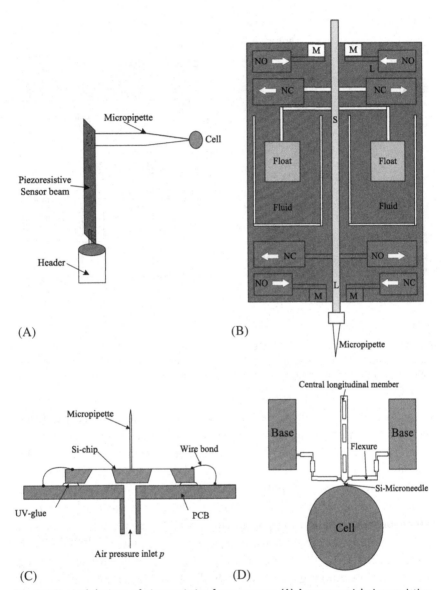

Fig. 2.8 Typical designs of piezoresistive force sensors. (A) A commercial piezoresistive force sensor (model AE801, from Kronex Technologies Corp.) is modified for use in penetration force sensing [21], (B) MEMS piezoresistive microforce sensor [61], (C) silicon-membrane-based piezoresistive force sensor [7], and (D) silicon-structured piezoresistive force sensor [62].

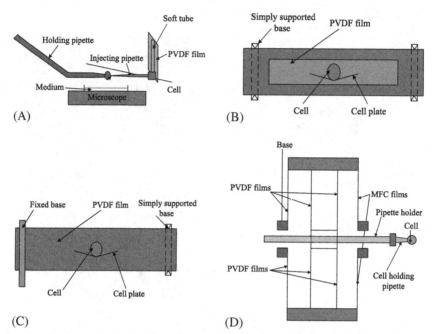

Fig. 2.9 Typical designs of PVDF force sensors. (A) Cantilever-PVDF force sensor [64], (B) simply supported PVDF force sensor [16], (C) fixed-simply supported PVDF force sensor [65], and (D) fixed-guided beam-type PVDF force sensor [66].

using a cantilever-PVDF force sensor and proved that the force feedback with resolution of μN can improve the success rate (see Fig. 2.9A). In addition, Shen et al. [67] employed closed-loop control to regulate the cantilever-PVDF force sensor to gain higher resolution of sub-μN. Furthermore, Chen et al. [68] built a model for a cantilever-PVDF force sensor and designed an electric circuit to collect the generated charges, which improved its sensing performance with a resolution of sub-μN in the microinjection of living Drosophila embryos. Huang et al. [69] designed a cantilever-PVDF force sensor to detect the force variations during microinjection. Although the cantilever-PVDF sensors are of high sensitivity, they will cause large curve deformation, and hence, cause significant damage to cells, which results in a low survival rate. Moreover, the micropipette is directly bonded with the cantilever, leading to difficulty in changing a broken micropipette and injecting materials.

As for a simply supported PVDF force sensor (see Fig. 2.9B), a PVDF film is fixed vertically with two ends bonded onto a supporting beam. A cell plate is perpendicularly placed at the center of the PVDF film [70].

Similarly, a fixed-simply supported PVDF force sensor was presented in [65] (see Fig. 2.9B). Nevertheless, these two force sensors cannot hold the cells stably.

Concerning a fixed-guided PVDF sensor (see Fig. 2.9C), PVDF films serve as flexure beams in multistage compound parallelogram flexure (MCPF). Nevertheless, the resolution of several hundred µN is relatively low [66].

The piezoelectric film has a simple structure, high sensitivity, high compliance, and high bandwidth. However, the output signals are dynamic and sensitive to environmental noise. Therefore, custom-designed signal collection electronic circuits and electromagnetic isolation of noise should be provided to gain the desired stable and accurate outputs.

4.6 Comparison of the force sensors

For different applications, the force sensor should be carefully selected according to their advantages and disadvantages. In consideration of the cost efficiency, vision-based force sensors are outstanding in handling many types of cells without adding extra equipment to an available microinjection system. It is noted that piezoresistive strain-gauge and piezoelectric film are also cost-effective, but they require additional signal collection or amplification equipment. In terms of resolution, the optical force sensors perform best (nN to sub-nN), and MEMS capacitive/piezoresistive force sensors can obtain resolution of nN to µN level. Regarding the response speed, the contact force sensors (e.g., capacitive, piezoresistive, and piezoelectric types) exhibit advantages over noncontact (e.g., optical and vision-based) ones. Concerning the measurement range, piezoelectric, piezoresistive, and capacitive force sensors can offer µN-mN force range. By contrast, optical force sensors can only provide nN-µN force range. Furthermore, a detailed comparison table can be referred to in our previous work [55].

5 Current challenges of cell microinjection and future development

Currently, there are still many challenges to realize automated microinjection with force control. The desired microinjection system should provide the properties of high-speed microinjection, high-level automation, high flexibility, abundant feedback, low cost, high success rate, and high survival rate. To realize these desired properties, six challenges and related promising methods are discussed as follows.

5.1 Microinjector design

The microinjector should complete the microinjection operation accurately and repeatably in a short time, and also ensure high survival rate. The force resolution of the microinjector should be of μN level at least. The integrated microinjector should not deteriorate the performance of motion control. These requirements should be collaboratively solved by both mechanical design and control design. Note that the piezoelectric actuator has proven its effectiveness in microinjection with rapid response and extremely high resolution of motion, which achieved a survival rate of 80% for injecting sperm into mouse oocytes [1]. Moreover, the vibration of piezoelectric actuator has also been utilized for microinjection with less damage to cells [13]. However, the piezoelectric actuator exhibits inherent nonlinearities, which calls for comprehensive control algorithms to achieve precise motion control [71]. Additionally, the motion control should not only accurately regulate the motion of the actuator, but also control the actuation performance of the microinjector. The motion control of the microinjector should also meet the requirement of transient response and steady state.

5.2 Injection control design

During the microinjection, visual-servo control and force control alternately dominate the entire process. First, the cells, microinjection pipette, and holding device are recognized and marked (with location and dimension information) through image processing. Then, the cell is translated to approach the micropipette (or the micropipette is moved to approach the cell) under the guide of visual-servo control. During the visual-servo control, a cooperative control of the micromanipulators for driving the microinjector and holding devices can improve the microinjection speed. Next, the visual-servo motion control is switched to force control [72]. The force control regulates the microinjection force to implement the microinjection with less damage. It is notable that the force control should tolerate certain disturbances, such as the varying stiffness of the same batch of cells or at different developing stages [73]. Concerning the switching process between visual-servo control and force control, there are mainly two methods, that is, impedance control method [24] and weight coefficient method [72]. In particular, the impedance control can smoothly complete the switch process, and control the position and microinjection force with the same impedance algorithm [74].

5.3 Cell holder design

Currently, there are three main methods to mechanically fix the cells, that is, microwells, hydrodynamic traps, and vacuum-based holder [75]. The microwells cannot ensure a stable immobilization, because the cells can still move inside the microwells. In hydrodynamic traps, it is hard for the micropipette to access the cell. The vacuum-based holder applies a vacuum to catch up the cells, which has been widely adopted in cell microinjection. Conventionally, a holding pipette can only fix one cell at a time, which limits the work efficiency. Thus, cell holders are needed to quickly fix multiple cells without damage. Regarding the materials used for cell holders, glass is preferred, because it does not influence the image under the most frequently used differential interference contrast (DIC) microscope [76].

In the literature, Fujisato et al. [77] presented a microporous glass-based cell holder, which can filter the cells out of a mixture of cells and liquid, and maintain the cells in holes (see Fig. 2.10A). Lu et al. [21] proposed a V-shaped groove-based cell holder, which is composed of gel. However, the cells may slip along the grooves (see Fig. 2.10B). Huang et al. [69]

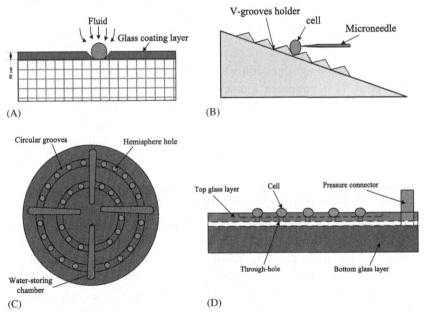

Fig. 2.10 Typical designs of mechanical confinement cell holders. (A) Microporous glass with hole-type cell holder [77], (B) parallel V-groove-based gel cell holder [21], (C) hemispherical hole and groove-based circular rotary cell holder [69], and (D) vacuum-based through-hole cell holder [45].

designed a hemispherical cell holder based on holes and grooves in a circular plate, which can fix the cells and maintain the wet environment (see Fig. 2.10C). Nevertheless, as the interval between holes is relatively big, the cells need to be manually placed in the holes, increasing the risk of damage. Liu and Sun [75] presented a vacuum hole array-based cell holder to quickly and evenly fix multiple holes, which improved the efficiency of fixing the cells (see Fig. 2.10D).

5.4 Penetration scheme design

Minimum microinjection force or deformation is very important to conduct high-throughput biology and genetics study. During penetration, the deformation of cells should be maintained within a reasonable range. When the penetration speed is constant, the cell will incur a large deformation. Thus, an accelerated and fast penetration motion created by an impact actuator is preferred to reduce the caused deformation [78]. Note that a piezoelectric actuator combined with compound flexure stage can generate a fast acceleration and variable speed with rapid response speed [35]. Moreover, a rapid deceleration is also important after the penetration process. Nevertheless, vibration occurs during the process of rapid acceleration and deceleration, which may cause significant damage to cells. Tradeoff should be made between the fast acceleration/deceleration and generated vibration to guarantee a successful microinjection with minimum damage. In the literature, Huang et al. [24] presented a trajectory of velocity and acceleration for microinjection of zebrafish embryos. Interestingly, vibration with small amplitude and high frequency can effectively reduce the microinjection force and deformation [13]. A combination of vibration and linear translation can greatly reduce the induced deformation or microinjection force.

5.5 Microinjection pipette maintenance

Conventionally, the microinjection pipette is fabricated using a glass tube, which is conducted under the combination of heating and pulling with a commercial pipette puller. The fabricated glass pipette has a closed front end, which should be carefully broken at a proper position to obtain the desired tip diameter. The process is not only labor intensive, but the obtained surface is usually rough, which produces irregular wounds and induces large deformation during the microinjection. The fabricated glass pipette is very fragile. Zhang et al. [57] demonstrated that the silicon nitride microinjector has a higher efficiency than the conventional glass needle.

Considering that the micropipette is very fragile and usually gets clogged, it is necessary to monitor its real condition. Machine vision and electrical method are potential techniques to monitor the condition. In the literature, Lukkari and Kallio [79] presented an electrical method with impedance measurement to detect multiple conditions, such as break, clogging, and injection solution. The broken and clogged micropipettes should be changed quickly, and hence, a fast exchanger is preferred. In the literature, Matsuoka et al. [14] suggested a modular design for the front end of the microinjector, which avoided influencing other components during micropipette replacement.

5.6 Issue of injection volume

To achieve a consistent volume of microinjection, a robotic microinjection system normally adopts the same uniform microinjection time and microinjection pressure. However, the microinjection volume is still different due to efflux or influx phenomena [80]. The efflux and influx phenomena cause less volume and more volume, respectively, leading to unreliable results. Moreover, the influx may narrow down the area of cross-section and even cause blocking. Thus, the efflux of solution in a micropipette is preferred [17]. Note that large efflux is a major cause of cell death, and carriers of alginate and nanodiamonds can reduce the efflux [81].

For controlling the microinjection volume, calibration methods (such as oil droplet) were adopted [16]. However, the calibration methods are affected by the adopted equipment and biological surroundings [17]. In the literature, Zhang et al. [13] equipped a microfabricated microinjection pipette with a piezoresistive sensor to monitor the microinjection volume. In summary, realizing a consistent microinjection volume and avoiding efflux phenomenon will improve the survival rate and transplantation efficiency.

6 Conclusion

Robotic cell injection with force sensing and control can achieve real-time force feedback. It is more flexible, more reliable, and more robust than the stand-alone visual-servo control. In this chapter, the state-of-the-art force sensing techniques for robotic cell microinjection have been extensively reviewed. Force control can benefit the robotic microinjection of both adherent and suspended cells. However, there are still some challenges to realize full automation of cell microinjection and the related promising solutions are also discussed. In the future, more autonomous robotic micro-injection systems with force control are expected to achieve higher success and survival rates.

Acknowledgments

This work was funded in part by the National Natural Science Foundation of China under Grant 52175556 and Grant 51575545; the Macao Science and Technology Development Fund under Grant 0045/2021/APD, Grant 0153/2019/A3, Grant 0022/2019/AKP, and Grant 0102/2022/A2; and the University of Macau under Grant MYRG2018-00034-FST, Grant MYRG2019-00133-FST, and Grant MYRG2022-00068-FST.

References

[1] Y. Kimura, R. Yanagimachi, Intracytoplasmic sperm injection in the mouse, Biol. Reprod. 52 (4) (1995) 709–720.

[2] Y. Zhang, K. Tan, S. Huang, Vision-servo system for automated cell injection, IEEE Trans. Ind. Electron. 56 (1) (2009) 231–238.

[3] W. Wang, X. Liu, D. Gelinas, B. Ciruna, Y. Sun, A fully automated robotic system for microinjection of zebrafish embryos, PLoS One 2 (9) (2007) e862.

[4] Y. Xie, D. Sun, C. Liu, H. Tse, S. Cheng, A force control approach to a robot-assisted cell microinjection system, Int. J. Robot. Res. 29 (9) (2010) 1222–1232.

[5] T. Nakayama, H. Fujiwara, K. Tastumi, K. Fujita, T. Higuchi, A new assisted hatching technique using a piezomicromanipulator, Fertil. Steril. 69 (4) (1998) 784–788.

[6] S. Li, D. Zheng, N. Li, X. Wang, Y. Liu, Size-adjustable microdroplets generation based on microinjection, Micromachines 8 (3) (2017) 88.

[7] T. Beutel, N. Ferreira, A. Balck, M. Leester-Schadel, S. Buttgenbach, Cell manipulation system based on a self-calibrating silicon micro force sensor providing capillary status monitoring, IEEE Sensors J. 12 (10) (2012) 3075–3081.

[8] W. Walther, U. Stein, Viral vectors for gene transfer a review of their use in the treatment of human diseases, Drugs 60 (2) (2000) 249–271.

[9] M. Rols, Electropermeabilization, a physical method for the delivery of therapeutic molecules into cells, Biochim. Biophys. Acta (BBA) Biomembranes 1758 (3) (2006) 423–428.

[10] A.K. Banga, S. Bose, T.K. Ghosh, Iontophoresis and electroporation: comparisons and contrasts, Int. J. Pharm. 179 (1) (1999) 1–19.

[11] J. Sundaram, B.R. Mellein, S. Mitragotri, An experimental and theoretical analysis of ultrasound-induced permeabilization of cell membranes, Biophys. J. 84 (5) (2003) 3084–3101.

[12] M.T. Lin, L. Pulkkinen, J. Uitto, K. Yoon, The gene gun: current applications in cutaneous gene therapy, Int. J. Dermatol. 39 (3) (2000) 161–170.

[13] X. Zhang, M.P. Scott, C.F. Quate, O. Solgaard, Microoptical characterization of piezoelectric vibratory microinjections in Drosophila embryos for genome-wide RNAi screen, J. Microelectromech. Syst. 15 (2) (2006) 277–286.

[14] H. Matsuoka, S. Shimoda, Y. Miwa, M. Saito, Automatic positioning of a microinjector in mouse ES cells and rice protoplasts, Bioelectrochemistry 69 (2) (2006) 187–192.

[15] H. Matsuoka, T. Komazaki, Y. Mukai, M. Shibusawa, H. Akane, High throughput easy microinjection with a single-cell manipulation supporting robot, J. Biotechnol. 116 (2) (2005) 185–194.

[16] A. Uchida, P. Monsma, J. Fenn, A. Brown, Live-cell imaging of neurofilament transport in cultured neurons, Methods Cell Biol. 131 (2016) 21–90.

[17] P. Kallio, J. Kuncova-Kallio, Capillary pressure microinjection of living adherent cells: challenges in automation, J. Micromechatron. 3 (3–4) (2006) 189220.

[18] A. Ghanbari, B. Horan, S. Nahavandi, X. Chen, W. Wang, Haptic microrobotic cell injection system, IEEE Syst. J. 8 (2) (2014) 371–383.

[19] P. Scherp, K.H. Hasenstein, Microinjection—a tool to study gravitropism, Adv. Space Res. 31 (10) (2003) 2221–2227.

[20] N. Tran, X. Liu, Z. Yan, D. Abbote, Q. Jiang, Efficiency of chimeraplast gene targeting by direct nuclear injection using a GFP recovery assay, Mol. Ther. 7 (2) (2003) 248–253.

[21] Z. Lu, P. Chen, J. Nam, R. Ge, W. Lin, A micromanipulation system with dynamic force-feedback for automatic batch microinjection, J. Micromech. Microeng. 17 (2) (2007) 314–321.

[22] S. Zappe, M. Fish, M.P. Scott, O. Solgaard, Automated MEMS-based Drosophila embryo injection system for high-throughput RNAi screens, Lab Chip 6 (8) (2006) 1012–1019.

[23] T.J. Chen, C.C. Wu, M.J. Tang, J.S. Huang, F.C. Su, Complexity of the tensegrity structure for dynamic Energy and force distribution of cytoskeleton during cell spreading, PLoS One 5 (12) (2010) e14392.

[24] H. Huang, D. Sun, J.K. Mills, W.J. Li, S.H. Cheng, Visual-based impedance control of out-of-plane cell injection systems, IEEE Trans. Autom. Sci. Eng. 6 (3) (2009) 565–571.

[25] Y. Sun, B.J. Nelson, Biological cell injection using an autonomous microrobotic system, Int. J. Robot. Res. 21 (10–11) (2002) 861–868.

[26] Q. Xu, Micromachines for Biological Micromanipulation, Springer, New York, NY, 2018.

[27] S. Permana, E. Grant, G.M. Walker, J.A. Yoder, A review of automated microinjection systems for single cells in the embryogenesis stage, IEEE/ASME Trans. Mechatron. 21 (5) (2016) 2391–2404.

[28] W. Zhang, A. Sobolevski, B. Li, Y. Rao, X. Liu, An automated force-controlled robotic micromanipulation system for mechanotransduction studies of Drosophila larvae, IEEE Trans. Autom. Sci. Eng. 13 (2) (2016) 789–797.

[29] X. Zhang, W. Wang, A. Nordin, F. Li, S. Jang, The influence of the electrode dimension on the detection sensitivity of electric cell-substrate impedance sensing (ECIS) and its mathematical modeling, Sens. Actuators B 247 (2017) 780–790.

[30] A. Meister, M. Gabi, P. Behr, P. Studer, J. Voros, FluidFM: combining atomic force microscopy and fluidics in a universal liquid delivery system for single cell applications and beyond, Nano Lett. 9 (6) (2009) 2501–2507.

[31] A. Laws-King, A. Trounson, H. Sathananthan, I. Kola, Fertilization of human oocytes by microinjection of a single spermatozoon under the zona pellucida, Fertil. Steril. 48 (4) (1987) 637–642.

[32] X. Li, G. Zong, S. Bi, W. Zhao, Automatic micromanipulating system for biological applications with visual servo control, J. Micromechatron. 1 (4) (2001) 345–363.

[33] X. Liu, K. Kim, Y. Zhang, Y. Sun, Nanonewton force sensing and control in microrobotic cell manipulation, Int. J. Robot. Res. 28 (8) (2009) 1065–1076.

[34] J. Kim, F. Janabi-Sharifi, J. Kim, A haptic interaction method using visual information and physically based modeling, IEEE/ASME Trans. Mechatron. 15 (4) (2010) 636–645.

[35] Z. Lu, P.C.Y. Chen, W. Lin, Force sensing and control in micromanipulation, IEEE Trans. Syst. Man Cybern. Part C Appl. Rev. 36 (6) (2006) 713–724.

[36] M. Kass, A. Witkin, D. Terzopoulos, Snakes: active contour models, Int. J. Comput. Vis. 1 (4) (1988) 321–331.

[37] F. Leymarie, M.D. Levine, Tracking deformable objects in the plane using an active contour model, IEEE Trans. Pattern Anal. Mach. Intell. 15 (6) (1993) 617–634.

[38] Y. Canny, A computational approach to edge detection, IEEE Trans. Pattern Anal. Mach. Intell. PAMI-8 (6) (1986) 679–698.

[39] M. Ammi, A. Ferreira, Biological cell injection visual and haptic interface, Adv. Robot. 20 (3) (2006) 283–304.

[40] C. Lim, E. Zhou, S. Quek, Mechanical models for living cells—a review, J. Biomech. 39 (2) (2006) 195–216.

[41] Y. Tan, D. Sun, W. Huang, S.H. Cheng, Characterizing mechanical properties of biological cells by microinjection, IEEE Trans. Nanobiosci. 9 (3) (2010) 171–180.

[42] E.G. Rens, L. Edelstein-Keshet, From energy to cellular forces in the cellular Potts model: an algorithmic approach, PLoS Comput. Biol. 15 (12) (2019) e1007459.

[43] M. Dao, J. Li, S. Suresh, Molecularly based analysis of deformation of spectrin network and human erythrocyte, Mater. Sci. Eng. C 26 (8) (2006) 1232–1244.

[44] M. Mehrbod, M.R. Mofrad, On the significance of microtubule flexural behavior in cytoskeletal mechanics, PLoS One 6 (10) (2011) e25627.

[45] X. Liu, Y. Sun, B.M. Lansdorp, Vision-based cellular force measurement using an elastic microfabricated device, J. Micromech. Microeng. 17 (7) (2007) 1281–1288.

[46] F. Karimirad, S. Chauhan, B. Shirinzadeh, Vision-based force measurement using neural networks for biological cell microinjection, J. Biomech. 47 (5) (2014) 1157–1163.

[47] D. Desmaele, M. Boukallel, S. Regnier, A planar structure sensitive to out-of-plane forces for the force-controlled injection of suspended and adherent cells, in: Proceedings of 2011 Annual International Conference of the IEEE Engineering in Medicine and Biology Society, Boston, MA, USA, August 30–September 1, 2011, pp. 8420–8423.

[48] R. Li, Z. Mahmoud, M. Rasras, I. Elfadel, D. Choi, Design, modelling and characterization of comb drive MEMS gap-changeable differential capacitive accelerometer, Measurement 169 (2021) 108377.

[49] M. Bao, Analysis and Design Principles of MEMS Devices 175-212, Elsevier, 2005.

[50] S. Yang, Q. Xu, Design of a microelectromechanical systems microgripper with integrated electrothermal actuator and force sensor, Int. J. Adv. Robot. Syst. 13 (5) (2016) 17298.

[51] Q. Xu, Design, fabrication, and testing of an MEMS microgripper with dual-axis force sensor, IEEE Sensors J. 15 (10) (2015) 6017–6026.

[52] Y. Sun, B.J. Nelson, MEMS capacitive force sensors for cellular and flight biomechanics, Biomed. Mater. 2 (1) (2007) S16–S22.

[53] K. Hammond, M.G. Ryadno, B.W. Hoogenboom, Atomic force microscopy to elucidate how peptides disrupt membranes, Biochim. Biophys. Acta Biomembr. 1863 (1) (2021) 183447.

[54] S. Muntwyler, F. Beyeler, B.J. Nelson, Three-axis micro-force sensor with sub-micro-Newton measurement uncertainty and tunable force range, J. Micromech. Microeng. 20 (2) (2010) 025011.

[55] R. Ungai-Salánki, B. Peter, T. Gerecsei, N. Orgovan, B. Horvath, A practical review on the measurement tools for cellular adhesion force, Adv. Colloid Interface Sci. 269 (2019) 309–333.

[56] Z.H. Shan, E. Zhang, D. Pi, H. Gu, W. Cao, Controlled rotation of cells using a single-beam anisotropic optical trap, Opt. Commun. 475 (2020) 126169.

[57] X.J. Zhang, S. Zappe, R.W. Bernstein, O. Sahin, C.C. Chen, Micromachined silicon force sensor based on diffractive optical encoders for characterization of microinjection, Sens. Actuators A Phys. 26 (12) (1990) 2148–2157.

[58] G. Meyer, N.M. Amer, Novel optical approach to atomic force microscopy, Appl. Phys. Lett. 53 (12) (1988) 1045–1047.

[59] O. Loh, R. Lam, M. Chen, N. Moldovan, H. Huang, Nanofountain-probe-based high-resolution patterning and single-cell injection of functionalized nanodiamonds, Small 5 (14) (2009) 1667–1674.

[60] R. Jumpertz, A. Hart, O. Ohlsson, F. Saurenbach, J. Schelten, Piezoresistive sensors on AFM cantilevers with atomic resolution, Microelectron. Eng. 41–42 (1998) 441–444.

[61] A. Shulev, I. Roussev, K. Kostadinov, Force sensor for cell injection and characterization, in: Proceedings of 2012 International Conference on Manipulation, Manufacturing and Measurement on the Nanoscale (3M-NANO), Xi'an, China, August 29–September 1, 2012, pp. 335–338.

[62] V.T. Stavrov, A.A. Shulev, C.M. Hardalov, V.M. Todorov, I.R. Roussev, All-silicon microforce sensor for bio applications, in: Proceedings of Society of Photo-Optical Instrumentation Engineers (SPIE), Grenoble, France, May 17, 2013, p. 87630Y.

[63] K. Kim, X. Liu, Y. Zhang, Y. Sun, Nanonewton force-controlled manipulation of biological cells using a monolithic MEMS microgripper with two-axis force feedback, Microelectron. Eng. 18 (5) (2008) 055013.

[64] A. Pillarisetti, M. Pekarev, A.D. Brooks, J.P. Desai, Evaluating the effect of force feedback in cell injection, IEEE Trans. Autom. Sci. Eng. 4 (3) (2007) 322–331.

[65] Y. Zhou, S. Liu, C. Yang, C. Liu, Y. Xie, Design of a microforce sensor based on fixed-simply supported beam: towards realtime cell microinjection, in: Proceedings of 2015 IEEE International Conference on Cyber Technology in Automation, Control, and Intelligent Systems (CYBER), Shenyang, China, June, 8–12, 2015, pp. 1080–1084.

[66] Y. Wei, Q. Xu, Design of a PVDF-MFC force sensor for robot-assisted single cell microinjection, IEEE Sens. J. 17 (13) (2017) 3975–3982.

[67] Y. Shen, E. Winder, N. Xi, C.A. Pomeroy, U.C. Wejinya, Closed-loop optimal control-enabled piezoelectric microforce sensors, IEEE/ASME Trans. Mechatron. 11 (4) (2006) 420–427.

[68] Z. Chen, Y. Shen, N. Xi, X. Tan, Integrated sensing for ionic polymer-metal composite actuators using PVDF thin films, Smart Mater. Struct. 16 (2) (2007) S262–S271.

[69] H.B. Huang, D. Sun, J.K. Mills, S.H. Cheng, Robotic cell injection system with position and force control: toward automatic batch biomanipulation, IEEE Trans. Robot. 25 (3) (2009) 727–737.

[70] Y. Xie, D. Sun, C. Liu, Penetration force measurement and control in robotic cell microinjection, in: Proceedings of 2009 IEEE/RSJ International Conference on Intelligent Robots and Systems, St. Louis, MO, USA, October 10–15, 2009, pp. 4701–4706.

[71] Q. Xu, K.K. Tan, Advanced Control of Piezoelectric Micro-/Nano-Positioning Systems, Springer, New York, NY, 2015.

[72] G. Wang, Q. Xu, Design and precision position/force control of a piezo-driven microinjection system, IEEE/ASME Trans. Mechatron. 22 (4) (2017) 1744–1754.

[73] J. Abadie, C. Roux, E. Piat, C. Filiatre, C. Amiot, Experimental measurement of human oocyte mechanical properties on a micro and nanoforce sensing platform based on magnetic springs, Sens. Actuators B 190 (2014) 429–438.

[74] N. Hogan, Impedance control: an approach to manipulation: part I—theory, J. Dyn. Syst. Meas. Control 107 (1) (1985) 1–7.

[75] X. Liu, Y. Sun, Microfabricated glass devices for rapid single cell immobilization in mouse zygote microinjection, Biomed. Microdevices 11 (6) (2009) 1169.

[76] D.B. Murphy, Fundamentals of Light Microscopy and Electronic Imaging, Wiley, Hoboken, NJ, 2002.

[77] T. Fujisato, S. Abe, T. Tsuji, M. Sada, F. Miyawaki, The development of an OVA holding device made of microporous glass plate for genetic engineering, in: Proceedings of the 20th Annual International Conference of the IEEE Engineering in Medicine and Biology Society, Hong Kong, China, June 8–12, 1998, pp. 2981–2982.

[78] T. Tanikawa, T. Arai, Development of a micro-manipulation system having a two-fingered micro-hand, Biomed. Microdevices 15 (1) (1999) 152–162.

[79] M. Lukkari, P. Kallio, Multi-purpose impedance-based measurement system to automate microinjection of adherent cells, in: Proceedings of 2005 International Symposium on Computational Intelligence in Robotics and Automation, Espoo, Finland, June 27–30, 2005, pp. 701–706.

[80] G. Minaschek, G. Bereiter-Hahn, G. Bertholdt, Quantitation of the volume of liquid injected into cells by means of pressure, Exp. Cell Res. 183 (2) (1989) 434–442.

[81] B.A. Aguado, W. Mulyasasmita, J. Su, K.J. Lampe, S.C. Heilshorn, Improving viability of stem cells during syringe needle flow through the design of hydrogel cell carriers, Tissue Eng. Part A 18 (7–8) (2011) 806–815.

CHAPTER 3

Robotic orientation control and enucleation of cells

Lin Feng[a,b], Wei Zhang[a], Chunyuan Gan[a], Chutian Wang[a], Hongyan Sun[a], Yiming Ji[a], and Luyao Wang[a]
[a]School of Mechanical Engineering & Automation, Beihang University, Beijing, China
[b]Beijing Advanced Innovation Center for Biomedical Engineering, Beihang University, Beijing, China

1 Introduction

As the basic structural and functional unit of life, cells play an essential role in life development and health states [1], attracting worldwide research interest and corresponding biomedical applications over the years. Many biological and medical applications, such as in vitro fertilization [2,3], cell surgery [4,5], somatic cell nuclear transfer [6,7], and so on, have involved in vitro cell manipulations in their procedures; in particular, cell orientation and enucleation. In the past two decades, the rise of the era of biotechnology and robotics has provided a unique platform for promoting efficient and precise cell orientation and enucleation, exhibiting many significant advantages over conventional manual cell operations. Numerous field-based technical methods reported in recent years, such as magnetic [8,9], optical [10,11], hydrodynamic [12,13], acoustic [14,15], and electrical [16,17] methods, have led to significant advances in robotic cell rotation and enucleation in various biomedical applications.

In this chapter, we focus on the existing micro- and nanorobotic technologies for cell orientation and enucleation, and report various research endeavors on cell manipulations involving different external fields. In the cell orientation section, these field-controlled cell manipulations are reported at the manipulating method level, including magnetic, acoustic, and optoelectronic fields, which exhibit different control mechanisms and manipulating features. In the second section, the combination of microfluidics and microrobotics for cell enucleation is summarized, followed with the related technological challenges in enucleation. Lastly, we discuss some potential trends of field-controlled cell orientation and enucleation for future development.

Robotics for Cell Manipulation and Characterization
https://doi.org/10.1016/B978-0-323-95213-2.00015-6

2 Robotic orientation control of cells

As introduced in the Introduction, cell orientation, especially cell rotation, plays an important role in cell microsurgical operations for many biomedical applications and, therefore, has received particular attention in recent years, with numerous study innovations based on different techniques. Herein, robotic rotation or orientation manipulations of cells are overviewed within three categories, according to controlling fields: magnetic, acoustic, and optoelectronic methods.

2.1 Cell rotation based on magnetic fields

2.1.1 Noninvasive manipulation

Magnetic field-based manipulation is a reliable tool for studying biological cells. A magnetic medium placed in a magnetic field is subjected to a magnetic force or magnetic moment, which causes motion, and this property of the magnetic medium can be used to guide its manipulation of cells. Compared to optical techniques, magnetic manipulation allows precise control of the intensity and direction of the force and causes minimal damage to living cells. The main magnetic media that have been demonstrated to be usable are the following: magnetic spheres, magnetic wires, magnetic special structures, magnetically controlled micromechanical arms, etc.

As shown in Fig. 3.1A, Lee et al. [18] achieved remote control of individual T cell activation using the unique magnetic response of Janus magnetic spheres, which are coated with a film of magnetically responsive material on one side and display stimulatory ligands to activate T cells on the other side. By manipulating the rotation and motion of the Janus spheres through an external magnetic field, the platform can control the particle-cell recognition direction, and thus, activate and control single cells.

In addition to utilizing ligands for cell rotation, Petit et al. [19] exploited low Reynolds number moving microvortices generated by self-assembled magnetic dipoles and nickel nanowires spinning in a fluid (Fig. 3.1B), where the amplitude and position of the vortex can be precisely controlled for selective capture and transport of individual objects such as polystyrene microspheres, *E. coli*, etc. Zhang et al. [20] reported a method for selective capture and transport of single objects by rotating magnetic field-driven nickel nanowires to perform basic contact and noncontact manipulation tasks on cells (see Fig. 3.1C). The fluid flow generated around the rotating nickel nanowires was used to manipulate micro-objects such as individual flagellated microorganisms and human blood cells in a noncontact manner.

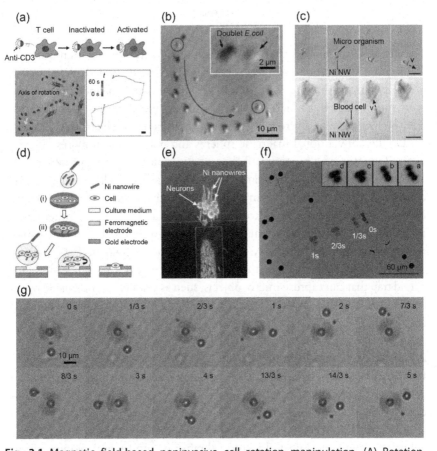

Fig. 3.1 Magnetic field-based noninvasive cell rotation manipulation. (A) Rotation movement of T cells using magnetic Janus sphere [18]. Both scale bars are 5 μm. (B) Manipulation of individual *E. coli* cells with micro-vortices generated during rotation of self-assembled microspheres [19]. (C) Nickel nanowires manipulate flagellated microorganisms and blood cells to make rotational movements [20]. Both scale bars are 30 μm. (D) Schematic of transport of living cells internalized into nickel nanowires using magnetic localization technique [21]. (E) Cells are transported by an external magnetic field into the gap between the ferromagnetic electrode and the gold electrode [21]. (F) Trapping and rotating magnetic particles using dumbbell-shaped magnetic actuators [22]. (G) Dumbbell-shaped magnetic actuators manipulate yeast cells [22]. *(Source: Panel (A) Reproduced and adapted with permission K. Lee, et al., Remote control of T cell activation using magnetic Janus particles, Angew. Chem. Int. Ed. 55(26) (2016) 7384–7387. Copyright 2016, John Wiley and Sons. Panel (B) Reproduced and adapted with permission T. Petit, et al., Selective trapping and manipulation of microscale objects using mobile microvortices, Nano Lett. 12(1) (2012) 156–160. Copyright 2012, American Chemical Society. Panel (C) Reproduced and adapted with permission L. Zhang, et al., Targeted cargo delivery using a rotating nickel nanowire, Nanomedicine. 8(7) (2012) 1074–1080. Copyright 2012, Elsevier. Panels (D–E) Reproduced and adapted with permission D. Choi, et al. Transport of living cells with magnetically assembled nanowires, Biomed. Microdevices. 9(2) (2007) 143–148. Copyright 2007, Springer Nature. Panels (F–G) Reproduced and adapted with permission Q. Zhou, et al., Dumbbell fluidic tweezers for dynamical trapping and selective transport of microobjects, Adv. Funct. Mater. 27(1) (2017) 1604571. Copyright 2017, John Wiley and Sons.)*

Choi et al. [21], on the other hand, presented a technique for transport and localization of internalized cells using magnetic field-guided nickel nanowires, as illustrated in Fig. 3.1D and E, where nanoscale magnetic nanowires were internalized by rat neuroblastoma and cells were transported and localized by magnetic fields from magnetic material-coated electrodes.

Unlike conventional magnetic spheres and magnetic nanowires, Steager et al. [23] designed a biocompatible magnetic structure with a groove in the middle that enables fully automated manipulation of cells, such as long-distance transport of rat hippocampal neurons, with a visual tracking algorithm that makes this process fully automated. Zhou et al. [22], on the other hand, prepared a dumbbell-shaped magnetic actuator assembled from a rotating nickel nanowire and two polystyrene microspheres (Fig. 3.1F–G), where the micro-vortex generated by the rotation of the nickel nanowire can induce a fluid trap that can capture micro-objects, such as yeast cells, and enable operations such as their rotation and transport as long as the rotation frequency is high enough.

2.1.2 Invasive manipulation

In addition to the manipulation of objects such as magnetic spheres to achieve magnetically controlled cell rotation by contacting cells, cell rotation by cell injection or endocytosis of magnetic nanoparticles has also been widely studied in recent years. Berndt et al. [9] produced a magnetic hydrogel microsphere and injected it into zebrafish or mouse embryonic cells to achieve three-dimensional cell rotation by means of a gradient field control system. As shown in Fig. 3.2A, three-dimensional observation of embryonic cells and their growth and development process was achieved. Liu et al. [24] obtained gastric cancer cells endocytosed with magnetic particles after incubating 170 ± 50 nm ferric tetroxide nanoparticles with human gastric cancer cells at 37°C for 24 h. As shown in Fig. 3.2B, the gastric cancer cells could be clearly observed to rotate by adjusting the orientation of the two-dimensional plane of the magnetic field.

Vieira et al. [25] designed specific structures to achieve magnetic particle and cellular motion and rotation. As shown in Fig. 3.2C, this chip utilizes a highly localized permanent magnetic field gradient to assemble labeled cells or microspheres onto a designed array on a vertex pattern of ferromagnetic herringbone lines. By combining this platform with an externally controlled weak (~60 Oe) field, cells can then be transported across the surface by programmable directional forces that are gentle enough not to damage the cells. Dai et al. [26,28] endocytosed the ferric oxide magnetic nanoparticles by

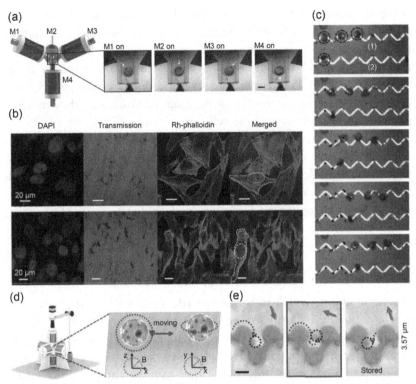

Fig. 3.2 Cell rotation with invasive magnetic particles. (A) Three-dimensional rotation of embryonic cells by injecting magnetic gel spheres [9]. Scale bar: 500 μm. (B)–(D) Cellular orientation by internalized magnetic particles [24–26]. (E) Rotational manipulation of cells by magnetophoretic forces [27]. Scale bar: 10 μm. *(Source: Panel (A) Reproduced and adapted with permission F. Berndt, et al., Dynamic and non-contact 3D sample rotation for microscopy, Nat. Commun. 9(1) (2018) 5025. Copyright 2018, Springer Nature. Panel (B) Reproduced and adapted with permission J. Liu, et al., Manipulation of cellular orientation and migration by internalized magnetic particles, Mat. Chem. Front. 1(5) (2017) 933–936. Copyright 2017, Royal Society of Chemistry. Panel (C) Reproduced and adapted with permission G. Vieira, et al., Magnetic wire traps and programmable manipulation of biological cells, Phys. Rev. Lett. 103(12) (2009) 128101. Copyright 2009, American Physical Society. Panel (D) Reproduced and adapted with permission Y. Dai, et al., Magnetically actuated cell-robot system: precise control, manipulation, and multimode conversion, Small 18(15) (2022) 2105414. Copyright 2022, John Wiley and Sons. Panel (E) Reproduced and adapted with permission S. R. Goudu, et al., Mattertronics for programmable manipulation and multiplex storage of pseudo-diamagnetic holes and label-free cells, Nat. Commun. 12(1) (2021) 3024. Copyright 2021, Springer Nature.)*

macrophages, and, as shown in Fig. 3.2D, the magnetic particles would be aggregated inside the cells after being endocytosed by the cells, at which time the cells as a whole came to have a specific magnetic field orientation. When a three-dimensional Helmholtz coil is applied with a uniform field in different directions, the cell will turn to the magnetic field direction generated by the magnetic control system due to the magnetic moment, thus realizing three-dimensional rotation of the cell.

Rotational manipulation of cells by magnetophoretic forces was achieved by Goudu et al. [27], who designed a special structure and a biocompatible magnetic solution that is essentially undamaged when the cells are in this line of solution. Through changes in the external uniform magnetic field, magnetophoretic forces are generated at the special structure, causing the cells to rotate (Fig. 3.2E). However, this method does not require the cells to be magnetic.

2.2 Cell rotation based on acoustic fields

Acoustic methods, which integrate acoustic wave with medium systems, provide a potent solution for noninvasive cell manipulation. Basically, through employing acoustic vibration in fluid platforms, the acoustofluidic systems utilize diverse physical effects of acoustic waves (such as wave potential wells, acoustic radiation forces, streaming patterns, and so on) to contactlessly and efficiently manipulate different target cells in real time [29,30]. One significant property of acoustic methods is their broad applicability in controlling multiple objects with a wide operating size and multiple shapes [31]. For cell rotation, while some acoustic systems utilize single or compound interdigital electrodes (IDT) to form vortex acoustic pressure fields based on surface acoustic waves (SAWs) [32–34]. Other acoustic methods are commonly coupled with microstructures to generate vortical streaming patterns actuated from bulk acoustic waves (BAWs), such as sharp edges [35–37], micropillars [38,39], trapped bubbles [40–44], and so on. Herein, the research endeavors of acoustically driven cell rotation are categorized into two parts in terms of the working mechanisms: SAW-based cell rotation and BAW-based cell rotation.

2.2.1 Cell rotation in SAW-based devices

Through converting electrical signals into surface vibration, SAW-based devices commonly integrate IDTs with a piezoelectric substrate [45]. Apart from patterning, screening, and other manipulations described in the existing research literature, SAW acoustofluidic devices could employ several pairs or more compound designs of IDTs to generate different SAW fields for object rotational operations.

Bernard et al. [32] described a kind of SAW-based acoustofluidic platform with two pairs of IDTs for controllable rotation and translation manipulations, as shown in Fig. 3.3A. Working at high driving frequencies of 36.3 MHz, two perpendicular standing waves were generated in a microfluidic channel (Fig. 3.3B). As demonstrated in Fig. 3.3C, when adjusting the phase lag between the two IDTs pairs into $\frac{\pi}{2}$, red blood cells (RBCs) could be trapped in the potential minimum nodes, and further rotate in the trapped site, which agrees well with the simulated acoustic potential distribution. Furthermore, the distribution feature of pressure nodes shows a potential probability for simultaneous control of translation and rotation of numerous cells.

By designing a more complex configuration of IDT pairs, Wang et al. [33] proposed a kind of tri-directional symmetrical acoustic tweezers, which enable programmable transportation and rotation of objects (Fig. 3.3D). The platform modulated multiple excitation combinations of IDTs to generate traveling surface acoustic waves (TSAWs) and reshape the acoustic pressure fields as shown. In this way, apart from linear motion control, the platform achieved cell rotation in both clockwise and anticlockwise manners, as demonstrated in the experimental and simulated results in Fig. 3.3E and F. By utilizing the modulation of driving electric signals and fluid dynamic viscosities, the SAW-based device could further accomplish programmable motion of cells, showing a promising solution for cell reorientation in various biomedical applications.

2.2.2 Cell rotation in BAW-based devices

Commonly, with different microstructures in the microfluidic platform, BAW-based acoustic methods can rotate targeted cells at lower driving conditions. Furthermore, considering the changeable features (like sizes, locations, numbers, and so on) of structures, the manipulation areas in BAW-based acoustofluidic devices could be customized and made suitable to be integrated with other microfluidic systems.

For acoustically driven solid microstructures, one typical design is sharp edge, which features high reliability and robustness in generating acoustic microstreaming and further manipulate cells efficiently and conveniently [46]. Ozcelik et al. described a sidewall sharp-edge acoustofluidic platform for biological rotation, as shown in Fig. 3.4A. Owing to the viscous dissipation in the microfluidic channel, symmetric vortical streaming was

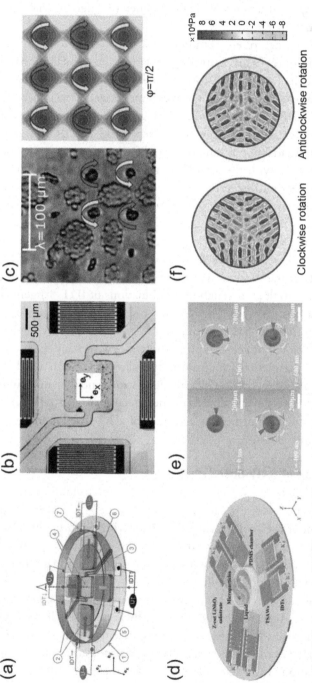

Fig. 3.3 Cell rotation in SAW-based platforms. (A)–(B) Schematic of the SAW-based manipulating platform with two orthogonal IDTs [32]. (C) Experiment results of parallel cell rotation and simulation of acoustic potentials in the platform [32]. (D) Schematic diagram of SAW-based tweezers with two sets of IDTs [33]. (E) Clockwise rotation of a single egg cell [33]. (F) Simulated acoustic pressure patterns of rotating manipulation in the platform [33]. *(Source: Panels (A–C) Reproduced and adapted with permission I. Bernard, et al., Controlled rotation and translation of spherical particles or living cells by surface acoustic waves, Lab Chip. 17(14) (2017) 2470–2480. Copyright 2017, Royal Society of Chemistry. Panels (D–F) Reproduced and adapted with permission Y. Wang, et al., Programmable motion control and trajectory manipulation of microparticles through tri-directional symmetrical acoustic tweezers, Lab Chip. 22(6) (2022) 1149–1161. Copyright 2022, Royal Society of Chemistry.)*

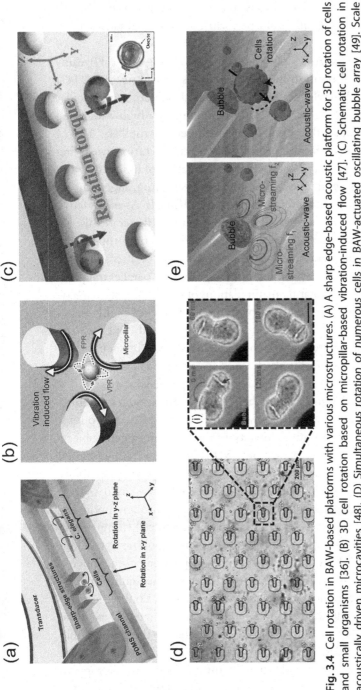

Fig. 3.4 Cell rotation in BAW-based platforms with various microstructures. (A) A sharp edge-based acoustic platform for 3D rotation of cells and small organisms [36]. (B) 3D cell rotation based on micropillar-based vibration-induced flow [47]. (C) Schematic cell rotation in acoustically driven microcavities [48]. (D) Simultaneous rotation of numerous cells in BAW-actuated oscillating bubble array [49]. Scale bar: 10μm. (E) Schematic of the bubble-based microgripper in the acoustic field [50]. *(Source: Panel (A) Reproduced and adapted with permission A. Ozcelik, et al., Acoustofluidic rotational manipulation of cells and organisms using oscillating solid structures. Small 12(37) (2016) 5120–5125. Copyright 2016, John Wiley and Sons. Panel (B) Reproduced and adapted with permission T. Hayakawa, et al., On-chip 3D rotation of oocyte based on a vibration-induced local whirling flow. Microsyst. Nanoeng. 1(1) (2015) 15001. Copyright 2015, Springer Nature. Panel (C) Reproduced and adapted with permission X. Bai, et al., Parallel trapping, patterning, separating and rotating of micro-objects with various sizes and shapes using acoustic microstreaming, Sens. Actuator A. Phys. 315 (2020) 112340. Copyright 2020, Elsevier. Panel (D) Reproduced and adapted with permission Q. Tang, et al., On-chip simultaneous rotation of large-scale cells by acoustically oscillating bubble array, Biomed. Microdevices 22(1) (2020) 13. Copyright 2020, Springer Nature. Panel (E) Reproduced and adapted with permission Y. Zhou, et al., Soft-contact acoustic microgripper based on a controllable gas–liquid interface for biomicromanipulations, Small 17(49) (2021) 2104579. Copyright 2021, John Wiley and Sons.)*

generated by the oscillating sharp edges under acoustic actuation. In this way, HeLa cells and *C. elegans* could perform continuous rotation along the streaming patterns.

Similarly, micropillars are widely studied for generating vortical streaming patterns for 3D cell rotation in fluid conditions. Hayakawa et al. [47] designed an on-chip 3D cell rotation platform with vibrating micropillar patterns as shown in Fig. 3.4B. The platform generated a highly localized whirling streaming around the three vibration-driven micropillars. Therefore, at the center of the micropillar group, the integrated whirling streaming will further induce rotational flow to achieve 3D rotation of mouse oocytes. Such a method was also employed on an open chip structure, suggesting its stability and allowing for convenient observation of 3D cells. As an attractive tool for cell culture in the microfluidic systems, microcavities could also be actuated acoustically for vortical streaming generation to trap, rotate, and even separate cells, which enriches their functionalities in biomedical areas. Song et al. [48] developed a bottom microcavity array for parallel trapping, patterning, separating, and rotating manipulations of micro-objects in a biocompatible manner (Fig. 3.4C). In this platform, two kinds of highly localized microvortexes were generated, both inside and outside the microcavities, under a low frequency acoustic field. Affected by such microvortices, cells and small organisms like *E. gracilis* suspended in the microfluid were then trapped and rotated in the microcavities parallelly.

Another common design of BAW-based device utilizes microbubbles. Compared with solid microstructures, oscillating bubbles show large output forces, effectiveness, and multifunctionality in manipulating micro-objects and achieving corresponding applications [51]. Tang et al. [49] proposed a cell manipulation strategy of horseshoe-structured bubble array, enabling simultaneous capture and rotation of large-scale cells, as shown in Fig. 3.4D. High homogeneity of bubble shapes was achieved through modifying the horseshoe structures, steadily rotating single cells, and cell clumps along consistent acoustic streaming patterns among microbubble arrays. Zhou et al. [50] described a soft-contact acoustic microgripper based on a microbubble for cell manipulations (Fig. 3.4E). After setting in the liquid environment in the acoustic field, the microcapillary trapping with a bubble at its tip oscillated to enhance adhesion of its interface and generate 3D acoustic microstreaming vortexes. In this way, cells suspended in the liquid could not only be captured, transported, and rotated around different axes under acoustic actuation, but be released when acoustic excitation was stopped. Such designs equipped with a pressure controller control the bubble volume and manipulation stability well. The movable microgripper also achieved controllable and precise cell transportation during the rotating manipulation, providing a convenient solution to orientate cells for biological and medical applications.

2.3 Cell rotation based on AC E-field in optoelectronic tweezers

Optical tweezers can handle anything from simple micron cell manipulation to single-molecule and submicrometer precision measurement [52]. However, the heat damage induced by the focused laser beam, which causes light pressure [53], cannot be ignored. Chiu et al. [54], from UC Berkeley, first introduced optoelectronic tweezers in 2005. In an optoelectronic tweezers (OET) system, light pattern filmed on a photoconductive layer leads to local electric-field gradients, and dielectrophoresis (DEP) forces enable control and movement of particles. Then OET system can manipulate different targets, including cells [55], protists [56], microrobots [57], metal beads [58], and carbon nanoparticles [59]. Liang et al. build a cross-platform involving a graphical user interface (G)UI) and cloud server to help simplify the operation process [60]. The effect of cell rotation based on DEP forces can be divided into two categories: (1) the cells rotating within a rotational AC electric field; and (2) certain types of cells rotating within an irrotational AC electric field.

2.3.1 Cell rotation in a rotational AC electric field

In 1896, Quincke observed the rotation of small solid spherical particles in dielectric liquids when exposed to an electrostatic field. Cell rotation in rotational E-field was also noticed by Arnold and Zimmermann in 1982 [61]. A theory to explain the cell rotation in a rotational AC electric field has been presented by researchers [61,62].

The cell experiences a torque Γ_{DEP} when exposed in a rotational AC E-field:

$$\Gamma_{DEP} = -4\pi R^3 \varepsilon_m Im[K_{cm}]E^2 \qquad (3.1)$$

Here K_{CM} is Clausius-Mossotti (CM) coefficient [63]. The torque is related to cell size, cell dielectric parameters, and solution dielectric parameters (ε_m). The basic structure consists of four electrodes, which are placed in a crisscross pattern, and four signals with a 90 phase difference from each other are applied to the four electrodes.

To trap the cells, Han et al. [64] constructed a three-dimensional (3D) electrode, which was placed on the top and bottom of the microchannel, as shown in Fig. 3.5A. The device measured electrorotation (ROT) spectra of human leukocyte subpopulations (T and B lymphocytes, granulocytes, and monocytes) and metastatic human breast (SkBr3) and lung (A549) cancer cell lines and avoided disturbance by lifting the cells. Lannin et al. [69] inferred the electrical properties of pancreatic cancer cells and blood cells by their observed rotation rates at each field frequency. Elkeles et al. [65] used ROT to characterize the properties of "dynamic" microcapsules that are sensitive to PH conditions (Fig. 3.5B). Huang et al. [66] fabricated a low-cost ITO microelectrode using a laser etching technique to characterize yeast cells, as shown in Fig. 3.5C. Wang et al. [67] focused on 3D cell rotation and designed a platform with an open-top sub-mm square chamber enclosed by four sidewall electrodes and two bottom electrodes (see Fig. 3.5D). They later added a PDMS-based capture part onto the electrodes, leading to a steady cell trapping [70]. They continued developing the device structure, which used carbon-black-nanoparticle-PDMS (C-PDMS) to build four vertical electrodes instead of brass, as shown in Fig. 3.5E [68]. They recorded stacked images of cells, reconstructed the 3D cell morphology and revealed the subtle difference of cell geometric parameters.

Besides pure rotation, electrorotation is an efficient tool to measure the dielectric properties of biological cells [71]. In decades, various cellular dielectric properties have been recorded to facilitate cell identification and improve cell manipulation effect based on DEP forces.

2.3.2 Cell rotation in an irrotational AC electric field
The local electric field gradients and electric field caused by light pattern in an OET system is relatively static and is irrotational.

Cell rotation within an irrotational AC electric field is an uncommon phenomenon. Liang et al. [72] demonstrated that yeast cells experience translation and rotation when reacting to the light pattern (Fig. 3.6A). Through simulation, they explained that the edge between the light part and dark part induces a rotating E-field. The rotating E-fields around the boundaries of virtual electrodes caused the illuminated part to make yeast cells rotate. Lee et al. [73] reported that Melan-a cells, lymphocytes, and white blood cells start to rotate without translation in an OET system. The applied voltage, frequency, and the distance from light pattern define the rotational speed. They hypothesized that self-rotation of specific cells is the result of a torque caused by uneven mass distribution and DEP force. Pigment cells have an obvious self-rotation phenomenon, so they used an OET system to measure the melanin in pigment

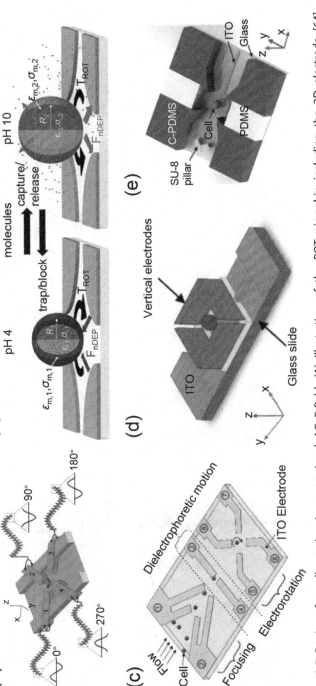

Fig. 3.5 Devices for cell rotation in a rotational AC E-field. (A) Illustration of the ROT-microchip including the 3D electrode [64]. (B) Representative turning directions of microcapsules using a crisscross electrode array at 1 MHz in low- and high-pH solutions [65]. (C) ITO etched low-cost crisscross electrode [66]. (D) A 3D cell rotation platform including four sidewall electrodes and two bottom electrodes [67]. (E) Carbon-black-nanoparticle-PDMS-based electrodes for cell rotation [68]. (*Source: Panel (A) Reproduced and adapted with permission S. Han, et al., An electrorotation technique for measuring the dielectric properties of cells with simultaneous use of negative quadrupolar dielectrophoresis and electrorotation, Analyst 138(5) (2013) 1529–1537. Copyright 2013, Royal Society of Chemistry. Panel (B) Reproduced and adapted with permission T. Elkeles, et al., Dielectrophoretic characterization of dynamic microcapsules and their magnetophoretic manipulation. ACS Appl. Energ. Mater. 14(13) (2022) 15765–15773. Copyright 2022, American Chemical Society. Panel (C) Reproduced and adapted with permission L. Huang, et al., Electrical properties characterization of single yeast cells by dielectrophoretic motion and electro-rotation, Biomed. Microdevices. 23(1) (2021) 11. Copyright 2021, Springer Nature. Panel (D) Reproduced and adapted with permission P. Benhal, et al., AC electric field induced dipole-based on-chip 3D cell rotation, Lab Chip 14(15) (2014) 2717–2727. Copyright 2014, Royal Society of Chemistry. Panel (E) Reproduced and adapted with permission L. Huang, et al., 3D cell electrorotation and imaging for measuring multiple cellular biophysical properties, Lab Chip 8(16) (2018) 2359–2368. Copyright 2018, Royal Society of Chemistry.*)

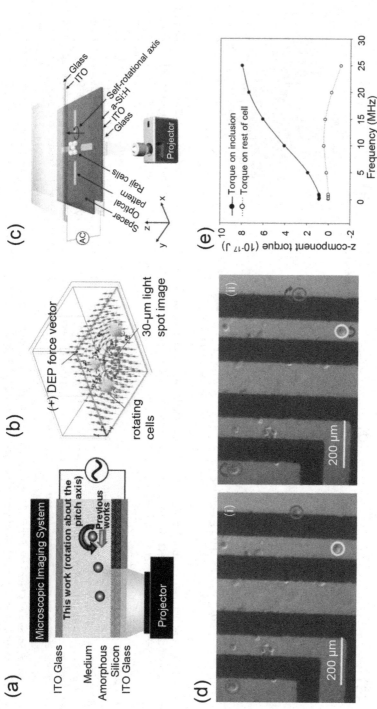

Fig. 3.6 Devices for cell rotation in an irrotational AC E-field. (A) The structure of a typical OET device and rotation of yeast cells [72]. (B) Time lapse for the Melan-a cells rotation with the applied frequency 100 kHz at 20 Vpp [73]. (C) Illustration of the OET chip and the rotation of Raji cells in four orthogonal light patterns [74]. (D) Rat adipose stem cells rotation with the applied frequency of 300 kHz at 200 Vpp [75]. (E) The Z-component torque on rotated cell inclusion and the rest of cell varying with AC frequencies [75]. (Source: Panel (A) Reproduced and adapted with permission Y. Liang, et al., Cell rotation using optoelectronic tweezers, Biomicrofluidics 4(4) (2010) 043003. Copyright 2010, AIP Publishing. Panel (B) Reproduced and adapted with permission L.-H. Chau, et al., Self-Rotation of cells in an irrotational AC E-field in an opto-electrokinetics chip. PLoS One 8(1) (2013) e51577. Copyright 2013, PLOS. Panel (C) Reproduced and adapted with permission W. Liang, et al., Distinctive translational and self-rotational motion of lymphoma cells in an optically induced non-rotational alternating current electric field, Biomicrofluidics 9(1) (2015) 014121. Copyright 2015, AIP Publishing. Panels (D–E) Reproduced and adapted with permission Y. Zhao, et al., Elucidating the mechanism governing cell rotation under DEP using the volumetric polarization and integration method, Biomed. Microdevices 20(3) (2018) 81. Copyright 2018, Springer Nature.)

cells, and sort the pigment and nonpigment cells, as shown in Fig. 3.6B [76]. Liu et al. [74] later filmed a light pattern of four orthogonal electrodes to manipulate Raji cells (see Fig. 3.6C). The Raji cells perform self-rotation when they deviate from the center part where the E-field is uneven.

Combining simulation with Turcu's theory [77], nonparallel polarization between specific cells and the electric field induces an initial electrical torque, which contributes to growing oscillations, resulting in the dynamic rotation. The self-rotational behaviors of red blood cells (RBCs) are noticed in OET chip, and the rotation speed is a biomarker to measure the qualities of RBCs. Liu et al. [78] figured out RBCs that are stored for a long time (3 weeks) are not suitable for the positive blood transfusion application by measuring the rotational speed. The self-rotational speed measurement is also applied in detecting different cancer cells and drug concentrations [79]. Zhang et al. [75] used a volumetric polarization and integration (VPI) method to research cell rotation, as demonstrated in Fig. 3.6D. They pointed out that eccentric inclusions with low conductivity inside the cell (Fig. 3.6E) are able to generate sufficient torque to initiate cell rotation and keep the rotation going continuously.

3 Robotic enucleation of cells

3.1 Introduction of robotic enucleation

Studies on cell surgery have been developed rapidly over decades in order to satisfy the great demand of assisted reproductive technology (ART), cytobiology, and novel drug delivery. Focusing on cell surgery demonstrates that the scale of the subjects has been shrunk from the organ level to the cellular level, and even to the subcellular level. Enucleation of cells is a kind of micromanipulation of cell surgery, together with cell injecting and cell cutting. It can be divided into two categories: enucleation of reproductive cells, that is, oocytes, and nonreproductive cells for different purposes, respectively. Generally, complete and purified nuclei are isolated from nonreproductive cells, such as stable cell lines or tissue samples (liver and spleen), by rupturing their cell membranes with chemical treatment and then collecting nuclei [80]. This conventional form of nuclear extraction is suitable for a number of cell biology applications, such as providing nuclear sources for protein, chromatin, genomic DNA, preparing nuclei for apoptosis detection, cellular transcriptional evaluation, and generating functional hybrid cells in vitro [81]. The protocol that utilizes this method isolates very abundant yields and without protease or ribozyme.

However, when it comes to the enucleation of oocytes, which is particularly fundamental and essential in the cloning process, this method is not appropriate due to the strict quality requirements for cloning. The outer membrane of an oocyte has higher elasticity and, with the existence of zona pellucida (ZP), this increases the difficulty of enucleation of oocytes [82,83]. It is necessary that both donor nuclei and cytoplast recipients should retain their intrinsic structure and high viability and suffer no exposure to contamination. What's more, the dexterity and proficiency of the operator is also a critical factor, because the growth of the embryo is greatly influenced even by minimal damage in the initial phase. This has raised the demand for shorter operation time, nondestructive positioning, and greater cutting precision.

There are some other techniques of enucleation, including manipulator operation (single or batch) and femtosecond laser pulses techniques [84]. Huang et al. presented an automatic cell-injection system, particularly for batch of suspended cells, which utilizes cells arranged and held in circular arrays for downstream insertion. A polyvinylidene fluoride microforce sensor was equipped on the micropipette to determine the real-time injection force [85]. Hogg et al. reported an automated system capable of injecting up to 600 *Xenopus* oocytes per hour [86]. Kai et al. demonstrated a noninvasive automated enucleation of porcine oocytes in combination with imaging by using femtosecond laser pulses [87]. These methods show defects in maintaining cell vitality and assured success rate, and low repeatability. These operation platforms consist of sophisticated and expensive equipment that is usually bulky, making it inconvenient and uneconomical for lab studies; furthermore, laser irradiation might cause DNA damage and delay of meiosis anaphase at the irradiation site.

To overcome all the barriers above, robots on a microfluidic chip have attracted many studies due to much lower cost, and high repeatability and throughput. Conventional methods of cell enucleation are highly dependent on manual manipulations under microscopes by experienced operators. The low repeatability and the high risk of contamination make these methods inefficient and unreliable [88]. Microfluidic chip with microactuators, also known as on-chip microrobots, have been demonstrated to achieve complicated cell manipulations with high quality and repeatability. These automated micromanipulators, normally driven by fluidic force, magnetic field, acoustic field, and electrostatic force, are of great potential in reducing the risk of contamination and increasing the success rate in the field of cell enucleation.

3.2 Microrobots for enucleation

Magnetically driven microrobot is one of the most popular types of micro-robot that can be applied to the confined space in a microfluidic chip (micro-tube/microchannel). Arai et al. [89] developed an on-chip microrobot system with the ability of achieving on-chip enucleation surgery of oocyte. The on-chip microrobot system consisted of a microfluidic chip and two magnetically driven microtool (MMT). The 2DOF-MMTs system with a 5 μm positioning precision can cut the oocyte in half with a 2mN applied force. This low-cost system significantly reduced the size of the mechanical micromanipulator compared to conventional tools. However, the incision of the cell was not smooth and the cutting volume was not controlled, making it impossible for the viability of the enucleated oocyte to be guaranteed.

To maintain high precision of the removal part and guarantee high viability of the enucleated oocyte, Ichikawa et al. proposed an on-chip micro-robot enucleation system using a combination of untethered microrobots [90]. The microrobot system consisted of a microknife and a microgripper, both driven by external magnetic force generated by permanent magnets. The positional accuracy was less than 5 μm and the production rate of the enucleated oocyte was 100%, suggesting the method to be minimally invasive. However, the manufacturing of the microrobots could be complex.

Based on the advantages of fluidic force enucleation over traditional blade cutting shown by the results of Ichikawa et al. [91], Feng et al. developed a fluidic driven on-chip microrobot enucleation system [84,92]. By adjusting the interface between the MMT and the microchannel, the system can adjust the speed of the flow. The oocyte was delivered by the flow and was stopped at the interface. The orientation of the cell can be adjusted by the position of MMT. When the nucleus was sucked into the opposite side of the interface, the MMT would be actuated to close the interface, thus, the portion with nucleus would be separated from the enucleated part. In this method, the cut-off volume of oocyte could be controlled minimally and the average removal proportion is reduced to 20%. The smooth incision of the enucleation part guaranteed the viability of enucleated oocyte with the maximum possibility. The average enucleation time is 2.5 s for one oocyte, demonstrating that the microrobot system could accomplish enucleation at a relatively higher speed. By changing the angle of the corner, the enucleation time could also be adjusted. However, the orientation control of the cell was not precise and fast enough. The operation time of properly orienting the position of the cell can be several seconds to minutes. The development of a high-speed, precise cell orientation control method will assist the microfluidic on–chip robot system in achieving repeatable, automatic, precise cell micromanipulation (Fig. 3.7).

In order to implement the 3D rotation orientation control of single cells and reduce the enucleation proportion, Feng et al. proposed a

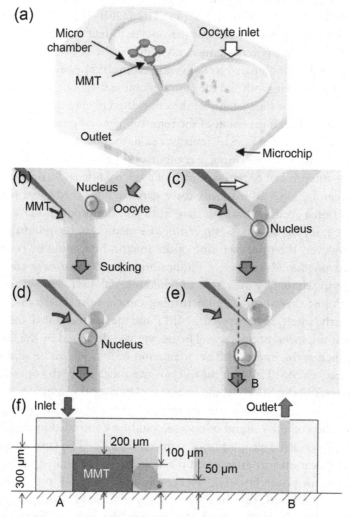

Fig. 3.7 Scheme of enucleation of bovine oocyte by using an acoustically levitated microrobot in the microchannel [92]. (A) Overview of the enucleation microchip. (B–E) Concept of the oocyte enucleation process by the use of magnetically driven MMTs in a microfluidic chip. *Blue arrows (gray* in print version) show the flow direction. (F) Height differences in the microchannel design. *White arrows* show the movement of the MMT. *(Source: Panels (A–F) Reproduced and adapted with permission L. Feng, et al., On-chip enucleation of bovine oocytes using microrobot-assisted flow-speed control. Micromachines 4(2) (2013) 272–285. Copyright 2013, MDPI.)*

high-precision motion control with ultrasonic levitation [93]. The ultrasonic levitation generated a noncontact suspension, which increased the speed of the rotation and reduced the affection of contact friction. With a visualized tracking algorithm and a comparatively high output force of 6.5 to 60mN, the robot system could capture single cells to achieve vertical and horizontal 3D rotation as well as in-plane translational motions (at a speed of 5 mm/s). The system achieved high precision single cell orientation control with an accuracy of 1°. Compared to conventional systems, the rotational speed (3 rad/s) was faster, making it possible to fulfill single-cell control within seconds. With the help of ultrasonic levitation, the system is relatively more robust than conventional methods. Combining this system with an on-chip microrobot enucleation system, automated high-speed cell enucleation with high repeatability will be possible (Fig. 3.8).

Furthermore, aiming at the orientation control problem of a single cell, Shin et al. presented a microfluidic system that was able to realize the orientation control of mouse embryos by employing hydrodynamic forces [94]. A special microchannel with cell trapping structures and micropipettes was designed in the experiment. With the orientation detection and rolling behavior of the cell being carried out through visual tracking, the reorientation of embryos could be realized within a significantly shorter time and with

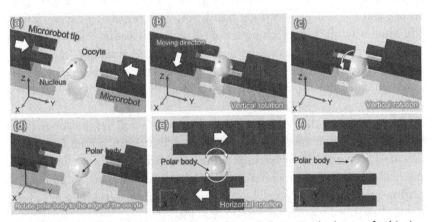

Fig. 3.8 Schematic of 3D oocyte rotation [93]. (A–C) Conceptual scheme of achieving oocyte vertical rotation control with the microrobot system. (D–F) Conceptual scheme of achieving oocyte horizontal rotation control with the microrobot system. *(Source: Panels (A–F) Reproduced and adapted with permission L. Feng, et al., High-precision motion of magnetic microrobot with ultrasonic levitation for 3-D rotation of single oocyte, Int. J. Robot. Res. 35(12) (2016) 1445–1458. Copyright 2016, SAGE Publications.)*

adequate accuracy. The system provides an orientation control method with a contact-free manipulation approach.

Since noncontact method can be less invasive to the cell, Park et al. presented an on-chip enucleation method incorporated with an acoustically oscillating bubble [95]. The oscillating bubble can induce cavitational microstreaming. When the bubble approached the embryo, the nucleus of the embryo was extracted into the bubble due to the effect of cavitational microstreaming. Although the generation and control of the acoustic field could be complicated, this noninvasive technique is of significant potential in the field of cell enucleation.

3.3 Future directions of robotic cell enucleation

The enucleation of cells has a wide range of applications such as ART, cloning, membrane physiology, apoptosis, signaling, metabolism, and proteomics. Thanks to the development of microfluidics techniques, various untethered actuation forms can be used to achieve single cell localization in a confined environment. Although many emerging techniques have been reported, we still need to improve them to meet the demands for high flexibility and mobility in the future. Challenges still exist for further research: (i) electromagnetic coils should be equipped to replace the permanent magnets for multimode magnetic field during the absorption and ejection operations; and (ii) bench-to-bedside translation with respect to enucleation is still some way off. Most techniques of robotic enucleation end with nuclei extraction rather than successful animal cloning. Researchers should pay more attention to the fusion process of the donor nuclei and cytoplast recipient, which is the most convincing method to investigate the effect of enucleation.

In conclusion, the pursuit of diversified single-cell surgical operations will focus on high repeatability and throughput with minimal invasion and ideal precision. The ever-improving enucleation platform will broaden the field of medical microrobots and could be adapted to other cell manipulation applications, eventually greatly promoting the development of biomedical engineering and precision medicine.

4 Conclusion and perspectives

Thanks to the development of microfabrication and robotic technologies, many efficient technical systems have been developed in recent years for potent robotic orientation and enucleation of cells, providing promising solutions to facilitate developments in biomedical research and therapeutic applications. In this chapter, we overviewed multiple state-of-the-art robotic controlling systems for cell rotation based on different external fields, including magnetic, acoustic, and optoelectronic fields. Moreover, we summarized the importance of robotic enucleation of cells in comparison with conventional manual methods from a biological standpoint, followed by the introduction of several typical robotic enucleation systems developed so far.

Although many advances have been made in recent years regarding related supporting techniques, we should realize that there is still a gap to replace traditional manual methods practically and commercially, requiring more effort in the future. For example, parallel precise control of numerous cells could be a valuable trend for practical biological and biomedical applications. Next, before being put into clinical practice, biocompatibility and safety of external field control in long-term cellular activity should be further validated. Last but not least, reliable automation of robotic cell orientation and enucleation should be another future direction, integrating with various potential supporting techniques, such as artificial neural network, fuzzy logic, and so on.

References

[1] B. Alberts, et al., Essential Cell Biology, Garland Science Pub, New York, 2004.
[2] N. Kashaninejad, M.J.A. Shiddiky, N.-T. Nguyen, Advances in microfluidics-based assisted reproductive technology: from sperm sorter to reproductive system-on-a-Chip, Adv. Biosys. 2 (3) (2018) 1700197.
[3] L. Weng, IVF-on-a-Chip: recent advances in microfluidics technology for in vitro fertilization, SLAS Technol. 24 (4) (2019) 373–385.
[4] L. Feng, et al., Cell injection millirobot development and evaluation in microfluidic chip, Micromachines 9 (11) (2018) 590.
[5] M.Y. Xie, et al., Out-of-plane rotation control of biological cells with a robot-tweezers manipulation system for orientation-based cell surgery, IEEE Trans. Biomed. Eng. 66 (1) (2019) 199–207.
[6] Y. Liu, et al., Robotic batch somatic cell nuclear transfer based on microfluidic groove, IEEE Trans. Autom. Sci. Eng. 17 (4) (2020) 2097–2106.
[7] H. Gong, et al., Automatic cell rotation based on real-time detection and tracking, IEEE Robot. Autom. Lett. 6 (4) (2021) 7909–7916.
[8] R. Elbez, et al., Nanoparticle induced cell magneto-rotation: monitoring morphology, stress and drug sensitivity of a suspended single cancer cell, PLoS One 6 (12) (2011), e28475.
[9] F. Berndt, et al., Dynamic and non-contact 3D sample rotation for microscopy, Nat. Commun. 9 (1) (2018) 5025.
[10] G. Carmon, M. Feingold, Rotation of single bacterial cells relative to the optical axis using optical tweezers, Opt. Lett. 36 (1) (2011) 40–42.
[11] X. Chen, et al., Observation of spin and orbital rotation of red blood cell in dual-beam fibre-optic trap with transverse offset, J. Opt. 19 (5) (2017), 055612.

[12] Y. Yalikun, et al., Hydrodynamic vertical rotation method for a single cell in an open space, Microfluid. Nanofluid. 20 (5) (2016) 1–10.

[13] J.P. Shelby, D.T. Chiu, Controlled rotation of biological micro-and nano-particles in microvortices, Lab Chip 4 (3) (2004) 168–170.

[14] A. Ozcelik, et al., Acoustic tweezers for the life sciences, Nat. Methods 15 (12) (2018) 1021–1028.

[15] O. Fuchiwaki, et al., Multi-axial non-contact in situ micromanipulation by steady streaming around two oscillating cylinders on holonomic miniature robots, Microfluid. Nanofluid. 22 (8) (2018) 80.

[16] A. Rohani, et al., Electrical tweezer for highly parallelized electrorotation measurements over a wide frequency bandwidth, Electrophoresis 35 (12–13) (2014) 1795–1802.

[17] Y.T. Chow, et al., Liquid metal-based multifunctional micropipette for 4D single cell manipulation, Adv. Sci. 5 (7) (2018) 1700711.

[18] K. Lee, Y. Yi, Y. Yu, Remote control of T cell activation using magnetic Janus particles, Angew. Chem. Int. Ed. 55 (26) (2016) 7384–7387.

[19] T. Petit, et al., Selective trapping and manipulation of microscale objects using mobile microvortices, Nano Lett. 12 (1) (2012) 156–160.

[20] L. Zhang, et al., Targeted cargo delivery using a rotating nickel nanowire, Nanomedicine 8 (7) (2012) 1074–1080.

[21] D. Choi, et al., Transport of living cells with magnetically assembled nanowires, Biomed. Microdevices 9 (2) (2007) 143–148.

[22] Q. Zhou, et al., Dumbbell fluidic tweezers for dynamical trapping and selective transport of microobjects, Adv. Funct. Mater. 27 (1) (2017) 1604571.

[23] E.B. Steager, et al., Automated biomanipulation of single cells using magnetic microrobots, Int. J. Robot. Res. 32 (3) (2013) 346–359.

[24] J. Liu, et al., Manipulation of cellular orientation and migration by internalized magnetic particles, Mat. Chem. Front. 1 (5) (2017) 933–936.

[25] G. Vieira, et al., Magnetic wire traps and programmable manipulation of biological cells, Phys. Rev. Lett. 103 (12) (2009), 128101.

[26] Y. Dai, et al., Magnetically actuated cell-robot system: precise control, manipulation, and multimode conversion, Small 18 (15) (2022) 2105414.

[27] S.R. Goudu, et al., Mattertronics for programmable manipulation and multiplex storage of pseudo-diamagnetic holes and label-free cells, Nat. Commun. 12 (1) (2021) 3024.

[28] Y. Dai, et al., Precise control of customized macrophage cell robot for targeted therapy of solid tumors with minimal invasion, Small 17 (41) (2021) 2103986.

[29] S. Mohanty, I.S.M. Khalil, S. Misra, Contactless acoustic micro/nano manipulation: a paradigm for next generation applications in life sciences, Proc. Math. Phys. Eng. Sci. 476 (2243) (2020) 20200621.

[30] F. Akkoyun, S. Gucluer, A. Ozcelik, Potential of the acoustic micromanipulation technologies for biomedical research, Biomicrofluidics 15 (6) (2021), 061301.

[31] K. Kolesnik, et al., Unconventional acoustic approaches for localized and designed micromanipulation, Lab Chip 21 (15) (2021) 2837–2856.

[32] I. Bernard, et al., Controlled rotation and translation of spherical particles or living cells by surface acoustic waves, Lab Chip 17 (14) (2017) 2470–2480.

[33] Y. Wang, et al., Programmable motion control and trajectory manipulation of microparticles through tri-directional symmetrical acoustic tweezers, Lab Chip 22 (6) (2022) 1149–1161.

[34] J. Zhang, et al., Surface acoustic waves enable rotational manipulation of *Caenorhabditis elegans*, Lab Chip 19 (6) (2019) 984–992.

[35] L. Feng, et al., On-chip tunable cell rotation using acoustically oscillating asymmetrical microstructures, Micromachines 9 (11) (2018) 596.

[36] A. Ozcelik, et al., Acoustofluidic rotational manipulation of cells and organisms using oscillating solid structures, Small 12 (37) (2016) 5120–5125.

[37] L. Feng, et al., On-chip rotational manipulation of microbeads and oocytes using acoustic microstreaming generated by oscillating asymmetrical microstructures, Biomicrofluidics 13 (6) (2019) 064103.

[38] Z. Ma, et al., Ultrasonic microstreaming for complex-trajectory transport and rotation of single particles and cells, Lab Chip 20 (16) (2020) 2947–2953.

[39] K. Kaneko, et al., Numerical and experimental analyses of three- dimensional unsteady flow around a micro-pillar subjected to rotational vibration, Micromachines 9 (12) (2018) 668.

[40] N.F. Läubli, et al., 3D mechanical characterization of single cells and small organisms using acoustic manipulation and force microscopy, Nat. Commun. 12 (1) (2021) 2583.

[41] N.F. Läubli, et al., Embedded microbubbles for acoustic manipulation of single cells and microfluidic applications, Anal. Chem. 93 (28) (2021) 9760–9770.

[42] W. Zhang, et al., Versatile acoustic manipulation of micro-objects using mode-switchable oscillating bubbles: transportation, trapping, rotation, and revolution, Lab Chip 21 (24) (2021) 4760–4771.

[43] D. Ahmed, et al., Rotational manipulation of single cells and organisms using acoustic waves, Nat. Commun. 7 (1) (2016) 11085.

[44] L. Meng, et al., Sonoporation of cells by a parallel stable cavitation microbubble array, Adv. Sci. 6 (17) (2019) 1900557.

[45] J. Shi, et al., Acoustic tweezers: patterning cells and microparticles using standing surface acoustic waves (SSAW), Lab Chip 9 (20) (2009) 2890–2895.

[46] Z. Chen, et al., Sharp-edge acoustic microfluidics: principles, structures, and applications, Appl. Mater. Today 25 (2021), 101239.

[47] T. Hayakawa, S. Sakuma, F. Arai, On-chip 3D rotation of oocyte based on a vibration-induced local whirling flow, Microsyst. Nanoeng. 1 (1) (2015) 15001.

[48] X. Bai, et al., Parallel trapping, patterning, separating and rotating of micro-objects with various sizes and shapes using acoustic microstreaming, Sens. Actuator A Phys. 315 (2020), 112340.

[49] Q. Tang, et al., On-chip simultaneous rotation of large-scale cells by acoustically oscillating bubble array, Biomed. Microdevices 22 (1) (2020) 13.

[50] Y. Zhou, et al., Soft-contact acoustic microgripper based on a controllable gas–liquid interface for biomicromanipulations, Small 17 (49) (2021) 2104579.

[51] Y. Li, et al., Bubbles in microfluidics: an all-purpose tool for micromanipulation, Lab Chip 21 (6) (2021) 1016–1035.

[52] Y. Chen, et al., Recent advances in field-controlled micro–nano manipulations and micro–nano robots, Adv. Intell. Syst. 4 (3) (2022) 2100116.

[53] M.L. Juan, M. Righini, R. Quidant, Plasmon nano-optical tweezers, Nat. Photon. 5 (6) (2011) 349–356.

[54] P.Y. Chiou, A.T. Ohta, M.C. Wu, Massively parallel manipulation of single cells and microparticles using optical images, Nature 436 (7049) (2005) 370–372.

[55] W. Hu, Q. Fan, A.T. Ohta, An opto-thermocapillary cell micromanipulator, Lab Chip 13 (12) (2013) 2285–2291.

[56] S. Liang, et al., Interaction between positive and negative dielectric microparticles/ microorganism in optoelectronic tweezers, Lab Chip 21 (22) (2021) 4379–4389.

[57] S. Zhang, et al., The optoelectronic microrobot: a versatile toolbox for micromanipulation, Proc. Natl. Acad. Sci. U. S. A. 116 (30) (2019) 14823–14828.

[58] S. Zhang, et al., Manipulating and assembling metallic beads with optoelectronic tweezers, Sci. Rep. 6 (1) (2016) 32840.

[59] S.E. Wang, et al., Deposition of carbon nanoparticles using optically induced dielectrophoretic force, in: Proceedings—2011 16th International Solid-State Sensors, Actuators and Microsystems Conference, 2011.

[60] S. Liang, et al., A versatile optoelectronic tweezer system for Micro-objects manipulation: transportation, patterning, sorting, rotating and storage, Micromachines 12 (3) (2021) 271.

[61] W.M. Arnold, U. Zimmermann, Rotating-field-induced rotation and measurement of the membrane capacitance of single mesophyll cells of *Avena sativa*, Z. Naturforsch. C 37 (10) (1982) 908–915.

[62] R.V.E. Lovelace, D.G. Stout, P.L. Steponkus, Protoplast rotation in a rotating electric field: the influence of cold acclimation, J. Membrane Biol. 82 (2) (1984) 157–166.

[63] R. Pethig, Review article—dielectrophoresis: status of the theory, technology, and applications, Biomicrofluidics 4 (2) (2010), 022811.

[64] S.-I. Han, Y.-D. Joo, K.-H. Han, An electrorotation technique for measuring the dielectric properties of cells with simultaneous use of negative quadrupolar dielectrophoresis and electrorotation, Analyst 138 (5) (2013) 1529–1537.

[65] T. Elkeles, et al., Dielectrophoretic characterization of dynamic microcapsules and their magnetophoretic manipulation, ACS Appl. Energ. Mater. 14 (13) (2022) 15765–15773.

[66] L. Huang, Q. Fang, Electrical properties characterization of single yeast cells by dielectrophoretic motion and electro-rotation, Biomed. Microdevices 23 (1) (2021) 11.

[67] P. Benhal, et al., AC electric field induced dipole-based on-chip 3D cell rotation, Lab Chip 14 (15) (2014) 2717–2727.

[68] L. Huang, P. Zhao, W. Wang, 3D cell electrorotation and imaging for measuring multiple cellular biophysical properties, Lab Chip 18 (16) (2018) 2359–2368.

[69] T. Lannin, et al., Automated electrorotation shows electrokinetic separation of pancreatic cancer cells is robust to acquired chemotherapy resistance, serum starvation, and EMT, Biomicrofluidics 10 (6) (2016), 064109.

[70] L. Huang, et al., Towards on-chip single cell manipulation of trap and rotation, in: Proceedings—2016 International Conference on Manipulation, Automation and Robotics at Small Scales (MARSS), 2016.

[71] J. Gimsa, et al., Dielectrophoresis and electrorotation of Neurospora slime and murine myeloma cells, Biophys. J. 60 (4) (1991) 749–760.

[72] Y.-L. Liang, et al., Cell rotation using optoelectronic tweezers, Biomicrofluidics 4 (4) (2010), 043003.

[73] L.-H. Chau, et al., Self-rotation of cells in an irrotational AC E-field in an optoelectrokinetics chip, PLoS One 8 (1) (2013), e51577.

[74] W. Liang, et al., Distinctive translational and self-rotational motion of lymphoma cells in an optically induced non-rotational alternating current electric field, Biomicrofluidics 9 (1) (2015), 014121.

[75] Y. Zhao, et al., Elucidating the mechanism governing cell rotation under DEP using the volumetric polarization and integration method, Biomed. Microdevices 20 (3) (2018) 81.

[76] L.H. Chau, et al., Inducing self-rotation of Melan-a cells by ODEP, in: Proceedings—2012 7th IEEE International Conference on Nano/Micro Engineered and Molecular Systems (NEMS), 2012.

[77] I. Turcu, Electric field induced rotation of spheres, J. Phys. A Math. Gen. 20 (11) (1987) 3301–3307.

[78] W. Liang, et al., Characterization of the self-rotational motion of stored red blood cells by using optically-induced electrokinetics, Opt. Lett. 41 (12) (2016) 2763–2766.

[79] W. Liang, et al., Label-free characterization of different kinds of cells using optoelectrokinetic-based microfluidics, Opt. Lett. 45 (8) (2020) 2454–2457.

[80] X. Li, et al., Nucleus-translocated ACSS2 promotes gene transcription for lysosomal biogenesis and autophagy, Mol. Cell 66 (5) (2017) 684–697.e9.

[81] S. Wang, et al., Macrophage-tumor chimeric exosomes accumulate in lymph node and tumor to activate the immune response and the tumor microenvironment, Sci. Transl. Med. 13 (615) (2021), eabb6981.

[82] W. Johnson, et al., A flexure-guided piezo drill for penetrating the zona pellucida of mammalian oocytes, IEEE T. Biomed. Eng. 65 (3) (2018) 678–686.

[83] C. Dai, et al., Design and control of a piezo drill for robotic piezo-driven cell penetration, IEEE Robot. Autom. Lett. 5 (2) (2020) 339–345.

[84] L. Feng, et al., Smooth enucleation of bovine oocyte by microrobot with local flow speed control in microchannel, in: Proceedings—2012 IEEE/RSJ International Conference on Intelligent Robots and Systems, 2012.

[85] H.B. Huang, et al., Robotic cell injection system with position and force control: toward automatic batch biomanipulation, IEEE Trans. Robot. 25 (3) (2009) 727–737.

[86] R.C. Hogg, et al., An automated system for intracellular and intranuclear injection, J. Neurosci. Methods 169 (1) (2008) 65–75.

[87] K. Kai, et al., Combined multiphoton imaging and automated functional enucleation of porcine oocytes using femtosecond laser pulses, J. Biomed. Opt. 15 (4) (2010) 1–8.

[88] B. Ahmad, et al., Mobile microrobots for in vitro biomedical applications: a survey, IEEE Trans. Robot. 38 (1) (2022) 646–663.

[89] F. Arai, et al., Omni-directional actuation of magnetically driven microtool for enucleation of oocyte, in: Proceedings—2010 IEEE 23rd International Conference on Micro Electro Mechanical Systems (MEMS), 2010.

[90] A. Ichikawa, et al., On-chip enucleation of an oocyte by untethered microrobots, J. Micromech. Microeng. 24 (9) (2014), 095004.

[91] T. Tamio, A. Satoshi, O. Kohtaro, Automatic cell cutting by high-precision microfluidic control, J. Robot. Mechatron. 23 (1) (2011) 13–18.

[92] L. Feng, et al., On-Chip enucleation of bovine oocytes using microrobot-assisted flow-speed control, Micromachines 4 (2) (2013) 272–285.

[93] L. Feng, P. Di, F. Arai, High-precision motion of magnetic microrobot with ultrasonic levitation for 3-D rotation of single oocyte, Int. J. Robot. Res. 35 (12) (2016) 1445–1458.

[94] Y.K. Shin, Y. Kim, J. Kim, Automated microfluidic system for orientation control of mouse embryos, in: Proceedings—2013 IEEE/RSJ International Conference on Intelligent Robots and Systems, 2013.

[95] I.S. Park, et al., On-chip enucleation using an untethered microrobot incorporated with an acoustically oscillating bubble, in: Proceedings—2015 28th IEEE International Conference on Micro Electro Mechanical Systems (MEMS), 2015.

CHAPTER 4

Robotic cell manipulation for in vitro fertilization

Changsheng Dai[a], Guanqiao Shan[b], and Yu Sun[b]
[a]School of Mechanical Engineering, Dalian University of Technology, Ganjingzi District, Dalian, Liaoning, China
[b]Department of Mechanical and Industrial Engineering, University of Toronto, Toronto, ON, Canada

1 Introduction

Infertility affects over 48 million couples, amounting to one in six couples worldwide, according to a World Health Organization report [1]. In vitro fertilization (IVF) was invented for infertility treatment, with more than 8 million IVF babies born since 1978 [2]. With the great advances in IVF in the past decades, there are still several challenges: (1) manual operation requires lengthy training and suffers from inconsistency and (2) cell surgeries in IVF induce excessive damage to cells, leading to low live birth rates of around 30% globally. Robotics holds the potential to overcome these difficulties by using design and automation to standardize IVF procedures and achieve minimally invasive cell surgeries. This chapter reviews the development of robotic cell manipulation for IVF.

Cell surgery in IVF involves the manipulation of human sperm and oocyte (i.e., egg cell). A swimming sperm needs to be immobilized by tapping its tail with a glass micropipette before it is injected into an oocyte. The success of sperm immobilization depends on precise alignment of end-effector, but manual alignment suffers from a high rate of error and inconsistency. Robotic end-effector alignment allows for accurate rotational positioning of end-effector and a high success rate of sperm immobilization, as discussed in Section 2.

After immobilization, sperm rotation is performed to make sperm orientation parallel with the micropipette. If the sperm is not parallel, a higher flow rate is required to aspirate the sperm into the micropipette, which risks losing the sperm when it enters the micropipette at speed. Manual sperm rotation is a trial-and-error process, and may damage the sperm head, which contains the DNA. Robotic sperm rotation was developed based on modeling and path planning, as discussed in Section 3.

Before injecting sperm into an oocyte, the oocyte's orientation needs to be controlled so that its polar body aligns in the 12 o'clock direction. The size and shape variances of polar bodies pose difficulties for detection. The deformability of cells also needs to be taken into account in robotic control and path planning. Robotic oocyte orientation control has been developed as discussed in Section 4.

To inject sperm into an oocyte, the oocyte membrane needs to be penetrated; however, this leads to significant cell deformation and potential damage. The piezo drill was developed by using a piezoelectric actuator to generate micropipette vibration so that the penetration energy was increased. However, existing piezo drills require the use of a damping fluid such as mercury to attenuate the lateral vibration, which raises biosafety concerns. A new form of piezo drill without the use of damping fluid and related robotic techniques is discussed in Section 5.

The final section of this chapter provides an outlook for the future development of robotic cell manipulation in IVF. Opportunities and challenges for technological innovation are discussed.

2 Robotic end-effector alignment for sperm immobilization

Different end-effectors are used in the manipulation of cells, such as atomic force microscope (AFM) tips and glass micropipettes. Improper end-effector alignment can cause manipulation failure. In intracytoplasmic sperm injection (ICSI) for infertility treatment, an angled micropipette must be aligned before sperm immobilization can be performed, with its bent tip parallel with the substrate (Fig. 4.1A). Sperm immobilization is conducted by tapping a sperm's tail with the micropipette tip to disrupt the motor proteins in the tail and immobilize the sperm [3]. When the micropipette is aligned, its tip and body are collinear under the microscope (Fig. 4.1B). When the micropipette is misaligned (Fig. 4.1C), misalignment error is defined as the angle between the micropipette's tip and body in the microscopy image, as shown in Fig. 4.1D. Since the diameter of the sperm tail is about 1 μm [4], a gap larger than the sperm tail allows the sperm to escape, resulting in immobilization failure.

To achieve robotic end-effector alignment, a rotational degree of freedom (DOF) along the end-effector axis is required. Most micromanipulators lack rotational DOF, since rotation-induced translation can easily move the end-effector out of the microscope's field of view and focus [5]. For a

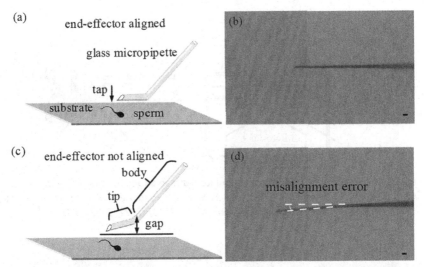

Fig. 4.1 (A) Micropipette alignment is required before sperm immobilization, aligning the micropipette tip parallel with the substrate. (B) The micropipette tip and body are collinear under the microscope when the micropipette is aligned. (C) If the micropipette is not aligned, there is a gap between the micropipette tip and substrate, and the sperm can escape through the gap, causing immobilization failure. (D) When the micropipette is not aligned, there is an angle between the micropipette tip and body under the microscope. Scale bar: 10 μm. *Source: C. Dai, S. Zhuang, G. Shan, C. Ru, Z. Zhang, Y. Sun, Automated end-effector alignment in robotic micromanipulation. IEEE/ASME Trans. Mech. 27 (5) (2022) 3932–3941.*

robotic system to perform sperm immobilization (Fig. 4.2A), a design was developed to add rotation DOF by connecting the end-effector with the motor shaft without intermediate transmission (Fig. 4.2B), making the structure compact and immune from transmission errors [6]. Considering that a micropipette holder is connected to a plastic tube for fluid aspiration/dispensing, a hollow shaft motor is used, such that the plastic tube passes through the hollow shaft to be connected with a hydraulic pump. This enables compatibility with standard setups in clinics and biomedical labs.

A kinematic model was established to describe the end-effector's motion during alignment. To calibrate unknown model parameters, the end-effector needs to be rotated for an angle, but an angled micropipette is easily rotated out of the microscope's field of view. The calibration points can only be collected from a limited local range and may introduce large model uncertainty. To achieve robustness in model uncertainty, a sliding mode controller was designed. Using the developed model and controller, a micropipette was automatically aligned, as shown in Fig. 4.3.

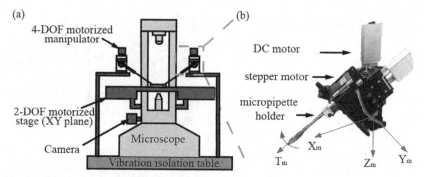

Fig. 4.2 (A) System setup. (B) Micromanipulator with an added rotational DOF. *Source: C. Dai, S. Zhuang, G. Shan, C. Ru, Z. Zhang, Y. Sun, Automated end-effector alignment in robotic micromanipulation. IEEE/ASME Trans. Mech. 27 (5) (2022) 3932–3941.*

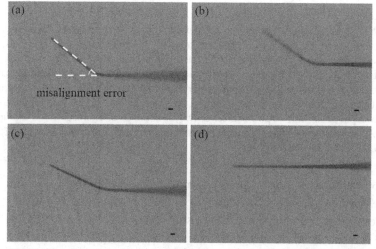

Fig. 4.3 (A) The injection micropipette had a misalignment error between the micropipette tip and body. (B) The micropipette was rotated by 5 degree to calibrate the unknown parameters of the kinematic model. The calibration from this limited range only provided a rough estimate of model parameters. (C) The micropipette was rotated for an angle based on misalignment error, and the kinematic model generated the required movement to bring the micropipette tip back to the field of view and in focus. (D) The control stopped when the misalignment error was less than 1 degree. Scale bar: 10 μm. *Source: C. Dai, S. Zhuang, G. Shan, C. Ru, Z. Zhang, Y. Sun, Automated end-effector alignment in robotic micromanipulation. IEEE/ASME Trans. Mech. 27 (5) (2022) 3932–3941.*

The misalignment error was 0.5 ± 0.3 degree for robotic alignment, significantly less than the error of 4.2 ± 3.7 degree made by an experienced operator ($P < .05$, $n = 50$ for each group). When comparing the time cost, the robotic alignment significantly reduced the time cost to 17.9 ± 7.3 s (vs 54.5 ± 23.1 s by manual alignment, $P < .05$, $n = 50$ for each group). Robotic micropipette alignment achieved a success rate of 98% for sperm immobilization ($n = 300$ sperm). The few failure cases occurred when the sperm swam away from the substrate (focal plane) and were lost in computer vision tracking. In comparison, manual micropipette alignment only achieved a success rate of 90% for sperm immobilization ($n = 300$ sperm).

3 Robotic rotation of sperm as deformable linear objects

Biological entities are deformable, and a number of cells and organisms exhibit linear shapes, such as sperm, worms, and zebrafish. The linear shape offers these cells and organisms hydrodynamic advantages for movement [7]. Manipulating cells and organisms with a linear shape has many biomedical and clinical applications. For instance, in clinical infertility treatment, a sperm needs to be rotated to be parallel with the micropipette before aspirating into the micropipette and depositing into an oocyte for fertilization.

At present, sperm tail rotation is performed manually by pushing the sperm tail with the micropipette, which is a trial-and-error process. Manual manipulation also decreases the fertilization potential of sperm. After immobilization, the tail membrane is broken to release PLCζ for oocyte activation [8], and the higher time cost in manual manipulation reduces the amount of PLCζ left within the tail.

To achieve robotic sperm rotation, robust sperm detection is required. A Kalman filter was integrated into a maximum intensity region algorithm to detect sperm tail [9] and fuzzy C means clustering was used to segment the sperm [10]. However, these methods rely on prior knowledge of the sperm tail, such as shape and length, and cannot accommodate the variance of tails among different sperm (Fig. 4.4A and B). To overcome these issues, deep

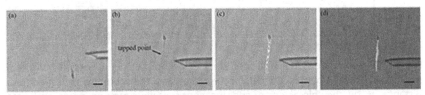

Fig. 4.4 (A and B) Tails of different sperm vary in shape and length. (C) A sperm tail is manually labeled as ground truth by drawing a curve along the sperm tail. (D) Using the trained neural networks, the contour of the sperm tail is detected. Scale bar: 10 μm. *Source: C. Dai, G. Shan, H. Liu, C. Ru, Y. Sun, Robotic manipulation of sperm as a deformable linear object. IEEE Trans. Robot. 38 (5) (2022) 2799–2811.*

neural networks were developed with robustness to tail variances [11]. Compared to conventional image processing techniques, neural networks can learn features from raw images [12]. To achieve robust sperm detection, U-Net was used as the network architecture for its effectiveness in image segmentation [13]. It is composed of a contracting path and an expansive path, forming a U-shape structure. The sperm tails are manually labeled as ground truth by drawing a curve along the sperm tail (see dashed curve in Fig. 4.4C). Using the trained neural networks, the contour of the detected sperm tail is shown in Fig. 4.4D. The networks achieved 98% in accuracy, 96% in precision, 94% in sensitivity, 99% in specificity, and 94% in dice coefficient.

To perform robotic sperm manipulation, a geometric model was established to describe the deformation of a sperm tail. In Fig. 4.5A, the tail is discretized into multiple control points, $c(s)$, to determine tail curvedness and to be used as the candidates for manipulation points in path planning. Force and kinematic models were further developed to describe the sperm rotation process. As shown in Fig. 4.5B, to provide a clockwise moment, M_p, to rotate the sperm tail, a micropipette is controlled to push the sperm tail at the manipulation point, X_p, which is above X_c. The sperm is in culture medium, and it encounters a friction moment, M_f, caused by the fluid during rotation. Since the micropipette moves at a constant speed, quasistatic dynamics are assumed, and the driving moment is balanced by the friction moment.

To rotate the object, the manipulation path can be derived by connecting the manipulation point of each rotation step. However, since the sperm

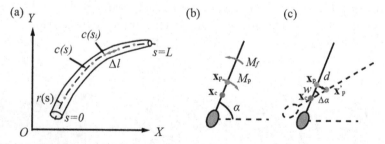

Fig. 4.5 (A) Geometric model of the deformable linear object. (B) Force model of sperm rotation. (C) In the kinematic model, for a rotation step, the sperm tail is pushed for a distance, d, by the micropipette, and the corresponding rotation angle and next manipulation point after pushing can be determined. *Source: C. Dai, G. Shan, H. Liu, C. Ru, Y. Sun, Robotic manipulation of sperm as a deformable linear object. IEEE Trans. Robot. 38 (5) (2022) 2799–2811.*

tail is deformable, this approach only manipulates a small part of the sperm tail, while the remaining part bends under the friction. To control the deformation along the sperm tail, the manipulation point is updated in each rotation step, as shown in Fig. 4.5C. Path planning is based on the tail's deformation behavior. A configuration of sperm tail is represented by its centerline curve and curvature along the curve. After each rotation step, the current configuration of the sperm tail is obtained by visual feedback. A state transition function is defined to determine the updated manipulation point based on the current configuration and target configuration.

Experiments were performed to compare the performance of robotic and manual sperm rotation. After sperm immobilization (Fig. 4.6A), 100 sperm were robotically rotated (Fig. 4.6B) and the other 100 sperm were manually rotated by an experienced operator (Fig. 4.6C). The orientation error of robotic sperm rotation was 0.8 ± 0.3 degree, significantly less than the error of 4.1 ± 2.4 degree incurred by manual rotation ($P < .05$, $n = 100$ in each group).

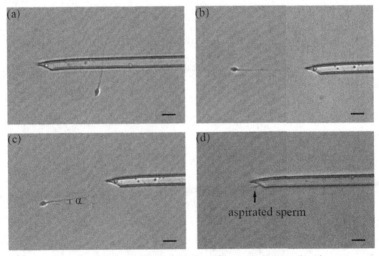

Fig. 4.6 (A) A sperm was robotically immobilized by tapping its tail with a micropipette. (B) Robotic sperm manipulation was performed to rotate the sperm tail to be coaxial with the micropipette and minimize tail curvedness. (C) Manual sperm rotation suffered from large orientation error, α, and tail curvedness. (D) After robotic rotation the sperm was aspirated into the micropipette at a lower speed due to less orientation error and tail curvedness, comparing to manual operation. *Source: C. Dai, G. Shan, H. Liu, C. Ru, Y. Sun, Robotic manipulation of sperm as a deformable linear object. IEEE Trans. Robot. 38 (5) (2022) 2799–2811.*

The purpose of rotating a sperm was to facilitate subsequent sperm aspiration in cell surgery (Fig. 4.6D). Therefore, experiments were further performed to compare the success rates of sperm aspiration after robotic and manual rotation. Robotic sperm rotation achieved a success rate of 97%, higher than that of 76% by manual rotation. With lower orientation error and tail curvedness after robotic rotation, the sperm was aspirated into the micropipette at a lower speed and successfully positioned within the micropipette and the field of view.

4 Robotic orientation control of deformable oocyte

In cell surgery tasks, such as in intracytoplasmic sperm injection (ICSI), embryo biopsy, and pronuclear transfer, cell orientation control is a crucial procedure. For instance, in mammalian oocyte manipulation, the polar body of the oocyte must be rotated away from the injection site (3 o'clock in Fig. 4.7A and B) to prevent damage to the spindle, which is in the proximity of the polar body inside the oocyte. The accuracy of rotation of the polar body is important for achieving a high fertilization rate and preserving embryo development potential [14].

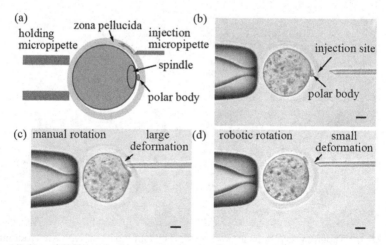

Fig. 4.7 (A and B) The polar body in the mammalian oocyte must be rotated away from the 3 o'clock position (i.e., injection site) toward the 6 or 12 o'clock position to protect the spindle, which is close to the polar body inside the oocyte. (C) In manual rotation, oocyte damage can arise from undesired large deformations. (D) Robotic orientation control achieves oocyte rotation with minimal oocyte deformation. Scale bar: 10 μm. *Source: C. Dai, Z. Zhang, Y. Lu, G. Shan, X. Wang, Q. Zhao, et al., Robotic manipulation of deformable cells for orientation control. IEEE Trans. Robot. 36 (1) (2019) 271–283.*

Manual oocyte orientation control is performed using a holding micropipette to gently hold the oocyte and an injection micropipette to push the oocyte and empirically rotate it. However, it is difficult to control oocyte deformation during manual rotation (see Fig. 4.7C). Cells are fragile and prone to damage under large deformation, thus, robotic cell orientation control was developed to minimize cell deformation [13], as shown in Fig. 4.7D.

Robotic orientation control of an oocyte requires path planning of the injection micropipette. The path for rotating the oocyte is divided into multiple steps, each corresponding to an indentation position of the injection micropipette. The path is designed to rotate an oocyte with minimal oocyte deformation, thus reducing oocyte damage. The process of path planning is shown in Fig. 4.8. The unfixed constraint between the oocyte and the holding micropipette is included in the force modeling and path planning. For each orientation step (control cycle), $\Delta\theta$, a force model determines the required minimal force, F, to apply to the oocyte. The force is then translated into the micropipette indentation, d, on the oocyte by contact mechanics. The indentation is updated to, $\hat{d}$, by the optimal controller based on the orientation error, e. The manipulation path is planned by connecting the indentation positions, P, of the micropipette in each orientation step.

Oocytes' geometrical and mechanical properties must be considered in path planning. The contour of an oocyte's zona pellucida (outer membrane of the oocyte) is detected by thresholding the local standard deviations of the image. In force analysis, an ellipsoidal oocyte model is established from the fitted ellipse to accommodate shape differences among oocytes. Oocytes have different Young's modulus values and friction coefficients to

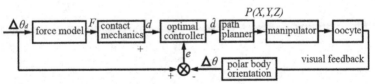

Fig. 4.8 Path planning for orientation control of deformable oocytes. For each orientation step, a force model is used to determine the required minimal force to rotate the oocyte. The force is translated into the injection micropipette's indentation on the oocyte by contact mechanics. An optimal controller has been designed to compensate for the variations of oocytes' mechanical parameters; the manipulation path is formed by connecting the indentation positions of the injection micropipette. *Source: C. Dai, Z. Zhang, Y. Lu, G. Shan, X. Wang, Q. Zhao, et al., Robotic manipulation of deformable cells for orientation control. IEEE Trans. Robot. 36 (1) (2019) 271–283.*

micropipettes. The variations of mechanical parameters can cause the force applied by the injection micropipette to be insufficient for rotating the oocyte. Young's modulus and friction coefficient can be experimentally calibrated [15], but both calibration procedures are time consuming. It is, therefore, desirable to minimize the time cost of each cell manipulation procedure, and an optimal controller (Fig. 4.8) was designed to compensate for the variations of oocytes' mechanical parameters without conducting calibrations.

Robotic orientation control of an ellipsoidal oocyte is shown in Fig. 4.9. When the polar body was not present in the microscope focal plane,

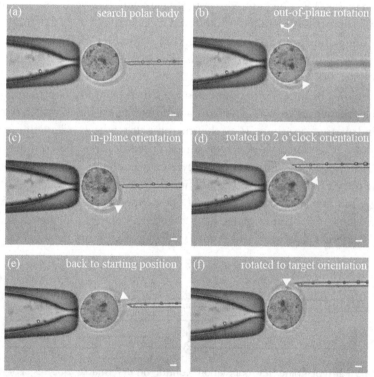

Fig. 4.9 Robotic orientation of an ellipsoidal oocyte. The polar body is labeled with a triangle. (A) Polar body was not present in the focal plane. (B) System performed out-of-plane orientation control and rotated the polar body into the focal plane. (C and D) In-plane orientation control followed to rotate the polar body toward the target orientation of 12 o'clock. (E and F) Second cycle of in-plane orientation control. Scale bar: 10 μm. *Source: C. Dai, Z. Zhang, Y. Lu, G. Shan, X. Wang, Q. Zhao, et al., Robotic manipulation of deformable cells for orientation control. IEEE Trans. Robot. 36 (1) (2019) 271–283.*

out-of-plane orientation control was performed to search for the polar body (Fig. 4.9A). Once the polar body was detected by deep neural networks, it was rotated to the focal plane in Fig. 4.9B. The polar body was at around 5 o'clock orientation and the robotic system performed in-plane orientation control to rotate the polar body toward the target orientation of 12 o'clock (Fig. 4.9C). Limited by the range of rotation in one cycle, after the first cycle of orientation control (Fig. 4.9D) the polar body was rotated to around 2 o'clock orientation. Then, the micropipette moved back to its starting position to initiate the second cycle of orientation control to rotate the polar body to the target orientation (Fig. 4.9E and F).

Maximum deformation of robotic cell orientation control was tested on 15 oocytes ($n=10$ tests for each oocyte). Robotic orientation control strategy achieved 2.70 ± 0.37 μm in maximum deformation, and oocyte deformation was consistently below 4 μm for each oocyte. Orientation error was defined as the difference between the final orientation of the polar body and the target orientation, and the robotic system achieved an orientation error of 0.7 ± 0.3 degree ($n=150$).

Robotic oocyte orientation control was compared with manual rotation performed by experienced embryologists at Toronto CReATe Fertility Centre. The time cost for rotating an oocyte was comparable between manual rotation and robotic control (10–15 s); however, the robotic system, due to modeling, optimal control, and path planning, achieved significantly less oocyte deformation (2.70 vs >10 μm) and orientation errors (0.7 vs >2 degree).

5 Robotic cell penetration with piezo drill

Cell penetration is required in many cell surgery tasks, and is achieved by using a sharp micropipette to passively indent and pierce the cell membrane. The penetration of mammalian oocytes (egg cells) is particularly challenging due to the surrounding zona pellucida, a protective layer made of glycoproteins. Passive penetration of an oocyte induces large deformations (Fig. 4.10A). To reduce cell deformation during penetration, piezo drills were developed by using piezoelectric actuators to generate micropipette vibration [16], which greatly facilitates cell penetration with less deformation, as shown in Fig. 4.10B.

However, existing piezo drills suffer from large lateral vibration and cause a large area of damage on the cell membrane. To reduce the undesired lateral vibration, a damping fluid (mercury [16] or Fluorinert (3M) [17]) is

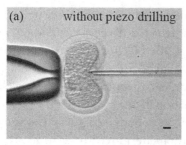 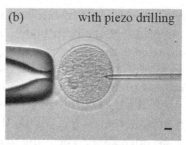

Fig. 4.10 (A) Cell penetration without piezo drilling suffers from large cell deformation. (B) Cell penetration with piezo drilling induces less cell deformation. Scale bar: 10 µm. *Source: C. Dai, L. Xin, Z. Zhang, G. Shan, T. Wang, K. Zhang, et al., Design and control of a piezo drill for robotic piezo-driven cell penetration. IEEE Robot. Autom. Lett. 5 (2) (2019) 339–345.*

filled into the micropipette, but this raises biosafety concerns. Although new piezo drill designs were also developed to circumvent the use of damping fluid [18], piezoelectric actuators were placed concentrically behind the micropipette in these designs. Such designs alter the standard setup in clinics and biology labs, where the micropipette is mounted on a standard micropipette holder.

A new design was developed with flexure beams to guide vibration along the axial direction and constrain lateral vibration [19]. The piezo drill is eccentrically mounted beside a standard micropipette holder, which does not disturb the standard clinical setup (Fig. 4.11A). In Fig. 4.11B, the piezoelectric actuator is placed behind the central beam and multiple pairs of flexure beams are connected to the central beam by flexure hinges. The flexure beams and hinges allow lower stiffness along the X-axis (3.5 N/µm) than along the Y- and Z-axes (>200 N/µm). Thus, the vibration from the piezoelectric actuator is guided along the X-axis (axial direction), resulting in minimal lateral vibration. Using the dynamic model in Fig. 4.11C, the resonant frequencies of the glass micropipette are also determined for the design of driving signals to prevent resonant vibration.

The driving signal is used to actuate the piezoelectric actuator to induce micropipette vibration. This is a pulse wave, comprising multiple pulses with intervals among them. Compared to continuous pulses without intervals, the pulse wave would transfer less power to the cell and reduce damage [20]. Because the frequency components of a pulse wave cover the full frequency spectrum, including the resonant frequencies of the glass micropipette, a bandpass filter was designed to remove resonant frequency components from the driving signal.

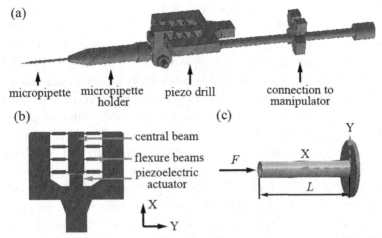

Fig. 4.11 (A) Eccentric configuration of piezo drill mounted on a standard micropipette holder. (B) Schematic of piezo drill design. Piezoelectric actuator is placed behind the central beam, and multiple pairs of flexure beams are connected to the central beam. (C) Dynamic models of the micropipette and holder. *Source: C. Dai, L. Xin, Z. Zhang, G. Shan, T. Wang, K. Zhang, et al., Design and control of a piezo drill for robotic piezo-driven cell penetration. IEEE Robot. Autom. Lett. 5 (2) (2019) 339–345.*

To achieve robotic cell penetration, micropipette tracking is required to provide visual feedback. When the micropipette indents the cell, the cell membrane and cytoplasm interfere with the micropipette tip during tracking and cause data association uncertainty. Most tracking algorithms utilize the nearest neighbor method by associating the nearest measurement to the target, and cannot effectively deal with data association uncertainty. A probabilistic data association filter (PDAF) is advantageous for tracking under interference because it performs measurement validation and uses the association probability of each valid measurement for state update [3].

The standard PDAF method only uses kinematics information (position and velocity), but during cell penetration, the cell membrane and cytoplasm have similar positions and velocities to the micropipette and can cause mismatch. Since the sharp micropipette has a corner feature but the cell membrane and cytoplasm lack this, a corner-feature probabilistic data association filter was formulated to distinguish the micropipette from interferences.

Experimental characterization of micropipette vibrations was performed, demonstrating maximum axial and lateral vibration amplitude of 1.44 and 0.36 μm, respectively. When the driving signal was not bandpass filtered, the micropipette tip's maximum axial and lateral vibration amplitudes were measured to be 1.26 and 1.80 μm, respectively.

The success rates for micropipette tracking were compared among the nearest neighbor method, standard PDAF, and developed corner-feature PDAF. The nearest neighbor method had a low success rate of 65.0%, because it does not account for false association. The standard PDAF had a success rate of 82.5%, since it only relies on kinematics information for object tracking. The corner-feature PDAF method achieved a success rate of 97.5%, significantly higher than the nearest neighbor method and standard PDAF. The inclusion of the corner feature was effective in distinguishing the micropipette tip from interfering cell membrane and cytoplasm.

To quantitatively evaluate cell deformation and damage during penetration with and without piezo drilling, experiments were performed on 80 hamster oocytes. Cell deformation by piezo-driven penetration was measured to be $5.68 \pm 2.74\,\mu m$, significantly less than the cell deformation of $54.29 \pm 10.21\,\mu m$ without piezo drilling ($P < .05$, $n = 40$ for each group), as shown in Fig. 4.12A.

To evaluate cell damage, after penetration, the oocytes were incubated for 12 h and classified as survived or degenerated, based on the standard morphology criteria reported in [21], as shown in Fig. 4.12B and C. Those shrunken oocytes with condensed cytoplasm were regarded as degenerated. The oocytes penetrated by piezo drilling had a survival rate of 82.5% (33/40), which is higher than the survival rate of 77.5% (31/40) without piezo drilling. These results suggest penetration with piezo drilling reduced cell damage during penetration.

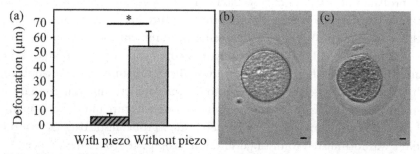

Fig. 4.12 (A) Cell deformation by piezo-driven penetration was significantly less than that without piezo drilling ($P < .05$, $n = 40$ for each group). (B and C) Hamster oocytes were classified as (B) survived and (C) degenerated, based on morphology criteria after incubation. Scale bar: 10 μm. *Source: C. Dai, L. Xin, Z. Zhang, G. Shan, T. Wang, K. Zhang, et al., Design and control of a piezo drill for robotic piezo-driven cell penetration. IEEE Robot. Autom. Lett. 5 (2) (2019) 339–345.*

6 Summary

Robotic cell manipulation has achieved significant progress in the area of in vitro fertilization. The developed techniques can relieve the burden of manual operation and achieve accurate and consistent cell surgeries. Meanwhile, emerging techniques offer new opportunities for robotic cell manipulation. The use of artificial intelligence can aid in image analysis and decision making, potentially improving the outcome of IVF treatment. Noncontact manipulation methods such as using magnetic and optical tweezers can further reduce the cell damage during robotic manipulation. Lastly, clinical compatibility must be taken into account in the development of robotic cell surgeries, to lower the adoption barrier into the clinical setting.

References

[1] M.N. Mascarenhas, S.R. Flaxman, T. Boerma, S. Vanderpoel, G.A. Stevens, National, regional, and global trends in infertility prevalence since 1990: a systematic analysis of 277 health surveys, PLoS Med. 9 (12) (2012), e1001356.

[2] G.D. Adamson, J. de Mouzon, G.M. Chambers, F. Zegers-Hochschild, R. Mansour, O. Ishihara, et al., International Committee for Monitoring Assisted Reproductive Technology: world report on assisted reproductive technology, 2011, Fertil. Steril. 110 (6) (2018) 1067–1080.

[3] Z. Zhang, C. Dai, J. Huang, X. Wang, J. Liu, C. Ru, et al., Robotic immobilization of motile sperm for clinical intracytoplasmic sperm injection, I.E.E.E. Trans. Biomed. Eng. 66 (2) (2018) 444–452.

[4] C. Serres, D. Escalier, G. David, Ultrastructural morphometry of the human sperm flagellum with a stereological analysis of the lengths of the dense fibres, Biol. Cell 49 (2) (1984) 153–161.

[5] Z. Zhang, X. Wang, J. Liu, C. Dai, Y. Sun, Robotic micromanipulation: fundamentals and applications, Annu. Rev. Contr. Robot. Auton. Syst. 2 (2019) 181–203.

[6] C. Dai, S. Zhuang, G. Shan, C. Ru, Z. Zhang, Y. Sun, Automated end-effector alignment in robotic micromanipulation, IEEE/ASME Trans. Mechatron. 27 (5) (2022) 3932–3941.

[7] E.A. Gaffney, H. Gadêlha, D. Smith, J. Blake, J.C. Kirkman-Brown, Mammalian sperm motility: observation and theory, Annu. Rev. Fluid Mech. 43 (2011) 501–528.

[8] K. Swann, M. Larman, C. Saunders, F. Lai, The cytosolic sperm factor that triggers ca2 + oscillations and egg activation in mammals is a novel phospholipase c: PLCζ, Reproduction 127 (4) (2004) 431–439.

[9] J. Liu, C. Leung, Z. Lu, Y. Sun, Quantitative analysis of locomotive behavior of human sperm head and tail, I.E.E.E. Trans. Biomed. Eng. 60 (2) (2012) 390–396.

[10] C. Dai, Z. Zhang, J. Huang, X. Wang, C. Ru, H. Pu, et al., Automated non-invasive measurement of single sperms motility and morphology, IEEE Trans. Med. Imaging 37 (10) (2018) 2257–2265.

[11] C. Dai, G. Shan, H. Liu, C. Ru, Y. Sun, Robotic manipulation of sperm as a deformable linear object, IEEE Trans. Robot. 38 (5) (2022) 2799–2811.

[12] G. Litjens, T. Kooi, B.E. Bejnordi, A. Setio, F. Ciompi, M. Ghafoorian, et al., A survey on deep learning in medical image analysis, Med. Image Anal. 42 (2017) 60–88.

[13] T. Falk, D. Mai, R. Bensch, O. Cicek, A. Abdulkadir, Y. Marrakchi, et al., U-net: deep learning for cell counting, detection, and morphometry, Nat. Methods 16 (1) (2019) 67–70.

[14] P. Rubino, P. Vigano, A. Luddi, P. Piomboni, The ICSI procedure from past to future: a systematic review of the more controversial aspects, Hum. Reprod. Update 22 (2) (2016) 194–227.

[15] Q. Zhao, M. Sun, M. Cui, J. Yu, Y. Qin, X. Zhao, X., Robotic cell rotation based on the minimum rotation force, IEEE Trans. Autom. Sci. Eng. 12 (4) (2015) 1504–1515.

[16] N. Yoshida, A.C. Perry, Piezo-actuated mouse intracytoplasmic sperm injection ICSI, Nat. Protoc. 2 (2) (2007) 296–304.

[17] K. Hiraoka, S. Kitamura, Clinical efficiency of piezo-ICSI using micropipettes with a wall thickness of 0.625 μm, J. Assist. Reprod. Genet. 32 (12) (2015) 1827–1833.

[18] C. Ru, P. Pan, R. Chen, The development of piezo-driven tools for cellular piercing, Appl. Sci. 6 (11) (2016) 314.

[19] C. Dai, L. Xin, Z. Zhang, G. Shan, T. Wang, K. Zhang, et al., Design and control of a piezo drill for robotic piezo-driven cell penetration, IEEE Robot. Autom. Lett. 5 (2) (2019) 339–345.

[20] K. Ediz, N. Olgac, Microdynamics of the piezo-driven pipettes in ICSI, I.E.E.E. Trans. Biomed. Eng. 51 (7) (2004) 1262–1268.

[21] J. Wu, L. Zhang, X. Wang, Maturation and apoptosis of human oocytes in vitro are age-related, Fertil. Steril. 74 (6) (2000) 1137–1141.

CHAPTER 5

Robotic cell transport for tissue engineering

Jiaxin Liu[a] and Huaping Wang[b]
[a]Key Laboratory of Biomimetic Robots and Systems, Beijing Institute of Technology, Ministry of Education, Beijing, China
[b]Intelligent Robotics Institute, School of Mechatronical Engineering, Beijing Institute of Technology, Beijing, China

1 Introduction

The clinical issue of repairing impaired, congenitally defective, or failed tissues and organs, coupled with donor shortage for organ transplantation [1], creates an urgent demand for the development of biological substitutes with native functions. As an emerging treatment technique, tissue engineering provides less-restricted sources to reconstruct native-mimicking tissues by combining engineering methodologies with the principles of biology.

Through conventional tissue engineering methods, such as electrospinning, cellular constructs with relatively simple architectures can be generated by seeding cells and inducing factors on biocompatible scaffolds and stimulating cells for gene expression. In native tissues, the distribution of cells and extracellular matrix (ECM) tends to be characterized by regionality, directionality, and inhomogeneity, such as the radial-like morphology of hepatic lobules [2], the anisotropic packing of myofibers [3], and the gradient changes of chondrocytes in articular cartilage [4]. However, it is quite challenging to seed different types of cells on 3D scaffolds with considerable depth according to densities and distributions that mimic native tissues. Thus, conventional methods cannot fully satisfy fabricating larger-scale artificial tissues or cellular constructs of more intricate structures (aligned fibers, neural junctions, musculoskeletal interfaces, etc.) with specific arrangement of cells or ECM components [5].

In the human body, most tissues are composed of reduplicated cellular units with structural integrity and direct biological boundaries (e.g., islets in the pancreas and nephrons in the kidney). As microscale scaffolds, these basic units with specific architecture are formed by different types of cells and

Robotics for Cell Manipulation and Characterization
https://doi.org/10.1016/B978-0-323-95213-2.00004-1
89

specific extracellular matrices (ECM) in a special arrangement. The structural features, chemical composition, and mechanical properties of these basic functional cell units have effects on the proliferation, differentiation, and gene expression of cells through cell-cell interactions and cell-ECM interactions, which, in turn, regulate the biological functions of tissues [6,7]. These reduplicated cellular units with similar compositions and morphologies are then self-assembled into tissues or organs with multiscale and hierarchical structures through cell-cell junctions [8,9]. Through self-assembly, native tissues are constructed with multiscale, hierarchical structures of cell- and ECM-specific distribution, which is inextricably linked to specific expression of biofunctions.

Inspired by the bottom-up assembly of native tissues, an innovative strategy for tissue engineering was proposed by researchers in which cellular modules are fabricated and then spatially assembled into a sophisticated architecture with biomimetic morphology [10]. Composed of encapsulated cells and natural or artificial biomaterials, basic cellular modules are produced through polymerization of hydrogels [11], self-aggregation [12], bioprinting [13], and other processes. This strategy facilities the regeneration of more intricate structures and physiological functions in artificial tissues by focusing on both microscale units and the macroscale constructs. However, the assembly of cells into 3D constructs within the constraints of spatial and temporal resolution is a highly orchestrated set of tasks. There are still many technical challenges in reconstructing highly biomimetic and functional tissues, such as creating even smaller, more precise modules to better mimic native cellular units, while also assembling larger, more robust tissues without sacrificing the desired microstructural features.

With high flexibility and accuracy, robotic cell transport holds great promise for boosting the construction of artificial tissues by participating in multiple processes from cell positioning to modular assembly. With trapping, moving, and immobilization as the basic process, cell transport can be expanded to a series of biologically significant micromanipulations of cells or cellular modules, including ordering, screening, isolation, aggregation, and assembly. Micro- and nanorobotic manipulation techniques have constantly developed to be versatile and adaptable over the past decades, spanning from contact-actuated methods to wireless-actuated methods (including magnetophoresis, dielectrophoresis, optical tweezers, etc.). By fusing vision/force feedback, all these systems can be controlled and automated to manipulate cells and cellular modules in a reproducible manner with high temporal and spatial resolution. The advances of micro- and nanorobotics applied to cell transport have significantly increased the feasibility of both high architectural complexity and high biofunctional completeness. In this chapter, we introduce a comprehensive account of the progress and contributions of robotic cell transport techniques in tissue engineering, with a focus on fabricating and assembling cellular modules, as well as cell isolation and positioning in preparation for the fabrication of cellular modules.

2 Robotic transport for cell isolating and positioning

As the smallest unit of living things, representing life and its associated consciousness, the single cell with highly complex and hierarchical structures formed by interconnected molecular networks is the fundamental building block for constructing biological substitutes [14]. To construct artificial tissues, the first step for tissue engineering is to obtain the desired cells according to application scenarios from various cell sources. Autologous cells from the patient, allogeneic cells from a human donor, and xenogeneic cells from a different species are the main cell sources to reconstruct functional living tissues. Furthermore, embryonic stem (ES) cells are the most potent source of cells due to their strong proliferative capacity and the diversity of induced differentiation [15]. Robotic cell transport techniques applied for collecting and isolating ideal cells from a cellular mixture liquid have been verified to be accurate and flexible.

In addition to isolating desired cells, immobilizing multiple cells at target positions to mimic the distribution of cells in native tissues is necessary before fabricating cellular modules. In vivo, cells always subsist in a special arrangement within tissues, according to biological regularity, including cell density, shape of cell aggregation, distribution of different types of cells, etc. The property of individual cells and the function of the tissues are regulated by cell–cell interactions through direct contact or exchange of soluble factors [16]. For example, hepatic lobules with hexagonal morphology are composed of two main types of cells that are distributed radially: parenchymal cells and nonparenchymal cells [17]. Since the surrounding environment, including the ECM and neighboring cells, has a significant impact on the health of cells, gene expression, and the secretion of functional biomolecules [18], it is necessary to precisely arrange positions of cells in artificial constructs according to the structural features of native tissues to maintain the normal cell phenotype and reproduce ideal biological functions. Novel micro-/nanorobotic manipulation techniques (such as optical tweezers (OTs), dielectrophoresis (DEP), etc.) with high throughput are promising solutions for the task of mass parallel transport, including multicell positioning.

2.1 Pick and place strategy

At the microscale, the reliable positioning of cells based on robotic micro-manipulation systems is not trivial due to the existence of scale effects and Brownian motion. One of the most widely exploited solutions for collecting and transporting cells to specific positions is the "pick and place" strategy based on mechanical contacts [19], as shown in Fig. 5.1A. Actuated by various mechatronic principles (such as piezoelectric actuation, motor actuation, hydraulic actuation, and so on), precise manipulation platforms equipped with end effectors are able to grasp, move and release cells. The assistance of visual feedback [20,21] and force feedback [22] in robot micro-manipulation systems is significant to achieve "pick and place" manipulation more efficiently and accurately.

As representative end effectors with diverse characteristics [23], micro-pipettes (Fig. 5.1B), atomic force microscopy (AFM) probes (Fig. 5.1C), and microgrippers (Fig. 5.1D) have been developed to conduct pick and place tasks. Powered by negative pressure provided by pneumatic or hydraulic pumps, micropipettes are able to aspirate [24] and move [25] cells in a less destructive and easily controlled manner. Aspiration processes are dynamically controlled to ensure that cells remain in a constant position in the micropipette during transportation [24,26]. To solve the issues of insensitive force feedback and liable damage to cells caused by the high stiffness of micropipettes, Xie et al. [27,28] proposed a precisely force-controlled methodology for cells transport based on a cantilevered micropipette probe (CMP) system. In this micro-/nanorobotic system, the frequency shift is measured to detect the interaction force between tip end and cell at pico-newton resolution, which provides sufficient manipulation force with minimal damage to cells.

AFM probes are uniquely effective in measuring and controlling manipulation force and position [29–31]. Cantilevers with hollow microfluidic channels have been fabricated [32] to provide controllable fluidic circuits, significantly broadening the capabilities and application scenarios of AFM probes, such as cellular isolation [33] and extraction [34]. Compared with single-ended effectors, such as AFM probes and micropipettes, microgrippers are mostly MEMS (micro-electro-mechanical systems)-based micro-tools and apply at least two fingers to grasp micro-objects [35], which provides more controllability and flexibility for micromanipulation [36,37]. The precision of manipulation force and position is realized and optimized by designing innovative mechanical structures [38] or employing

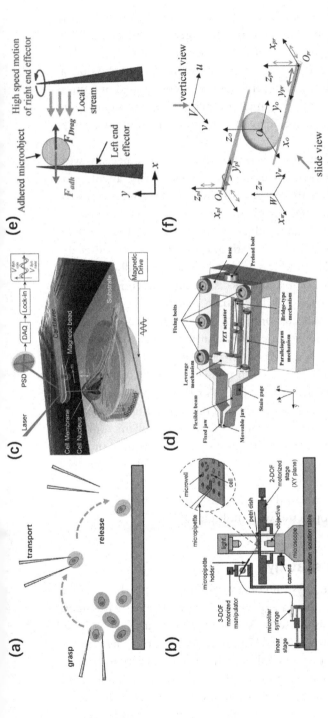

Fig. 5.1 Pick and place strategy based on mechanical contact for cell manipulation. (A) Schematic of pick and place strategy. (B) Diagram of the robotic manipulation system with a micropipette. (C) Schematic of the working mechanism of AFM probes to obtain force feedback information. (D) Diagram of a microgripper with a special structure. (E) Releasing strategies using the local stream caused by high speed motion of the right end-effector. (F) 2-Axis rotation based on two cooperative modes of micropipettes. *Images are reprinted with permission from: (B) from Zhe Lu, Christopher Moraes, George Ye, Craig A Simmons, Yu Sun, Single cell deposition and patterning with a robotic system. PLoS One 5 (10) (2010) e13542. Copyright 2010; (C) from Jianmin Song, Xianghe Meng, Hao Zhang, Kunming Zhao, Ying Hu, Hui Xie, Probing multidimensional mechanical phenotyping of intracellular structures by viscoelastic spectroscopy. ACS Appl. Mater. Interfaces 12 (1) (2019) 1913–1923. Copyright 2020 American Chemical Society; (D) Cunman Liang, Fujun Wang, Beichao Shi, Zhichen Huo, Kaihuan Zhou, Yanling Tian, Dawei Zhang, Design and control of a novel asymmetrical piezoelectric actuated microgripper for micromanipulation. Sensors Actuators A Phys. 269 (2018) 227–237. Copyright 2017 Elsevier B.V.; (E) Eunhye Kim, Masaru Kojima, Kazuto Kamiyama, Mitsuhiro Horade, Yasushi Mae, Tatsuo Arai, Accurate releasing of biological cells using two release methods generated by high speed motion of an end effector. In: 2016 IEEE/RSJ International Conference on Intelligent Robots and Systems (IROS), IEEE 2016, 2572–2577. Copyright 2016 IEEE; (F) Wanfeng Shang, Hao Ren, Mingjian Zhu, Tiantian Xu, Xinyu Wu, Dual rotating microsphere using robotic feedforward compensation control of cooperative flexible micropipettes. IEEE Trans. Autom. Sci. Eng. 17 (4) (2020) 2004–2013. Copyright 2020 IEEE.*

a controller with superior performance [39]. Faced with the issues that the microgrippers are not miniature enough and the fingers are always limited to two-dimensional distribution, a tethered microgripper [40] with a 120-degree distribution of the fingers is printed on the end face of the optical fiber by two-photon polymerization (2PP). A force sensor based on optical interferometry is integrated into this microgripper, and optical signals containing force information are reflected by the sensor and transmitted through optical fibers.

For transporting cells actuated by mechanical contact, especially with multiended microtools, an issue that requires extraordinary attention is how to release cells precisely in the presence of physical adhesion. The passive release methods mainly take effect by giving rise to easier fracture opportunities of the interface between target objects and microtools. For example, rolling resistance on a substrate coated with gold was utilized to offset the adhesion force [41], the pH value was adjusted to decrease adhesion [42], and the inertial force value was controlled by dispensing liquid microdroplets to overcome the adhesion force [43]. In contrast, vibration (Fig. 5.1E) [44], thrust [45], and vacuum-based pressure [46,47] are commonly used active release methods, which do not require specific surface properties of microtools. Active release is provided with better performance and potential in terms of controllability and efficiency than passive release strategies, which assists in automating the release process [48]. Apart from the transporting manipulation, the adjustment ability of robotic micromanipulation systems for cell posture will provide more abundant possibilities for cell micromanipulation [19]. Cell rotation tasks mostly require the cooperation of multiple end effectors [49,50] or the relative motion of the substrate and the microtool [51]. For more wide-angle and precise rotation, Xu et al. proposed a feedforward compensation control method to realize 2-axis rotation based on two cooperative modes of micropipettes with multiple DOFs (Fig. 5.1F) [52].

2.2 Wireless actuation strategy

Compared to the inevitable physical contact resulting from pick and place strategies, the wireless actuation strategy employs physical fields such as light, electric, or magnetic fields as the sources of energy for actuation. Cells are directly manipulated by a physical field [53], or indirectly driven by untethered field-actuated microrobots [54,55], which lifts the restriction that cell positioning has no alternative but to take place in an open environment. Furthermore, massively parallel manipulation is feasible for a wireless actuation strategy, as there is no limitation caused by the number of end effectors.

One of the most promoted wireless actuation methods is optical twee-zers (OTs), which ware first proposed by Arthur and Joseph in the late 1980s [56]. The scattering forces and the gradient forces are generated near the focused optical field in optical tweezers systems, and cells or other particles are trapped in an equilibrium point (Fig. 5.2A) under the effect of these two forces [57]. To control multiple targets synchro-nously for massively parallel micromanipulation, diverse methods [58–61] have been explored to generate multiple beams simultaneously, among which the computer-generated hologram (CGH) method based on holo-gram optical tweezers systems has received extensive attention due to its advantages of extending to 3-dimensional operations [62]. Tremendous efforts have been made in designing the control algorithm in terms of path planning [63,64] and path following (Fig. 5.2B) [65] for transporting multiple cells by optical tweezers, even with considerable performance in the multiobstacle environment and long-distance transportation tasks, which has highly improved the efficiency and accuracy of automation manipulation.

Dielectrophoresis (DEP) is the phenomenon in which a dielectrophore-tic force is exerted on the cell when it is subjected to a nonuniform electric field [66,67]. When an AC electric field is applied, cells can be trapped or guided to specific positions in the microfluidic chip by changing the field strength, frequency, and electrode geometries. Furthermore, DEP holds great potential for mass manipulation of multicells into arbitrarily shaped structures by fabricating microelectrodes with diverse geometries [17,68]. Resorting to programmable light projected on photoconductive materials, real-time changeable virtual electrodes are generated without demands for premanufacturing [69], which is known as optically-induced dielectrophor-esis (ODEP), as shown in Fig. 5.2C.

With the satisfactory performance of virtual electrodes in terms of flexibility, throughput, and controllability, sophisticated strategies [70–72] have been explored for trapping, transporting, rotating, and storing cells based on ODEP systems. Among these strategies, Zhang et al. proposed a circular gear-like microrobot as a versatile toolbox for micromanipulation (Fig. 5.2D), which can perform cell movement, isolation, release, and even drive micro-objects to hop onto a micro-plateau or over a barrier wall in 3D space [73,74]. Since the crossover frequency between the positive and neg-ative DEP forces exerted on cells is affected by the size and membrane capac-itance of the cells, ODEP provides a promising solution for the detection and isolation of target cells in multicell mixtures utilizing this electrical prop-erty [75–77].

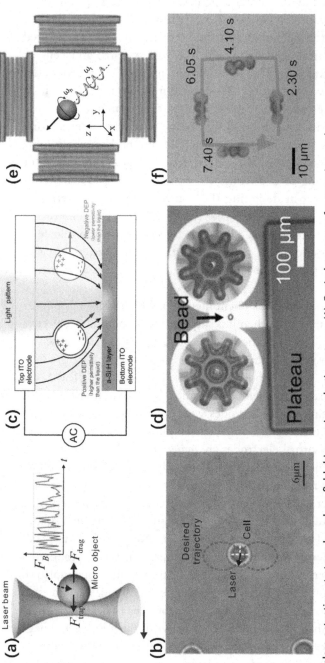

Fig. 5.2 Wireless actuation strategy based on field-driven microrobotic systems. (A) Optical tweezers for trapping and transporting micro-objects. (B) A cell actuated by optical tweezers to follow a curved path, like a figure "8." (C) Mechanism of positive and negative DEP force exerted on microparticles in an ODEP chip. (D) ODEP-based micro-feed-roller formed by two gear-like microrobots and virtual electrode for driving microbeads to hop onto a micro-plateau. (E) Magnetic field controlling the motion and posture of the magnetic microrobot. (F) Chlorella-templated biohybrid microrobot multimers actuated by magnetic field to follow a square trajectory. *Images are reproduced with permission from: (A) and (B) Xiang Li, Chien Chern Cheah, Stochastic optical trapping and manipulation of a micro object with neural-network adaptation. IEEE/ASME Trans. Mechatron. 22 (6) (2017) 2633–2642. Copyright 2017 IEEE; (C) Shuailong Zhang, Joan Juvert, Jonathan M Cooper, Steven L Neale, Manipulating and assembling metallic beads with optoelectronic tweezers. Sci. Rep. 6 (1) (2016) 1–10. Copyright 2016 The Author(s); (D) Shuailong Zhang, Mohamed Elsayed, Ran Peng, Yujie Chen, Yanfeng Zhang, Jiaxi Peng, Weizhen Li, M Dean Chamberlain, Adele Nikitina, Siyuan Yu, et al., Reconfigurable multi-component micromachines driven by optoelectronic tweezers. Nat. Commun. 12 (1) (2021) 1–9. 2021 The Author(s); (E) Huaijuan Zhou, Carmen C Mayorga-Martinez, Salvador Pane, Li Zhang, Martin Pumera, Magnetically driven micro and nanorobots. Chem. Rev. 121 (8) (2021) 4999–5041. Copyright 2021 American Chemical Society; (F) De Gong, Nuoer Celi, Deyuan Zhang, Jun Cai, Magnetic biohybrid microrobot multimers based on chlorella cells for enhanced targeted drug delivery. ACS Appl. Mater. Interfaces 14 (5) (2022) 6320–6330. Copyright 2022 American Chemical Society.*

Unlike optical or electric field-driven micromanipulations that are confined to 2D space, magnetic fields (Fig. 5.2E) are able to precisely control microrobots [29,78] and even cells [79] in 3D space [80]. Furthermore, magnetic fields have the ability to provide force and torque for transporting targets in relatively enclosed environments (blood vessels [81], brain [82], intestine [83], etc.) with the aid of nonoptical imaging techniques such as magnetic resonance imaging (MRI) [84], ultrasonic imaging [85,86], and computed tomography (CT). Wang applied the ionic sensitivity of alginate hydrogel for the first time to develop a shape-morphing microrobot responsive to environmental stimuli, driven by magnetic fields for sampling, releasing, and delivery of targets [83]. The deformation of this untethered microrobot occurs automatically when the physicochemical environment changes, which is of great significance to the intelligence of shape-morphing microrobots.

Motion control and biological applications of magnetotactic cells and biohybrid microrobots responsive to magnetic field actuation have been extensively studied by researchers due to their natural advantages in biocompatibility and drug loading ability [79,82]. Biohybrid microrobot based on Chlorella cells and Fe_3O_4 nanoparticles were assembled into multimers in a rotating magnetic field and actuated by precessing the magnetic field to follow a square trajectory, as shown in Fig. 5.2F [87]. However, the differentiated control of numerous magnetic microrobots has always been a challenging issue, which limits the throughput and efficiency of cellular micromanipulation based on magnetic actuation systems [88]. Some published research works may be instructive on this problem. For example, multiple collective modes of microrobotic swarms are flexibly programmed and transformed by applying alternating magnetic fields on hematite colloidal particles [89]. In addition, a decoupled independent control is realized by utilizing different phase responses of multiple microrobots with different magnetization directions to the oscillating magnetic field [90].

3 Robotic transport for fabrication and assembly of cellular modules

During embryo development and wound healing in vivo, cellular units are self-assembled into highly ordered tissues or organs with multiscale and hierarchical structures through the cell-cell junctions network including tight, adherens and gap junctions, etc. Biofunctions of native tissues are guided by the specific architectures and the self-organization mode of these units during cell proliferation, differentiation, and gene expression. Inspired by this natural process in vivo (self-assembly of cells from bottom up), fabricating heterogeneous building blocks with specific microarchitectures and assembling them into macroscale engineered tissues is one of the most promising solutions to faithfully regenerate intricate structures that mimic native tissues [3].

In the previous section, we highlighted the technical support provided by robotic transport methods for cell isolation and positioning, which serves to form cellular aggregates with complex structures and specific biofunctions. However, further manipulation (such as assembly and transformation) of these cellular aggregates in the absence of auxiliary material support is hard in terms of engineering implementation. Many types of cells are incapable of producing sufficient extracellular matrix and cell–cell junctions, leading to resultant cellular aggregates without tight internal traction and high cell densities [91]. Thus, constructing cellular modules in vitro requires encapsulating cells to establish direct biological boundaries and robust morphology, which provide appropriate microenvironments for specific functional expression of cells and the feasibility of subsequent 3D assembly. These micromodules encapsulated with cells are then assembled into tissues or organs with specific morphological- and biofunctions. With high scalability and multiple implementation routes [92], assembling cellular modules provides greater possibilities for architectural and compositional complexity of engineering tissue grafts. Robotic micromanipulation methods, especially the cell transport approaches, have been extensively involved in fabrication and assembly of cellular modules with high-resolution biostructures.

3.1 Cell encapsulation

To construct biofunctional cellular modules, various natural or artificial materials, including polymers, ceramics, and metals, are introduced into engineering scaffolds for specific cell types [93]. Among these materials, hydrogels are widely utilized due to their superior performance in terms of biocompatibility, transparency, and processability. The porous structure of the hydrogel provides cells with channels for transporting nutrients and metabolites while also protecting cells from mechanical damage. In addition to possessing similar chemical properties [94], mechanical characteristics [95,96], and topography as the extracellular matrix [97,98] in vivo, hydrogels are versatile in multiple application scenarios as they can be cross-linked into arbitrarily shaped micromodules as needed. However, limited by their lack of sufficient mechanical stiffness, microstructures fabricated by hydrogels that mimic native tissues are prone to occur distortion of geometry including rounding of corners or shrinkage of height [99].

By applying robotic techniques to micro-/nano processing, more intricate cellular modules with bionic geometries and related physiological functions have been developed with high fidelity. Programmable and automated methods facilitate the large scale production of cell-laden hydrogels with controllable size and shape. An integrated cellular module could be fabricated as a heterogeneous architecture with the regionalized distribution of multiple cells or components of solution by emerging transport techniques such as electrowetting and ODEP. Some two-dimensional cell-encapsulating modules could be transformed into 3D microtissues (e.g., reeled cell-laden fibers and rolled cell-laden sheets) by robotic manipulation methods. Below, we outline the common cell encapsulation methods including photolithography and microfluidics, and introduce the contribution of robotic micromanipulation techniques (especially cell transport) for encapsulating cells into modules.

3.1.1 Photolithography

Photolithography fabricates cell-laden hydrogels through photo-crosslinking a mixture of hydrogel precursor, photoinitiator, and cells. When UV or visible light is exposed to the hydrogel precursor, free radicals are released by the photoinitiator and attack the reactive double-bond group of the hydrogel precursor, which initiates the polymerization process [100]. In the photo-crosslinking process, the hydrogels are structured into arbitrary shapes by controlling the light exposure areas determined by the photo-mask with different patterns. By flexibly adjusting the micromirror array of DMD to change the light pattern in real-time and printing layer by layer, digital light processing (DLP) has received extensive attention for fabricating sophisticated micro-modules due to its relatively satisfactory performance in both printing efficiency and printing precision [101–104].

To further boost the printing resolution of morphology and mechanical stiffness, Wang et al. proposed a closed-loop control algorithm for cellular micromodule fabrication based on a DLP printing system integrated with digital holographic microscopy (DHM), which provides 3D imaging feedback information in real time [101]. The printing accuracy is demonstrated by the holographic reconstruction images (Fig. 5.3A) and microscope images (Fig. 5.3B) of the hydrogel microstructures. This work provides an innovative concept for high-precision control of hydrogel photopolymerization, which has great value in printing heterogeneous 3D scaffolds for tissue engineering.

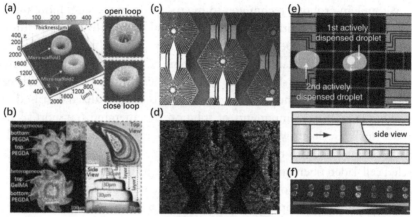

Fig. 5.3 Photolithography for cell encapsulation. (A) Holographic reconstruction of hollow circular micromodule and (B) 3D composite microscaffolds fabricated by DLP method based on closed-loop controller with DHM feedback to illustrate printing accuracy. (C) Image of the prefabricated three-layered electrodes on the chip to providing DEP force. (D) Fluorescent images of liver cells patterned actuated by DEP force between stellate microelectrode array and encapsulated with GelMA. (E) Image and schematic (side view) of applying electrowetting technology for on-demand formation of hydrogels by discretely manipulating droplets. (F) Array of 16 pillar-shaped microgels with differentiated contents formed by combining digital microfluidic methods with hydrogel polymerization methods. *Images are reproduced with permission from (A) and (B) Xin Li, Huaping Wang, Jiaxin Liu, Qing Shi, Zhe Wang, Pietro Ferraro, Qiang Huang, Toshio Fukuda, Holographic display-based control for high-accuracy photolithography of cellular micro-scaffold with heterogeneous architecture. IEEE/ASME Trans. Mechatron. 27 (2) (2021) 1117–1127. Copyright 2022 IEEE; (C) and (D) Yu-Shih Chen, Chen-Kuo Tung, Tzu-Hsuan Dai, Xiaohong Wang, Chau-Ting Yeh, Shih-Kang Fan, Cheng-Hsien Liu, Liver-lobule-mimicking patterning via dielectrophoresis and hydrogel photopolymerization. Sens. Actuators B 343 (2021) 130159. 2021 Elsevier B.V.; (E) and (F) Irwin A Eydelnant, Bingyu Betty Li, Aaron R Wheeler, Microgels on-demand. Nat. Commun. 5 (1) (2014) 1–9. Copyright 2014 Nature Publishing Group, a division of Macmillan Publishers Limited. All Rights Reserved.*

To maintain high cell viability for long-term cell culture, photolithography can flexibly adjust the molecular weight, the concentration of hydrogel, or the degree of crosslinking, to create an appropriate microenvironment including hardness, geometry, and ligand density [105–107]. For example, Cui et al. used photolithography integrated with a digital micromirror device (DMD) to construct recapitulated micromodules for cellular co-culture with physiological morphology, including radial architecture and lumen [2,108]. To avoid damage to cells caused by the biotoxicity of photoinitiators and the expensive costs of UV laser equipment,

Liu et al. developed a visible light-induced electropolymerization method for hydrogel fabrication based on an optically-induced electrophoretic (ODEP) microfluidic chip [109,110]. Cells are co-cultured with great viability in a multilayer 3D hydrogel scaffold, which is cured as a continuous polymerizing-delaminating-polymerizing mode.

In vivo, basic cellular units with specific biofunctions, such as lobules in the liver, are typically heterogeneous structures characterized by regionalized distribution of cells and the uneven composition of the ECM. Therefore, homogeneous modules fabricated by cell-encapsulating hydrogels are not adequate to regenerate structural complexity and functional completeness for tissue engineering. To fabricate cellular modules with heterogeneous architecture, photolithography is combined with robotic transport techniques, where electrowetting, optically-induced dielectrophoresis, and acoustics [111] are employed to flexibly control the components of the prepolymer solution and distribution of cells in the light exposure areas.

A novel method was explored by Zhang et al. to transport micro- and nano-objects to functional positions by ODEP and immobilize them via in situ photopolymerization [112]. Liu et al. designed a microchip with stellate microelectrode array (Fig. 5.3C), and 3T3 cells and HepG2 cells are trapped sequentially along the array based on DEP force. The spatially organized cells are encapsulated by the hydrogel with a radial shape mimicking the liver lobule (Fig. 5.3D) [17]. Eydelnant et al. [113] combined digital microfluidics methods with multiple hydrogel crosslinking methods to fabricate microgels (Fig. 5.3E) with arbitrary shape and content (Fig. 5.3F) on demand. To fabricate microstructures with heterogeneous hydrogels, Chiang et al. presented a multifunctional microchip to realize the micromanipulation of different kinds of prepolymer droplets in the same chamber based on electrowetting for the crosslinking of microgels. Through electrically moving different types of prepolymers through different exits, this system can realize the crosslinking of multitype micromodules with electrochemical deposition or photo-crosslinking [68].

3.1.2 Microfluidic formation methods

Since all life activities of cells are inseparable from the liquid environment, microfluidics is one of the most versatile techniques applied in biomedical studies, including cell analysis [114], pharmacology and pathology testing [115], construction of implantable artificial tissue [116], and so on. Actuated by fluid pumps and regulated by specific microchannels, microfluidics holds enough capacity to flexibly control the movement of cell-containing fluids, which makes it a robotic cell transport method in a broad sense. Microfluidics is ideal for mass production for hydrogel micromodule fabrication, benefiting from its high uniformity, reproducibility and design flexibility [9,117]. In contrast to the arbitrary shapes (even intricate 3D morphology) of micromodules fabricated by photolithography, three standardized shapes are mainly generated by microfluidics techniques: point [118] (cell-laden microbeads), line (cell-laden fibers) [119,120], and plane (cell-laden sheets) [121]. In this section, we mainly introduce the application of microfluidic techniques in fabricating the first two types of micromodules.

Hydrogel microbeads are formed by curing prepolymer droplets before the deformation and collapse of droplets occurs. A continuous aqueous stream is divided into discrete units by shear forces provided by the sudden squeeze of oil phase liquids, and these units become spherical due to hydrophobic interactions. 2D planar microfluidic devices (including T-junction microchannels in Fig. 5.4A, flow-focusing channels in Fig. 5.4B, 3D microfluidic setups (including axisymmetric flow-focusing devices (AFFDS)) in Fig. 5.4C, and micronozzle channels in Fig. 5.4D) are typical platforms for generating spherical droplets. The latter two categories avoid the deformation of the droplet caused by contact with the channels. The diameter of the hydrogel microbeads is strongly related to the flow rates of the aqueous phase and oil phase and the setup of the intersection [122], while the density of encapsulated cells in a single micromodule can be adjusted by controlling the concentration of cells in prepolymer solutions [123].

Tremendous efforts have been made for fabricating compartmentalized cellular modules, which broadens the application scenarios of cell-laden spherical hydrogels, such as loading of multiple drugs, hierarchical delivery, and controllable motion [124–126]. Complicated hydrogel beads, even with 20 compartments, are produced by controlling the plane divergent multiflows, which are decisive for the 3D structure of cellular micromodules [127]. To provide cells encapsulated in spherical micromodules with more favorable culture conditions for proliferation and differentiation, hydrogel microbeads of core-shell structure embedded with a medium droplet for cell culturing are generated based on a flow-focusing microfluidic device [128,129]. To facilitate the throughput of point-shaped cellular micromodules fabricated by microfluidic techniques, especially in fluids with low velocity for producing smaller-sized hydrogel beads, parallelized microfluidic devices with complex structures have been designed [130,131]. A two-layer PDMS microfluidic device with six flow-focusing nozzles was

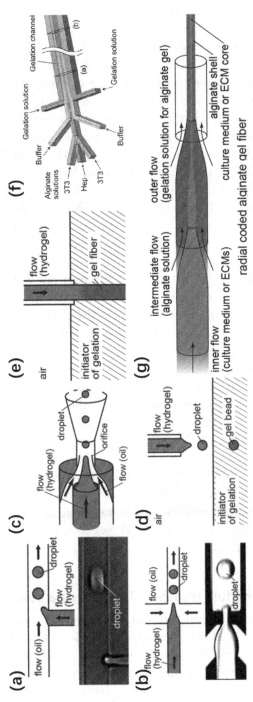

Fig. 5.4 Microfluidic techniques for fabrication of cellular micromodules. 2D planar microfluidic devices including (A) T-junction channels and (B) flow-focusing channels, 3D microfluidic setups such as (C) axisymmetric flow-focusing devices (AFFDS), and (D) micronozzle channels for producing hydrogel beads encapsulated with cells. (E) Extrusion method, (F) 2D laminar coaxial flow method, and (G) 3D coaxial flow method for fabricating cellular microfibers. *Images are reproduced with permission from: Yuya Morimoto, Amy Y Hsiao, Shoji Takeuchi, Point-, line-, and plane-shaped cellular constructs for 3D tissue assembly. Adv. Drug Deliv. Rev. 95 (2015) 29–39. Copyright 2015 Elsevier B.V.*

designed and applied for parallel fabrication of cell-laden hydrogel micro-spheres without significantly increased polydispersity [132].

Line-shaped cellular micromodules with a spindle-like shape, long and thin, commonly comprise cells and hydrogels [133]. Compared with con-ventional spinning methods, such as molding in tubes [134], with the lim-itation of fiber length related to the geometry of tubes, the microfluidics spinning technique is able to fabricate cell-laden fibers in a continuous mode ranging from a few microns to hundreds of microns in length [135,136]. Extrusion (Fig. 5.4E) [123], 2D laminar flow (Fig. 5.4F) [122], and 3D coax-ial flow (Fig. 5.4G) [135] are three representative methods based on micro-fluidic technologies for producing cellular microfibers. Flow rates and channel dimensions are two decisive parameters that influence the diameter of microfibers formed by microfluidic techniques, but the diameter of fiber produced by the extrusion method is almost independent of the flow rate [134].

Heterogeneous microfibers of multiple compartments with variable cross-sectional shapes and compositions have been developed by ameliorat-ing the structure of coaxial microfluidic devices [136,137]. Furthermore, programmable control methods have been applied for fabricating heteroge-neous hydrogel microfibers with volumetric encoding features in both radial and axial directions [138–140]. A digital fluid control method based on a 2D laminar flow device is proposed to form fibrous micromodules, along which morphologies, cell types, and compositions of liquid could be coded in serial, parallel or mixed manners, as required [138]. Faced with diverse requirements for the biofunction of microfibers, researchers are constantly enriching the palette of feasible structures, including grooved [141], hollow [142], core-shell [143], helical [144], and knotted [145,146] shapes. Hollow perfusible fibers with helical and straight channels have been constructed based on a novel coaxial microfluidic device to mimic blood vessel and to investigate the influence of vessel morphologies on nutrient and metab-olite transportation [147].

3.2 Cellular micromodule transport for 3D tissue assembly

It is widely appreciated that the application of robotic transport techniques in cellular assembly has received much attention for its flexible and high pre-cision [148,149]. Moreover, automated assembly with satisfactory efficiency and success rate is developed by combining microscopic observation tech-nologies with diverse actuating mechanisms including magnetic, acoustic, optical, and electric fields. In this section, we divided these assembly methods into three categories, mechanically actuated assembly, field-driven assembly [150–152], and field-driven self-assembly.

3.2.1 Mechanically actuated assembly

The mechanically actuated assembly method predominantly utilizes end effectors (microprobe, micropipettes, microgrippers, etc.) of micro-/nanorobotic systems, and has a relatively earlier start and more mature research than methods mentioned in Sections 3.2.2 and 3.2.3. As a contact micromanipulation technique, the mechanically actuated micro-/nanorobotic system assembles cellular modules into larger tissues through physical interaction [153]. In contrast to field-driven techniques, mechanically actuated microrobotics provides the most intuitive force in operation, which can be smoothly changed from micronewton to newton [154]. Unaffected by the interference of electrical or optical fields in the ambient environment [155,156], mechanically driven assembly takes place in the open manipulation environment, which provides a relatively broad three-dimensional workspace, enabling the assembly of micromodules to be more flexible and efficient.

In addition, the achievement of highly robotic control facilitates 3D assembly in the form of a continuous production line [154]. Avci et al. designed a two-fingered micromanipulation system, achieving high-speed multilayered cell assembly by pick and place manipulations, as shown in Fig. 5.5A and B [157,158]. Compared with the single pick and place strategy of mechanically robotic micromanipulation systems applied to single-cell transfer, more diverse strategies are employed for the assembly of cellular modules. To improve assembly efficiency while maintaining accuracy and repeatability, Liu developed a bubble-assisted assembly method (Fig. 5.5C) for constructing microvessel-like structures with annular modules (Fig. 5.5D) by introducing microfluidic forces into contact-based assembly manipulations [159,160]. A rail-guided micromanipulation system (Fig. 5.5E) with multiple microrobots was presented by Wang et al. and has been applied for assembling vascular-like microchannels [149,155,161,162]. The flexibility of the system was greatly improved by extending the attitude workspace of end effectors to an angle range of almost 360 degree and dynamically varying the coordinated mode of the microrobots (Fig. 5.5F). This microrobot system with variable configuration has high compatibility for various manipulation tasks and strong adaptability to different targets, allowing it be used as a universal platform of great significance for biological micromanipulation. Onoe et al. proposed a microfluidic weaving machine to create 3D macroscopic cellular structures to mimic

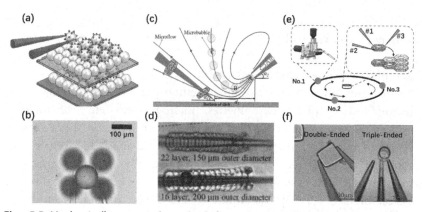

Fig. 5.5 Mechanically actuated method for assembly of cellular micromodules. (A) Schematic and (B) experimental result of high-speed multilayered assembly of NIH/3T3 cell spheroids based on a two-fingered micromanipulation system. (C) Schematic of bubble-assisted method for assembling annular modules to construct and (D) microvessel-like structure with multilayers. (E) Schematic of rail-guided micromanipulation system with multiple microrobots for modular assembly. (F) Manipulation of hydrogel micromodules via double-ended configuration and triple-ended configuration of rail-guided micromanipulation system, respectively. *Images are reproduced with permission from: (A) Ebubekir Avci, Kenichi Ohara, Chanh-Nghiem Nguyen, Chayooth Theeravithayangkura, Masaru Kojima, Tamio Tanikawa, Yasushi Mae, Tatsuo Arai, High-speed automated manipulation of microobjects using a two-fingered microhand. IEEE Trans. Industr. Electron. 62 (2) (2014) 1070–1079. Copyright 2015 IEEE; (B) Eunhye Kim, Masaru Kojima, Yasushi Mae, Tatsuo Arai, High-speed manipulation of microobjects using an automated two-fingered microhand for 3d microassembly. Micromachines 11 (5) (2020) 534. Copyright 2020 MDPI; (C) and (D) Xiaoming Liu, Qing Shi, Huaping Wang, Tao Sun, Ning Yu, Qiang Huang, Toshio Fukuda, Automated fluidic assembly of microvessel-like structures using a multimicromanipulator system. IEEE/ASME Trans. Mechatron. 23 (2) (2018) 667–678. Copyright 2018 IEEE; (F) Haojun Hu, Huaping Wang, Qing Shi, Han Tao, Qiang Huang, Toshio Fukuda, Rail-guided robotic system for multi-configuration cooperative micromanipulation based on formation control. In: 2021 IEEE International Conference on Real-time Computing and Robotics (RCAR), IEEE 2021, 165–170. Copyright 2021 IEEE.*

native tissues (including nerve networks and blood vessels) by mechanically higher-order assembly of various cell fibers [163]. Furthermore, double-striped helical tube structures were fabricated by reeling two different cell fibers on glass rods based on this weaving machine, which has great potential in artificial endothelial tubes.

3.2.2 Field-actuated assembly

Field-actuated assembly is a strategy to discretely assemble individual cell micromodules into artificial tissues or organoids through wireless manipulation methods such as magnetic field, optical field, and electric field actuation. The ODEP mechanism holds great potential in individual assembly due to its characteristics of both multithreaded workability and high throughput. The number, geometry, and size of the programmable projection can be modulated in real time, which means that there is enough flexibility to design customized "virtual electrodes," according to diverse features of the hydrogel. The dielectrophoretic force generated by light patterns can also be utilized as a frameless constraint to maintain the specific morphologies of the assembled structure [4]. Yang et al. combined ODEP and optofluidic maskless lithography for the high-throughput fabrication of hydrogel microstructures and rapid assembly of building blocks containing different cells into 3D heterogeneous microscale tissues [164].

In recent years, optothermally generated bubble robots with simple, rapid, and versatile properties have emerged as promising untethered microtools for modular assembly [165]. By studying the mechanism of bubble generation and disappearance, the researchers were able to flexibly adjust the movement and growth rate of the bubble microrobots by modulating the power of the laser. Based on the indirect actuation of the bubble robot, a vessel-like structure was constructed by adjusting the 3D pose of the hydrogel modules [166]. In addition to providing thrust and torque for the movement of the building blocks in the horizontal plane, the bubble robot can generate the rising convective flow to lift the hydrogel microstructure by growing between the substrate and the microstructure. The collaboration of multiple microrobots (Fig. 5.6A) is applied for efficient 3D assembly of hydrogel microstructures, for example, inserting the tail of a micropart into the socket of another micropart (Fig. 5.6B) [75].

Magnetic tweezers are also employed for the individual manipulation and assembly of cell-encapsulating hydrogels. Kim et al. proposed a novel method for fabricating a hepatic-lobule-like vascular network (Fig. 5.6C and D) in a 3D cellular structure by removing the steel rods and melting the assembled microfibers loaded with magnetic nanoparticles in lobule-like solidified fibrin gel to form perfusable channels [167]. The magnetic fibers are assembled into capillary-mimic networks with radial multilayered structures radiating from the "central vein" to the "portal vein" by applying a magnetic field. As an alternative to magnetic assembly based on encapsulated MNPs in hydrogels, untethered magnetic microrobots that provide temporal and spatial control have been developed to indirectly assemble heterogeneous cell-encapsulating hydrogels into 3D macrostructures [168]. Chung et al. demonstrated a magnetic microgripper robotic assembly system for heterogeneous microgel construction of multiple layers [169].

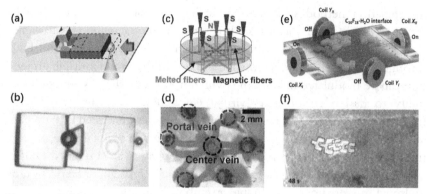

Fig. 5.6 Field-actuated assembly of cellular modules. (A) Schematic and (B) experimental result of three-dimensional assembly of hydrogels with pin and socket structures based on collaboration of two bubble microrobots. (C) A magnetic-tweezer system with seven poles for assembling microgels loaded with magnetic particles. (D) Hepatic-lobule-like 3D vascular network constructed by removing the steel rods and melting the assembled microfibers. (E) A magnetic microgripper for assembling microgels into heterogeneous microgel constructions of multiple layers. (F) Schematic and (G) experimental result of modular assembly based on magnetic microcapsule-robots (MMc-robots) of lock-and-key shape actuated by magnetic field. *Images are reprinted with permission from: (A) and (B) Liguo Dai, Daojing Lin, Xiaodong Wang, Niandong Jiao, Lianqing Liu, Integrated assembly and flexible movement of microparts using multifunctional bubble microrobots. ACS Appl. Mater. Interfaces 12 (51) (2020) 57587–57597. Copyright 2020 American Chemical Society; (C) and (D) Eunhye Kim, Masaru Takeuchi, Akiyuki Hasegawa, Akihiko Ichikawa, Yasuhisa Hasegawa, Qiang Huang, Toshio Fukuda, Construction of hepatic-lobule-like 3-D vascular network in cellular structure by manipulating magnetic fibers. IEEE/ASME Trans. Mechatron. 25 (1) (2019) 477–486. Copyright 2020 IEEE; (E) and (F) Yang Liu, Gen Li, Haojian Lu, Yuanyuan Yang, Zeyang Liu, Wanfeng Shang, Yajing Shen, Magnetically actuated heterogeneous microcapsule-robot for the construction of 3D bioartificial architectures. ACS Appl. Mater. Interfaces 11 (29) (2019) 25664–25673. Copyright 2019 American Chemical Society.*

The individual microgripper is opened and closed to grasp and release microgels by adjusting the magnitude of the magnetic torque. Shen et al. designed a magnetic microcapsule–robot (MMc–robot) with lock-and-key shape for modular assembly, which can transmit reliable forward force and steering torque while ensuring biocompatibility (Fig. 5.6E and F) [170,171].

3.2.3 Self-assembly

Self-assembly based on field actuation is a massively parallel and noninvasive manipulation method for assembling of cellular modules into large-scale structures such as artificial tissues or organoids. Instead of being controlled individually, large swarms of cellular modules respond simultaneously to forces or torques from external physical fields and are assembled randomly or under guidance (e.g., shape matching, magnetic interaction). These swarm self-assembly methods are inexpensive and highly efficient, despite sacrificing a certain degree of accuracy due to their inherent probabilistic properties. Acoustic fields have been successfully employed in self-assembly with efficient and noninvasive properties [150,172,173]. Xu et al. introduced an acoustic assembler (Fig. 5.7A) for fabricating large 3D complex cellular constructs of multiple layers (Fig. 5.7B) by assembling hydrogel micromodules [150].

Benefiting from the flexible modulation performance in terms of geometry [174], the acoustic standing wave has received extensive attention in assembling cellular modules into ordered structures. Chen et al. systematically explored the effects of chamber geometry, acoustic frequency, phase,

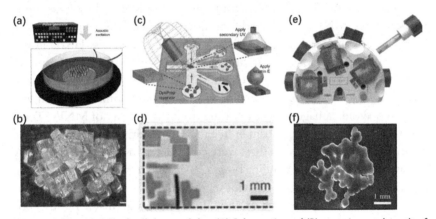

Fig. 5.7 Self-assembly of cellular modules. (A) Schematic and (B) experimental result of self-assembly of cell-encapsulating microscale hydrogels into multilayered structures based on acoustic field. (C) Schematic and (D) experimental result of self-assembly of hydrogels with different shapes and materials by utilizing the paramagnetism of free radicals. (E) A magnetic levitation bioassembler for assembling cellular spheroids into (f) 3D tissue constructs on the international space station. *Images are reproduced with permission from: (A) and (B) Feng Xu, Thomas D. Finley, Muge Turkaydin, Yuree Sung, Umut A. Gurkan, Ahmet S. Yavuz, Rasim O. Guldiken, Utkan Demirci, The assembly of cell-encapsulating microscale hydrogels using acoustic waves. Biomaterials 32 (31) (2011) 7847–7855. Copyright 2011 Elsevier Ltd.; (C) and (D) S. Tasoglu, C.H. Yu, H.I. Gungordu, S. Guven, T. Vural, U. Demirci, Guided and magnetic self-assembly of tunable magnetoceptive gels. Nat. Commun. 5 (1) (2014) 1–11. Copyright 2014, The Author(s); (E) and (F) Vladislav A. Parfenov, Yusef D. Khesuani, Stanislav V. Petrov, Pavel A. Karalkin, Elizaveta V. Koudan, Elizaveta K. Nezhurina, Frederico D.A.S. Pereira, Alisa A. Krokhmal, Anna A. Gryadunova, Elena A. Bulanova, et al., Magnetic levitational bioassembly of 3D tissue construct in space. Sci. Adv. 6 (29) (2020) eaba4174. Copyright 2020 The Authors, some rights reserved; exclusive licensee American Association for the Advancement of Science.*

and direction on the formation of standing waves [175]. Multiple modes have been developed for assembling cellular spheroids and cell-seeded microcarrier beads into diverse structures by flexibly tuning these above parameters. Long-term viable hepatic organoids have been constructed by assembling primary rat hepatocyte spheroids, and the capabilities of the hepatic organoids to form bile canaliculi, hepatic gap junctions, and extracellular matrix evaluated by fluorescent probes [176]. Through standing-wave acoustic assembly, Ren et al. not only fabricated ring-shaped cell micromodules with the assistance of hydrogel crosslinking, but also assembled them into a bracelet-shaped structure with heterogeneous composition [177].

Moreover, various magnetic self-assembly approaches of cell-encapsulating microgels into 3D structures have been reported for tissue engineering. Assembling cell-encapsulated microgels loaded with magnetic nanoparticles (MNPs) was an early solution that meets the requirements of both cell-carrying and magnetically actuated responsiveness [151]. However, cells were exposed to hydrogels mixed with toxic MNPs in this method, and there is no guarantee that MNPs can be entirely released from assembled large-scale structures, which is obviously detrimental to cell viability. To avoid damage to cells resulting from the toxicity of MNPs, Tasoglu et al. used the paramagnetism of free radicals as a driving mechanism, and complex heterogeneous structures were constructed in the magnetic field by self-assembly of different building blocks, as shown in Fig. 5.7C and D [178]. Magnetic levitation is also applied for self-assembly of cells and cellular modules without the addition of extra substances with high magnetic susceptibility [179–182]. Based on magnetic levitation assembly, Parfenov et al. developed a novel custom-designed magnetic bioassembler (Fig. 5.7E) for assembling tissue spheroids into 3D cellular constructs (Fig. 5.7F) on the international space station [183].

4 Summary

Recapitulating the structural complexity and physiological function of native tissues is a highly orchestrated set of tasks that requires manipulating cells in a reproducible manner and with high temporal and spatial resolution. Through tremendous efforts over the past decade, robotic cell transport techniques have been developed and applied for multiple key processes in tissue engineering, which facilitates the reproduction of living tissue becoming more efficient and adaptable.

This chapter highlights the recent progress and contributions of robotic cell transport techniques in tissue engineering, including single-cell manipulation, cell encapsulation, and assembly of cellular modules. In single-cell manipulation, robotic transport techniques are introduced to achieve efficient and accurate isolation and positioning of cells. The ideal cells are obtained and immobilized at target positions to mimic the distribution of cells in native tissues, which serves to form cellular patterns with specific morphology in preparation for the fabrication of cellular modules. In cell encapsulation, micro- and nanoprocessing techniques (photolithography and microfluidic formation methods) fused with robotics enable the efficient production of size- and shape-controllable cell-laden hydrogel micromodules with sophisticated structures, which serve as building units for subsequent bioassembly. In 3D cell assembly, inspired by the bottom-up assembly of native tissues, the combination of various applicable techniques (mechanically actuated assembly, field-driven assembly, and self-assembly) and microrobots can create spatially organized complex 3D structures composed of cellular building blocks, which provides the possibility of mimicking native tissues and shows great impact on regenerative medicine and biomedical research.

In the future, it can be expected that more advanced robotic cell transport techniques with cross-scale characteristic and multitask integrated adaptability will be developed to reproduce macroscale tissues with intricate microstructural features and highly biomimetic functions.

Acknowledgment

We thank the financial support from the National Natural Science Foundation under grant number 62073042 and 62222305.

References

[1] C. Van Blitterswijk, J. De Boer, P. Thomsen, J. Hubbell, R. Cancedda, J.D. De Bruijn, A. Lindahl, J. Sohier, D.F. Williams, Tissue Engineering, Elsevier, 2008.
[2] J. Cui, H. Wang, Z. Zheng, Q. Shi, T. Sun, Q. Huang, T. Fukuda, Fabrication of perfusable 3D hepatic lobule-like constructs through assembly of multiple cell type laden hydrogel microstructures, Biofabrication 11 (1) (2018), 015016.
[3] J.P.K. Armstrong, M.M. Stevens, Using remote fields for complex tissue engineering, Trends Biotechnol. 38 (3) (2020) 254–263.
[4] Y.-H. Lin, Y.-W. Yang, Y.-D. Chen, S.-S. Wang, Y.-H. Chang, W. Min-Hsien, The application of an optically switched dielectrophoretic (ODEP) force for the manipulation and assembly of cell-encapsulating alginate microbeads in a microfluidic perfusion cell culture system for bottom-up tissue engineering, Lab Chip 12 (6) (2012) 1164–1173.

[5] J. Laurent, G. Blin, F. Chatelain, V. Vanneaux, A. Fuchs, J. Larghero, M. Théry, Convergence of microengineering and cellular self-organization towards functional tissue manufacturing, Nat. Biomed. Eng. 1 (12) (2017) 939–956.

[6] J. Boublik, H. Park, M. Radisic, E. Tognana, F. Chen, M. Pei, G. Vunjak-Novakovic, L.E. Freed, Mechanical properties and remodeling of hybrid cardiac constructs made from heart cells, fibrin, and biodegradable, elastomeric knitted fabric, Tissue Eng. 11 (7–8) (2005) 1122–1132.

[7] S.M. Gopalan, C. Flaim, S.N. Bhatia, M. Hoshijima, R. Knoell, K.R. Chien, J.H. Omens, A.D. McCulloch, Anisotropic stretch-induced hypertrophy in neonatal ventricular myocytes micropatterned on deformable elastomers, Biotechnol. Bioeng. 81 (5) (2003) 578–587.

[8] V.L. Tsang, A.A. Chen, L.M. Cho, K.D. Jadin, R.L. Sah, S. DeLong, J.L. West, S.N. Bhatia, Fabrication of 3D hepatic tissues by additive photopatterning of cellular hydrogels, FASEB J. 21 (3) (2007) 790–801.

[9] M. Nie, S. Takeuchi, Bottom-up biofabrication using microfluidic techniques, Biofabrication 10 (4) (2018), 044103.

[10] D.L. Elbert, Bottom-up tissue engineering, Curr. Opin. Biotechnol. 22 (5) (2011) 674–680.

[11] J. Yeh, Y. Ling, J.M. Karp, J. Gantz, A. Chandawarkar, G. Eng, J.B. Iii, R. Langer, A. Khademhosseini, Micromolding of shape-controlled, harvestable cell-laden hydrogels, Biomaterials 27 (31) (2006) 5391–5398.

[12] D.M. Dean, A.P. Napolitano, J. Youssef, J.R. Morgan, Rods, tori, and honeycombs: the directed self-assembly of microtissues with prescribed microscale geometries, FASEB J. 21 (14) (2007) 4005–4012.

[13] V. Mironov, T. Boland, T. Trusk, G. Forgacs, R.R. Markwald, Organ printing: computer-aided jet-based 3D tissue engineering, Trends Biotechnol. 21 (4) (2003) 157–161.

[14] Y. Shen, T. Fukuda, State of the art: micro-nanorobotic manipulation in single cell analysis, Robot. Biomimet. 1 (1) (2014) 1–13.

[15] K. Takahashi, S. Yamanaka, Induction of pluripotent stem cells from mouse embryonic and adult fibroblast cultures by defined factors, Cell 126 (4) (2006) 663–676.

[16] E.E. Hui, S.N. Bhatia, Micromechanical control of cell–cell interactions, Proc. Natl. Acad. Sci. 104 (14) (2007) 5722–5726.

[17] Y.-S. Chen, C.-K. Tung, T.-H. Dai, X. Wang, C.-T. Yeh, S.-K. Fan, C.-H. Liu, Liver-lobule-mimicking patterning via dielectrophoresis and hydrogel photopolymerization, Sens. Actuators B 343 (2021), 130159.

[18] J. Cui, H. Wang, Q. Shi, P. Ferraro, T. Sun, P. Dario, Q. Huang, T. Fukuda, Permeable hollow 3D tissue-like constructs engineered by on-chip hydrodynamic-driven assembly of multicellular hierarchical micromodules, Acta Biomater. 113 (2020) 328–338.

[19] T. Tang, Y. Hosokawa, T. Hayakawa, Y. Tanaka, W. Li, M. Li, Y. Yalikun, Rotation of biological cells: fundamentals and applications, Engineering (2021).

[20] C. Dai, Z. Zhang, J. Huang, X. Wang, R. Changhai, P. Huayan, S. Xie, J. Zhang, S. Moskovtsev, C. Librach, et al., Automated non-invasive measurement of single sperm's motility and morphology, IEEE Trans. Med. Imaging 37 (10) (2018) 2257–2265.

[21] Y. Sun, S. Duthaler, B.J. Nelson, Autofocusing in computer microscopy: selecting the optimal focus algorithm, Microsc. Res. Tech. 65 (3) (2004) 139–149.

[22] W. Yinan, Z. Fan, Y. Fang, C. Liu, An intelligent AFM scanning strategy based on autonomous exploration, in: IEEE/ASME Transactions on Mechatronics, 2021.

[23] Z. Zhang, X. Wang, J. Liu, C. Dai, Y. Sun, Robotic micromanipulation: fundamentals and applications, Annu. Rev. Control, Robot. Auton. Syst. 2 (2019) 181–203.

[24] X.P. Zhang, C. Leung, Z. Lu, N. Esfandiari, R.F. Casper, Y. Sun, Controlled aspiration and positioning of biological cells in a micropipette, IEEE Trans. Biomed. Eng. 59 (4) (2012) 1032–1040.

[25] Y.H. Anis, M.R. Holl, D.R. Meldrum, Automated selection and placement of single cells using vision-based feedback control, IEEE Trans. Autom. Sci. Eng. 7 (3) (2010) 598–606.

[26] M. Sun, Y. Yao, L. Xiangfei Zhao, H.G. Li, J. Qiu, Y. Liu, X. Zhao, Precise aspiration and positioning control based on dynamic model inside and outside the micropipette, IEEE Trans. Autom. Sci. Eng. (2022).

[27] H. Xie, H. Zhang, J. Song, X. Meng, Y. Wen, L. Sun, High-precision automated micromanipulation and adhesive microbonding with cantilevered micropipette probes in the dynamic probing mode, IEEE/ASME Trans. Mechatron. 23 (3) (2018) 1425–1435.

[28] H. Xie, H. Zhang, J. Song, X. Meng, J. Geng, L. Sun, Living cell manipulation and in situ nanoinjection based on frequency shift feedback using cantilevered micropipette probes, IEEE Trans. Autom. Sci. Eng. 17 (1) (2019) 142–150.

[29] H. Xie, Y. Wen, X. Shen, H. Zhang, L. Sun, High-speed AFM imaging of nanopositioning stages using h and iterative learning control, IEEE Trans. Industr. Electron. 67 (3) (2019) 2430–2439.

[30] H. Xie, S. Régnier, High-efficiency automated nanomanipulation with parallel imaging/manipulation force microscopy, IEEE Trans. Nanotechnol. 11 (1) (2010) 21–33.

[31] S. Yuan, Z. Wang, N. Xi, Y. Wang, L. Liu, AFM tip position control in situ for effective nanomanipulation, IEEE/ASME Trans. Mechatron. 23 (6) (2018) 2825–2836.

[32] P. Saha, T. Duanis-Assaf, M. Reches, Fundamentals and applications of FluidFM technology in single-cell studies, Adv. Mater. Interfaces 7 (23) (2020) 2001115.

[33] O. Guillaume-Gentil, T. Zambelli, J.A. Vorholt, Isolation of single mammalian cells from adherent cultures by fluidic force microscopy, Lab Chip 14 (2) (2014) 402–414.

[34] O. Guillaume-Gentil, T. Rey, P. Kiefer, A.J. Ibanez, R. Steinhoff, R. Bronnimann, L. Dorwling-Carter, T. Zambelli, R. Zenobi, J.A. Vorholt, Single-cell mass spectrometry of metabolites extracted from live cells by fluidic force microscopy, Anal. Chem. 89 (9) (2017) 5017–5023.

[35] T. Chen, L. Sun, L. Chen, W. Rong, X. Li, A hybrid-type electrostatically driven microgripper with an integrated vacuum tool, Sensors Actuators A Phys. 158 (2) (2010) 320–327.

[36] S. Yang, X. Qingsong, A review on actuation and sensing techniques for MEMS-based microgrippers, J. Micro-Bio Robot. 13 (1) (2017) 1–14.

[37] M. Verotti, A. Dochshanov, N.P. Belfiore, A comprehensive survey on microgrippers design: mechanical structure, J. Mech. Des. 139 (6) (2017), 060801.

[38] Y. Liu, Y. Zhang, X. Qingsong, Design and control of a novel compliant constant-force gripper based on buckled fixed-guided beams, IEEE/ASME Trans. Mechatron. 22 (1) (2016) 476–486.

[39] F. Wang, C. Liang, Y. Tian, X. Zhao, D. Zhang, Design and control of a compliant microgripper with a large amplification ratio for high-speed micro manipulation, IEEE/ASME Trans. Mechatron. 21 (3) (2016) 1262–1271.

[40] M. Power, A.J. Thompson, S. Anastasova, G.-Z. Yang, A monolithic force-sensitive 3D microgripper fabricated on the tip of an optical fiber using 2-photon polymerization, Small 14 (16) (2018) 1703964.

[41] S. Saito, H.T. Miyazaki, T. Sato, K. Takahashi, Kinematics of mechanical and adhesional micromanipulation under a scanning electron microscope, J. Appl. Phys. 92 (9) (2002) 5140–5149.

[42] J. Dejeu, P. Rougeot, M. Gauthier, W. Boireau, Adhesion forces controlled by chemical self-assembly and pH: application to robotic microhandling, ACS Appl. Mater. Interfaces 1 (9) (2009) 1966–1973.

[43] Z. Fan, W. Rong, L. Wang, L. Sun, A single-probe capillary microgripper induced by dropwise condensation and inertial release, J. Micromech. Microeng. 25 (11) (2015), 115011.

[44] E. Kim, M. Kojima, K. Kamiyama, M. Horade, Y. Mae, T. Arai, Accurate releasing of biological cells using two release methods generated by high speed motion of an end effector, in: 2016 IEEE/RSJ International Conference on Intelligent Robots and Systems (IROS), IEEE, 2016, pp. 2572–2577.

[45] Y. Zhang, B.K. Chen, X. Liu, Y. Sun, Autonomous robotic pick-and-place of micro-objects, IEEE Trans. Robot. 26 (1) (2009) 200–207.

[46] W. Zesch, M. Brunner, A. Weber, Vacuum tool for handling microobjects with a nanorobot, in: Proceedings of International Conference on Robotics and Automation, vol. 2, IEEE, 1997, pp. 1761–1766.

[47] S. Ruggeri, G. Fontana, G. Legnani, I. Fassi, Design strategies for vacuum microgrippers with integrated release system, in: International Design Engineering Technical Conferences and Computers and Information in Engineering Conference, vol. 58165, American Society of Mechanical Engineers, 2017. V004T09A022.

[48] F. Wang, B. Shi, Z. Huo, Y. Tian, D. Zhang, Control and dynamic releasing method of a piezoelectric actuated microgripper, Precis. Eng. 68 (2021) 1–9.

[49] L. Huiying Gong, J.Q. Li, Y. Yao, Y. Liu, M. Cui, Q. Zhao, X. Zhao, M. Sun, Automatic cell rotation based on real-time detection and tracking, IEEE Robot. Autom. Lett. 6 (4) (2021) 7909–7916.

[50] Q. Zhao, M. Sun, M. Cui, Y. Jin, Y. Qin, X. Zhao, Robotic cell rotation based on the minimum rotation force, IEEE Trans. Autom. Sci. Eng. 12 (4) (2014) 1504–1515.

[51] Z. Wang, C. Feng, R. Muruganandam, W.T. Ang, S.Y.M. Tan, W.T. Latt, Three-dimensional cell rotation with fluidic flow-controlled cell manipulating device, IEEE/ASME Trans. Mechatron. 21 (4) (2016) 1995–2003.

[52] W. Shang, H. Ren, M. Zhu, X. Tiantian, W. Xinyu, Dual rotating microsphere using robotic feedforward compensation control of cooperative flexible micropipettes, IEEE Trans. Autom. Sci. Eng. 17 (4) (2020) 2004–2013.

[53] H. Xin, N. Zhao, Y. Wang, X. Zhao, T. Pan, Y. Shi, B. Li, Optically controlled living micromotors for the manipulation and disruption of biological targets, Nano Lett. 20 (10) (2020) 7177–7185.

[54] F. Qiu, S. Fujita, R. Mhanna, L. Zhang, B.R. Simona, B.J. Nelson, Magnetic helical microswimmers functionalized with lipoplexes for targeted gene delivery, Adv. Funct. Mater. 25 (11) (2015) 1666–1671.

[55] A. Ghosh, C.K. Yoon, F. Ongaro, S. Scheggi, F.M. Selaru, S. Misra, D.H. Gracias, Stimuli-responsive soft untethered grippers for drug delivery and robotic surgery, Front. Mech. Eng. 3 (2017) 7.

[56] A. Ashkin, J.M. Dziedzic, Optical trapping and manipulation of viruses and bacteria, Science 235 (4795) (1987) 1517–1520.

[57] K.R. Dhakal, V. Lakshminarayanan, Optical tweezers: fundamentals and some biophysical applications, in: Progress in Optics, 63, Elsevier, 2018, pp. 1–31.

[58] J. Gieseler, J.R. Gomez-Solano, A. Magazzu, I. Perez-Castillo, L. Perez-García, M. Gironella-Torrent, X. Viader-Godoy, F. Ritort, G. Pesce, A.V. Arzola, K. Volke-Sepulveda, G. Volpe, et al., Optical tweezers—from calibration to applications: a tutorial, Adv. Optics Photon. 13 (1) (2021) 74–241.

[59] K. Sasaki, M. Koshioka, H. Misawa, N. Kitamura, H. Masuhara, Pattern formation and flow control of fine particles by laser-scanning micromanipulation, Opt. Lett. 16 (19) (1991) 1463–1465.

[60] R.L. Eriksen, P.C. Mogensen, J. Glückstad, Multiple-beam optical tweezers generated by the generalized phase-contrast method, Opt. Lett. 27 (4) (2002) 267–269.

[61] J.E. Curtis, B.A. Koss, D.G. Grier, Dynamic holographic optical tweezers, Optics Commun. 207 (1-6) (2002) 169–175.

[62] K. Onda, F. Arai, Multi-beam bilateral teleoperation of holographic optical tweezers, Opt. Express 20 (4) (2012) 3633–3641.

[63] H. Yang, X. Li, X. Li, L. Sun, D. Sun, A virtual-assisted controller for biological cell transportation in a dynamic environment with variable field of view, IEEE/ASME Trans. Mechatron. 25 (3) (2020) 1255–1265.

[64] A.G. Banerjee, S. Chowdhury, W. Losert, S.K. Gupta, Real-time path planning for coordinated transport of multiple particles using optical tweezers, IEEE Trans. Autom. Sci. Eng. 9 (4) (2012) 669–678.

[65] X. Li, C.C. Cheah, Stochastic optical trapping and manipulation of a micro object with neural-network adaptation, IEEE/ASME Trans. Mechatron. 22 (6) (2017) 2633–2642.

[66] Z. Zhang, T. Zheng, R. Zhu, Characterization of single-cell biophysical properties and cell type classification using dielectrophoresis model reduction method, Sens. Actuators B 304 (2020), 127326.

[67] Y. Li, Y. Wang, K. Wan, W. Mingxue, L. Guo, X. Liu, G. Wei, On the design, functions, and biomedical applications of high-throughput dielectrophoretic micro-/nanoplatforms: a review, Nanoscale 13 (8) (2021) 4330–4358.

[68] M.-Y. Chiang, Y.-W. Hsu, H.-Y. Hsieh, S.-Y. Chen, S.-K. Fan, Constructing 3D heterogeneous hydrogels from electrically manipulated prepolymer droplets and cross-linked microgels, Sci. Adv. 2 (10) (2016), e1600964.

[69] P.Y. Chiou, A.T. Ohta, M.C. Wu, Massively parallel manipulation of single cells and microparticles using optical images, Nature 436 (7049) (2005) 370–372.

[70] S. Zhang, N. Shakiba, Y. Chen, Y. Zhang, P. Tian, M. Jastaranpreet Singh, D. Chamberlain, M. Satkauskas, A.G. Flood, N.P. Kherani, et al., Patterned optoelectronic tweezers: a new scheme for selecting, moving, and storing dielectric particles and cells, Small 14 (45) (2018) 1803342.

[71] W. Liang, Y. Wang, H. Zhang, L. Liu, Characterization of the self-rotational motion of stored red blood cells by using optically-induced electrokinetics, Opt. Lett. 41 (12) (2016) 2763–2766.

[72] T.-K. Chiu, A.C. Chao, W.-P. Chou, C.-J. Liao, H.-M. Wang, J.-H. Chang, P.-H. Chen, M.-H. Wu, Optically-induced-dielectrophoresis (ODEP)-based cell manipulation in a microfluidic system for high-purity isolation of integral circulating tumor cell (CTC) clusters based on their size characteristics, Sens. Actuators B 258 (2018) 1161–1173.

[73] S. Zhang, M. Elsayed, R. Peng, Y. Chen, Y. Zhang, J. Peng, M. Weizhen Li, D. Chamberlain, A. Nikitina, Y. Siyuan, et al., Reconfigurable multi-component micromachines driven by optoelectronic tweezers, Nat. Commun. 12 (1) (2021) 1–9.

[74] S. Zhang, E.Y. Scott, J. Singh, Y. Chen, Y. Zhang, M. El-sayed, M.D. Chamberlain, N. Shakiba, K. Adams, S. Yu, et al., The optoelectronic microrobot: a versatile toolbox for micromanipulation, Proc. Natl. Acad. Sci. 116 (30) (2019) 14823–14828.

[75] Y. Zhang, J. Zhao, Y. Haibo, P. Li, W. Liang, Z. Liu, G.-B. Lee, L. Liu, W.J. Li, Z. Wang, Detection and isolation of free cancer cells from ascites and peritoneal lavages using optically induced electrokinetics (OEK), Sci. Adv. 6 (32) (2020) eaba9628.

[76] S.-B. Huang, W. Min-Hsien, Y.-H. Lin, C.-H. Hsieh, C.-L. Yang, H.-C. Lin, C.-P. Tseng, G.-B. Lee, High-purity and label-free isolation of circulating tumor cells (CTCs) in a microfluidic platform by using optically-induced-dielectrophoretic (ODEP) force, Lab Chip 13 (7) (2013) 1371–1383.

[77] N. Liu, Y. Lin, Y. Peng, L. Xin, T. Yue, Y. Liu, R. Changhai, S. Xie, L. Dong, P. Huayan, et al., Automated parallel electrical characterization of cells using optically-induced dielectrophoresis, IEEE Trans. Autom. Sci. Eng. 17 (2) (2020) 1084–1092.

[78] Y. Jiangfan, B. Wang, D. Xingzhou, Q. Wang, L. Zhang, Ultra-extensible ribbon-like magnetic microswarm, Nat. Commun. 9 (1) (2018) 1–9.

[79] O. Felfoul, M. Mohammadi, S. Taherkhani, D. De Lanauze, X. Yong Zhong, D. Loghin, S. Essa, S. Jancik, D. Houle, M. Lafleur, et al., Magneto-aerotactic bacteria deliver drug-containing nanoliposomes to tumour hypoxic regions, Nat. Nanotechnol. 11 (11) (2016) 941–947.

[80] J. Liu, X. Tiantian, S.X. Yang, X. Wu, Navigation and visual feedback control for magnetically driven helical miniature swimmers, IEEE Trans. Industr. Inform. 16 (1) (2019) 477–487.

[81] Y. Alapan, U. Bozuyuk, P. Erkoc, A.C. Karacakol, M. Sitti, Multifunctional surface microrollers for targeted cargo delivery in physiological blood flow, Sci. Robot. 5 (42) (2020) eaba5726.

[82] H. Zhang, Z. Li, C. Gao, X. Fan, Y. Pang, T. Li, W. Zhiguang, H. Xie, Q. He, Dual-responsive biohybrid neutrobots for active target delivery, Sci. Robot. 6 (52) (2021) eaaz9519.

[83] Z. Zheng, H. Wang, L. Dong, Q. Shi, J. Li, T. Sun, Q. Huang, T. Fukuda, Ionic shape-morphing microrobotic end-effectors for environmentally adaptive targeting, releasing, and sampling, Nat. Commun. 12 (1) (2021) 1–12.

[84] M.R. Benoit, D. Mayer, Y. Barak, I.Y. Chen, H. Wei, Z. Cheng, S.X. Wang, D.M. Spielman, S.S. Gambhir, A. Matin, Visualizing implanted tumors in mice with magnetic resonance imaging using magnetotactic bacteria, Clin. Cancer Res. 15 (16) (2009) 5170–5177.

[85] D. Mahdy, R. Reda, N. Hamdi, I.S.M. Khalil, Ultrasound-guided minimally invasive grinding for clearing blood clots: promises and challenges, IEEE Instrumen. Measure. Mag. 21 (2) (2018) 10–14.

[86] Q. Wang, S. Yang, L. Zhang, Magnetic actuation of a dynamically reconfigurable microswarm for enhanced ultrasound imaging contrast, IEEE/ASME Trans. Mechatron. (2022) 1–11.

[87] D. Gong, N. Celi, D. Zhang, J. Cai, Magnetic biohybrid microrobot multimers based on chlorella cells for enhanced targeted drug delivery, ACS Appl. Mater. Interfaces 14 (5) (2022) 6320–6330.

[88] X.-Z. Chen, M. Hoop, F. Mushtaq, E. Siringil, H. Chengzhi, B.J. Nelson, S. Pane, Recent developments in magnetically driven micro-and nanorobots, Appl. Mater. Today 9 (2017) 37–48.

[89] H. Xie, M. Sun, X. Fan, Z. Lin, W. Chen, L. Wang, L. Dong, Q. He, Reconfigurable magnetic microrobot swarm: multimode transformation, locomotion, and manipulation, Sci. Robot. 4 (28) (2019) eaav8006.

[90] X. Tiantian, C. Huang, Z. Lai, W. Xinyu, Independent control strategy of multiple magnetic flexible millirobots for position control and path following, IEEE Trans. Robot. 38 (5) (2022) 2875–2887.

[91] J.W. Nichol, A. Khademhosseini, Modular tissue engineering: engineering biological tissues from the bottom up, Soft Matter 5 (7) (2009) 1312–1319.

[92] L. Ouyang, J.P.K. Armstrong, M. Salmeron-Sanchez, M.M. Stevens, Assembling living building blocks to engineer complex tissues, Adv. Funct. Mater. 30 (26) (2020) 1909009.

[93] H.N. Kim, D.-H. Kang, M.S. Kim, A. Jiao, D.-H. Kim, K.-Y. Suh, Patterning methods for polymers in cell and tissue engineering, Ann. Biomed. Eng. 40 (6) (2012) 1339–1355.

[94] S.H. Lee, H.N. Kim, K.Y. Suh, et al., Use of directly molded poly (methyl methacrylate) channels for microfluidic applications, Lab Chip 10 (23) (2010) 3300–3306.

[95] S. Nemir, J.L. West, Synthetic materials in the study of cell response to substrate rigidity, Ann. Biomed. Eng. 38 (1) (2010) 2–20.

[96] D.-H. Kim, P.K. Wong, J. Park, A. Levchenko, Y. Sun, Microengineered platforms for cell mechanobiology, Annu. Rev. Biomed. Eng. 11 (2009) 203–233.

[97] A. Dolatshahi-Pirouz, M. Nikkhah, K. Kolind, M.R. Dokmeci, A. Khademhosseini, Micro- and nanoengineering approaches to control stem cell-biomaterial interactions, J. Funct. Biomater. 2 (3) (2011) 88–106.

[98] A. Khademhosseini, R. Langer, J. Borenstein, J.P. Vacanti, Microscale technologies for tissue engineering and biology, Proc. Natl. Acad. Sci. 103 (8) (2006) 2480–2487.

[99] Youn Sang Kim, H.H. Lee, P.T. Hammond, High density nanostructure transfer in soft molding using polyurethane acrylate molds and polyelectrolyte multilayers, Nanotechnology 14 (10) (2003) 1140.

[100] J.W. Nichol, S.T. Koshy, H. Bae, C.M. Hwang, S. Yamanlar, A. Khademhosseini, Cell-laden microengineered gelatin methacrylate hydrogels, Biomaterials 31 (21) (2010) 5536–5544.

[101] X. Li, H. Wang, J. Liu, Q. Shi, Z. Wang, P. Ferraro, Q. Huang, T. Fukuda, Holographic display-based control for high-accuracy photolithography of cellular microscaffold with heterogeneous architecture, IEEE/ASME Trans. Mechatron. 27 (2) (2021) 1117–1127.

[102] M. Wang, W. Li, L.S. Mille, T. Ching, Z. Luo, G. Tang, C.E. Garciamendez, A. Lesha, M. Hashimoto, Y.S. Zhang, Digital light processing based bioprinting with composable gradients, Adv. Mater. 34 (1) (2022) 2107038.

[103] Y. Li, Q. Mao, X. Li, J. Yin, Y. Wang, F. Jianzhong, Y. Huang, High-fidelity and high-efficiency additive manufacturing using tunable pre-curing digital light processing, Addit. Manuf. 30 (2019), 100889.

[104] W. Yang, Y. Haibo, G. Li, F. Wei, Y. Wang, L. Liu, Mask-free fabrication of a versatile microwell chip for multidimensional cellular analysis and drug screening, Lab Chip 17 (24) (2017) 4243–4252.

[105] K. Dey, E. Roca, G. Ramorino, L. Sartore, Progress in the mechanical modulation of cell functions in tissue engineering, Biomater. Sci. 8 (24) (2020) 7033–7081.

[106] L.N. West-Livingston, J. Park, S.J. Lee, A. Atala, J.J. Yoo, The role of the microenvironment in controlling the fate of bioprinted stem cells, Chem. Rev. 120 (19) (2020) 11056–11092.

[107] A.N. Buxton, J. Zhu, R. Marchant, J.L. West, J.U. Yoo, B. Johnstone, Design and characterization of poly (ethylene glycol) photopolymerizable semi-interpenetrating networks for chondrogenesis of human mesenchymal stem cells, Tissue Eng. 13 (10) (2007) 2549–2560.

[108] Z. Zheng, H. Wang, J. Li, Q. Shi, J. Cui, T. Sun, Q. Huang, T. Fukuda, 3D construction of shape-controllable tissues through self-bonding of multicellular microcapsules, ACS Appl. Mater. Interfaces 11 (26) (2019) 22950–22961.

[109] P. Li, Y. Haibo, N. Liu, F. Wang, G.-B. Lee, Y. Wang, L. Liu, W.J. Li, Visible light induced electropolymerization of suspended hydrogel bioscaffolds in a microfluidic chip, Biomater. Sci. 6 (6) (2018) 1371–1378.

[110] N. Liu, W. Liang, L. Liu, Y. Wang, J.D. Mai, G.-B. Lee, W.J. Li, Extracellular-controlled breast cancer cell formation and growth using non-UV patterned hydrogels via optically-induced electrokinetics, Lab Chip 14 (7) (2014) 1367–1376.

[111] H. Xuejia, S. Zhao, Z. Luo, Y. Zuo, F. Wang, J. Zhu, L. Chen, D. Yang, Y. Zheng, Y. Zheng, et al., On-chip hydrogel arrays individually encapsulating acoustic formed multicellular aggregates for high throughput drug testing, Lab Chip 20 (12) (2020) 2228–2236.

[112] S. Zhang, W. Li, M. Elsayed, J. Peng, Y. Chen, Y. Zhang, Y. Zhang, M. Shayegannia, W. Dou, T. Wang, et al., Integrated assembly and photopreservation of topographical micropatterns, Small 17 (37) (2021) 2103702.

[113] I.A. Eydelnant, B.B. Li, A.R. Wheeler, Microgels on-demand, Nat. Commun. 5 (1) (2014) 1–9.

[114] J. Ma, S. Yan, C. Miao, L. Li, W. Shi, X. Liu, Y. Luo, T. Liu, B. Lin, W. Wenming, et al., Paper microfluidics for cell analysis, Adv. Healthc. Mater. 8 (1) (2019) 1801084.

[115] M. Boyd-Moss, S. Baratchi, M. Di Venere, K. Khoshmanesh, Self-contained microfluidic systems: a review, Lab Chip 16 (17) (2016) 3177–3192.

[116] N.W. Choi, M. Cabodi, B. Held, J.P. Gleghorn, L.J. Bonassar, A.D. Stroock, Microfluidic scaffolds for tissue engineering, Nat. Mater. 6 (11) (2007) 908–915.

[117] A.R. Kang, J.S. Park, J. Jongil, G.S. Jeong, S.-H. Lee, Cell encapsulation via microtechnologies, Biomaterials 35 (9) (2014) 2651–2663.

[118] A.C. Daly, L. Riley, T. Segura, J.A. Burdick, Hydrogel microparticles for biomedical applications, Nat. Rev. Mater. 5 (1) (2020) 20–43.

[119] L. Shang, Y. Yunru, Y. Liu, Z. Chen, T. Kong, Y. Zhao, Spinning and applications of bioinspired fiber systems, ACS Nano 13 (3) (2019) 2749–2772.

[120] M.C. McNamara, F. Sharifi, A.H. Wrede, D.F. Kimlinger, D.-G. Thomas, J.B. Vander, Y.C. Wiel, R. Montazami, N.N. Hashemi, Microfibers as physiologically relevant platforms for creation of 3D cell cultures, Macromol. Biosci. 17 (12) (2017) 1700279.

[121] M. Li, J. Ma, Y. Gao, L. Yang, Cell sheet technology: a promising strategy in regenerative medicine, Cytotherapy 21 (1) (2019) 3–16.

[122] Y. Jun, M.J. Kim, Y.H. Hwang, E.A. Jeon, A.R. Kang, S.-H. Lee, D.Y. Lee, Microfluidics-generated pancreatic islet microfibers for enhanced immunoprotection, Biomaterials 34 (33) (2013) 8122–8130.

[123] S. Mazzitelli, L. Capretto, D. Carugo, X. Zhang, R. Piva, C. Nastruzzi, Optimised production of multifunctional microfibres by microfluidic chip technology for tissue engineering applications, Lab Chip 11 (10) (2011) 1776–1785.

[124] L. Zhang, K. Chen, H. Zhang, B. Pang, C.-H. Choi, A.S. Mao, H. Liao, S. Utech, D.J. Mooney, H. Wang, et al., Microfluidic templated multicompartment microgels for 3D encapsulation and pairing of single cells, Small 14 (9) (2018) 1702955.

[125] G. Tang, R. Xiong, D. Lv, R.X. Xu, K. Braeckmans, C. Huang, S.C. De Smedt, Gas-shearing fabrication of multicompartmental microspheres: a one-step and oil-free approach, Adv. Sci. 6 (9) (2019) 1802342.

[126] S. Yoshida, M. Takinoue, H. Onoe, Compartmentalized spherical collagen microparticles for anisotropic cell culture microenvironments, Adv. Healthc. Mater. 6 (8) (2017) 1601463.

[127] W. Zengnan, Y. Zheng, L. Lin, S. Mao, Z. Li, J.-M. Lin, Controllable synthesis of multicompartmental particles using 3D microfluidics, Angew. Chem. Int. Ed. 59 (6) (2020) 2225–2229.

[128] Z. Liu, H. Zhang, Z. Zhan, H. Nan, N. Huang, X. Tao, X. Gong, H. Chengzhi, Mild formation of core–shell hydrogel microcapsules for cell encapsulation, Biofabrication 13 (2) (2021), 025002.

[129] K. Zhu, Y. Yunru, Y. Cheng, C. Tian, G. Zhao, Y. Zhao, All-aqueous-phase microfluidics for cell encapsulation, ACS Appl. Mater. Interfaces 11 (5) (2019) 4826–4832.

[130] J.M. de Rutte, J. Koh, D. Di Carlo, Scalable high-throughput production of modular microgels for in situ assembly of microporous tissue scaffolds, Adv. Funct. Mater. 29 (25) (2019) 1900071.

[131] T. Kamperman, L.M. Teixeira, S.S. Salehi, G. Kerckhofs, Y. Guyot, M. Geven, L. Geris, D. Grijpma, S. Blanquer, J. Leijten, Engineering 3D parallelized microfluidic droplet generators with equal flow profiles by computational fluid dynamics and stereolithographic printing, Lab Chip 20 (3) (2020) 490–495.

[132] D.M. Headen, J.R. García, A.J. García, Parallel droplet microfluidics for high throughput cell encapsulation and synthetic microgel generation, Microsyst. Nanoeng. 4 (1) (2018) 1–9.

[133] Y. Morimoto, A.Y. Hsiao, S. Takeuchi, Point-, line-, and plane-shaped cellular constructs for 3D tissue assembly, Adv. Drug Deliv. Rev. 95 (2015) 29–39.

[134] W. Zheng, R. Xie, X. Liang, Q. Liang, Fabrication of biomaterials and biostructures based on microfluidic manipulation, Small (2022) 2105867.

[135] A.Y. Hsiao, T. Okitsu, H. Onoe, M. Kiyosawa, H. Teramae, S. Iwanaga, T. Kazama, T. Matsumoto, S. Takeuchi, Smooth muscle-like tissue constructs with circumferentially oriented cells formed by the cell fiber technology, PLoS One 10 (3) (2015), e0119010.

[136] Y. Cheng, F. Zheng, L. Jie, L. Shang, Z. Xie, Y. Zhao, Y. Chen, G. Zhongze, Bioinspired multicompartmental microfibers from microfluidics, Adv. Mater. 26 (30) (2014) 5184–5190.

[137] D.H. Yoon, K. Kobayashi, D. Tanaka, T. Sekiguchi, S. Shoji, Simple microfluidic formation of highly heterogeneous microfibers using a combination of sheath units, Lab Chip 17 (8) (2017) 1481–1486.

[138] E. Kang, G.S. Jeong, Y.Y. Choi, K.H. Lee, A. Khademhosseini, S.-H. Lee, Digitally tunable physicochemical coding of material composition and topography in continuous microfibres, Nat. Mater. 10 (11) (2011) 877–883.

[139] M. Zhang, S. Wang, Y. Zhu, Z. Zhu, T. Si, R.X. Xu, Programmable dynamic interfacial spinning of bioinspired microfibers with volumetric encoding, Mater. Horiz. 8 (6) (2021) 1756–1768.

[140] C. Yang, Y. Yunru, X. Wang, L. Shang, Y. Zhao, Programmable knot microfibers from piezoelectric microfluidics, Small 18 (5) (2022) 2104309.

[141] M. Ebrahimi, S. Ostrovidov, S. Salehi, S.B. Kim, H. Bae, A. Khademhosseini, Enhanced skeletal muscle formation on microfluidic spun gelatin methacryloyl (GelMA) fibres using surface patterning and agrin treatment, J. Tissue Eng. Regen. Med. 12 (11) (2018) 2151–2163.

[142] R. Xie, A. Korolj, C. Liu, X. Song, L. Rick Xing Ze, B. Zhang, A. Ramachandran, Q. Liang, M. Radisic, h-FIBER: microfluidic topographical hollow fiber for studies of glomerular filtration barrier, ACS Central Sci. 6 (6) (2020) 903–912.

[143] Y. Yunru, G. Chen, J. Guo, Y. Liu, J. Ren, T. Kong, Y. Zhao, Vitamin metal-organic framework-laden microfibers from microfluidics for wound healing, Mater. Horiz. 5 (6) (2018) 1137–1142.

[144] J.-D. Liu, D. Xiang-Yun, S. Chen, A phase inversion-based microfluidic fabrication of helical microfibers towards versatile artificial abdominal skin, Angew. Chem. 133 (47) (2021) 25293–25300.

[145] R. Xie, X. Peidi, Y. Liu, L. Li, G. Luo, M. Ding, Q. Liang, Necklace-like microfibers with variable knots and perfusable channels fabricated by an oil-free microfluidic spinning process, Adv. Mater. 30 (14) (2018) 1705082.

[146] Y. Liu, N. Yang, X. Li, J. Li, W. Pei, X. Yiwen, Y. Hou, Y. Zheng, Water harvesting of bioinspired microfibers with rough spindle-knots from microfluidics, Small 16 (9) (2020) 1901819.

[147] X. Peidi, R. Xie, Y. Liu, G. Luo, M. Ding, Q. Liang, Bioinspired microfibers with embedded perfusable helical channels, Adv. Mater. 29 (34) (2017) 1701664.

[148] X. Qingsong, Y. Li, N. Xi, Design, fabrication, and visual servo control of an XY parallel micromanipulator with piezo-actuation, IEEE Trans. Autom. Sci. Eng. 6 (4) (2009) 710–719.

[149] H. Wang, Q. Shi, T. Yue, M. Nakajima, M. Takeuchi, Q. Huang, T. Fukuda, Microassembly of a vascular-like micro-channel with railed micro-robot team-coordinated manipulation, Int. J. Adv. Robot. Syst. 11 (7) (2014) 115.

[150] X. Feng, T.D. Finley, M. Turkaydin, Y. Sung, U.A. Gurkan, A.S. Yavuz, R.O. Guldiken, U. Demirci, The assembly of cell-encapsulating microscale hydrogels using acoustic waves, Biomaterials 32 (31) (2011) 7847–7855.

[151] X. Feng, C.-a.M. Wu, V. Rengarajan, T.D. Finley, H.O. Keles, Y. Sung, B. Li, U.A. Gurkan, U. Demirci, Three-dimensional magnetic assembly of microscale hydrogels, Adv. Mater. 23 (37) (2011) 4254–4260.

[152] T. Yue, M. Nakajima, H. Tajima, T. Fukuda, Fabrication of microstructures embedding controllable particles inside dielectrophoretic microfluidic devices, Int. J. Adv. Robot. Syst. 10 (2) (2013) 132.

[153] M. Savia, H.N. Koivo, Contact micromanipulation—survey of strategies, IEEE/ASME Trans. Mechatron. 14 (4) (2009) 504–514.

[154] A.A. Ramadan, T. Takubo, Y. Mae, K. Oohara, T. Arai, Developmental process of a chopstick-like hybrid-structure two-fingered micromanipulator hand for 3-D manipulation of microscopic objects, IEEE Trans. Industr. Electron. 56 (4) (2009) 1121–1135.

[155] H. Wang, Q. Huang, Q. Shi, T. Yue, S. Chen, M. Nakajima, M. Takeuchi, T. Fukuda, Automated assembly of vascular-like microtube with repetitive single-step contact manipulation, IEEE Trans. Biomed. Eng. 62 (11) (2015) 2620–2628.

[156] S. Fatikow, J. Seyfried, S.T. Fahlbusch, A. Buerkle, F. Schmoeckel, A flexible microrobot-based microassembly station, J. Intelli. Robot. Syst. 27 (1) (2000) 135–169.

[157] E. Avci, K. Ohara, C.-N. Nguyen, C. Theeravithayangkura, M. Kojima, T. Tanikawa, Y. Mae, T. Arai, High-speed automated manipulation of microobjects using a two-fingered microhand, IEEE Trans. Industr. Electron. 62 (2) (2014) 1070–1079.

[158] E. Kim, M. Kojima, Y. Mae, T. Arai, High-speed manipulation of microobjects using an automated two-fingered microhand for 3d microassembly, Micromachines 11 (5) (2020) 534.

[159] X. Liu, Q. Shi, H. Wang, T. Sun, Y. Ning, Q. Huang, T. Fukuda, Automated fluidic assembly of microvessel-like structures using a multimicromanipulator system, IEEE/ASME Trans. Mechatron. 23 (2) (2018) 667–678.

[160] H. Wang, J. Cui, Z. Zheng, Q. Shi, T. Sun, X. Liu, Q. Huang, T. Fukuda, Assembly of RGD-modified hydrogel micromodules into permeable three-dimensional hollow microtissues mimicking in vivo tissue structures, ACS Appl. Mater. Interfaces 9 (48) (2017) 41669–41679.

[161] H. Hu, H. Wang, Q. Shi, H. Tao, Q. Huang, T. Fukuda, Rail-guided robotic system for multi-configuration cooperative micromanipulation based on formation control, in: IEEE International Conference on Real-time Computing and Robotics (RCAR), 2021, IEEE, 2021, pp. 165–170.

[162] H. Wang, Q. Shi, T. Sun, X. Liu, M. Nakajima, Q. Huang, P. Dario, T. Fukuda, High-speed bioassembly of cellular microstructures with force characterization for repeating single-step contact manipulation, IEEE Robot. Autom. Lett. 1 (2) (2016) 1097–1102.

[163] H. Onoe, T. Okitsu, A. Itou, M. Kato-Negishi, R. Gojo, D. Kiriya, K. Sato, S. Miura, S. Iwanaga, K. Kuribayashi-Shigetomi, et al., Metre-long cell-laden microfibres exhibit tissue morphologies and functions, Nat. Mater. 12 (6) (2013) 584–590.

[164] W. Yang, Y. Haibo, G. Li, Y. Wang, L. Liu, High-throughput fabrication and modular assembly of 3D heterogeneous microscale tissues, Small 13 (5) (2017) 1602769.

[165] L. Dai, D. Lin, X. Wang, N. Jiao, L. Liu, Integrated assembly and flexible movement of microparts using multifunctional bubble microrobots, ACS Appl. Mater. Interfaces 12 (51) (2020) 57587–57597.

[166] Z. Ge, L. Dai, J. Zhao, Y. Haibo, W. Yang, X. Liao, W. Tan, N. Jiao, Z. Wang, L. Liu, Bubble-based microrobots enable digital assembly of heterogeneous microtissue modules, Biofabrication 14 (2) (2022), 025023.

[167] E. Kim, M. Takeuchi, A. Hasegawa, A. Ichikawa, Y. Hasegawa, Q. Huang, T. Fukuda, Construction of hepatic-lobule-like 3-D vascular network in cellular structure by manipulating magnetic fibers, IEEE/ASME Trans. Mechatron. 25 (1) (2019) 477–486.

[168] S. Tasoglu, E. Diller, S. Guven, M. Sitti, U. Demirci, Untethered micro-robotic coding of three-dimensional material composition, Nat. Commun. 5 (1) (2014) 1–9.

[169] S.E. Chung, X. Dong, M. Sitti, Three-dimensional heterogeneous assembly of coded microgels using an untethered mobile microgripper, Lab Chip 15 (7) (2015) 1667–1676.

[170] Y. Liu, G. Li, L. Haojian, Y. Yang, Z. Liu, W. Shang, Y. Shen, Magnetically actuated heterogeneous microcapsule-robot for the construction of 3D bioartificial architectures, ACS Appl. Mater. Interfaces 11 (29) (2019) 25664–25673.

[171] X. Yang, R. Tan, L. Haojian, Y. Shen, Magnetic-directed manipulation and assembly of fragile bioartificial architectures in the liquid–liquid interface, IEEE/ASME Trans. Mechatron. (2022).

[172] J.P.K. Armstrong, J.L. Puetzer, A. Serio, A.G. Guex, M. Kapnisi, A. Breant, Y. Zong, V. Assal, S.C. Skaalure, O. King, et al., Engineering nisotropic muscle tissue using acoustic cell patterning, Adv. Mater. 30 (43) (2018) 1802649.

[173] H. Cai, Z. Ao, H. Liya, Y. Moon, W. Zhuhao, L. Hui-Chen, J. Kim, F. Guo, Acoustofluidic assembly of 3D neurospheroids to model Alzheimer's disease, Analyst 145 (19) (2020) 6243–6253.

[174] M. Caleap, B.W. Drinkwater, Acoustically trapped colloidal crystals that are reconfigurable in real time, Proc. Natl. Acad. Sci. 111 (17) (2014) 6226–6230.

[175] P. Chen, Z. Luo, S. Guven, S. Tasoglu, A.V. Ganesan, A. Weng, U. Demirci, Microscale assembly directed by liquid-based template, Adv. Mater. 26 (34) (2014) 5936–5941.

[176] P. Chen, S. Guven, O.B. Usta, M.L. Yarmush, U. Demirci, Biotunable acoustic node assembly of organoids, Adv. Healthc. Mater. 4 (13) (2015) 1937–1943.

[177] T. Ren, P. Chen, L. Gu, M.G. Ogut, U. Demirci, Soft ring-shaped cellu-robots with simultaneous locomotion in batches, Adv. Mater. 32 (8) (2020) 1905713.

[178] S. Tasoglu, C.H. Yu, H.I. Gungordu, S. Guven, T. Vural, U. Demirci, Guided and magnetic self-assembly of tunable magnetoceptive gels, Nat. Commun. 5 (1) (2014) 1–11.

[179] N.G. Durmus, H.C. Tekin, S. Guven, K. Sridhar, A.A. Yildiz, G. Calibasi, I. Ghiran, R.W. Davis, L.M. Steinmetz, U. Demirci, Magnetic levitation of single cells, Proc. Natl. Acad. Sci. 112 (28) (2015) E3661–E3668.

[180] A. Tocchio, N.G. Durmus, K. Sridhar, V. Mani, B. Coskun, R. El Assal, U. Demirci, Magnetically guided self-assembly and coding of 3D living architectures, Adv. Mater. 30 (4) (2018) 1705034.

[181] S. Tasoglu, C.H. Yu, V. Liaudanskaya, S. Guven, C. Migliaresi, U. Demirci, Magnetic levitational assembly for living material fabrication, Adv. Healthc. Mater. 4 (10) (2015) 1469–1476.

[182] V.A. Parfenov, V.A. Mironov, E.V. Koudan, E.K. Nezhurina, P.A. Karalkin, F.D.A.S. Pereira, S.V. Petrov, A.A. Krokhmal, T. Aydemir, I.V. Vakhrushev, et al., Fabrication of calcium phosphate 3D scaffolds for bone repair using magnetic levitational assembly, Sci. Rep. 10 (1) (2020) 1–11.

[183] V.A. Parfenov, Y.D. Khesuani, S.V. Petrov, P.A. Karalkin, E.V. Koudan, E.K. Nezhurina, F.D.A.S. Pereira, A.A. Krokhmal, A.A. Gryadunova, E.A. Bulanova, et al., Magnetic levitational bioassembly of 3D tissue construct in space, Sci. Adv. 6 (29) (2020) eaba4174.

CHAPTER 6

Robotic cell biopsy for disease diagnosis

Mingyang Xie[a] and Adnan Shakoor[b]
[a]College of Automation Engineering, Nanjing University of Aeronautics and Astronautics, Nanjing, China
[b]Control and Instrumentation Engineering Department, King Fahd University of Petroleum and Minerals, Dhahran, Saudi Arabia

1 Introduction

Studies on bulk tissues only provide a statistical average of several actions taking place in different cells. Single cell investigation may disclose that genetic changes caused by tumorigenesis-related signaling pathways lead to the change of a healthy cell to a cancerous cell [1]. Single cell surgery at the subcellular level, such as the manipulation or removal of subcellular components or/and organelles [2] from a single cell, is increasingly needed to study diseases and their causes [3]. The techniques associated with single cell surgery include preimplantation and diagnosis [4], and understanding the organelle and subcellular activities [1], to name a few.

Several micro-/nanomanipulation tools for intracellular-level surgery have been developed over the past few decades. For instance, a scanning ion conductance microscopy-based nanobiopsy system utilized electro-wetting to uptake samples of cytoplasm into a glass nanopipette in a minimally invasive way, and then used high-throughput sequencing technology to analyze the extracted cellular material [5]. Atomic force microscopy (AFM) tips were utilized to extract specific mRNA from live cells, and nanorobots were fabricated for puncturing cells [6]. Similarly, carbon nanotubes placed at the tip of a glass pipette were used at the single cell level for transferring fluids, interrogating cells, and performing optical and electrochemical diagnostics [7].

These subcellular-level cell surgery tasks are currently performed by highly skilled operators. Depending on the operator's skill, various cell surgery tasks, such as positioning micropipette tip relative to the position of cell, inserting the tip at a specific location inside the cell after membrane perforation, and extracting the micropipette from the cell, are very tedious and time consuming. Several robotic micromanipulation systems for

Robotics for Cell Manipulation and Characterization
https://doi.org/10.1016/B978-0-323-95213-2.00001-6

123

manipulating biological cells have been reported in the literature [8–11]. Several studies have examined the automation of embryo biopsy, which include force measurements during micropipette insertion [12], contact detection [13], automatic position selection for zona pellucida dissection [14], and automated micropipette control and embryo rotation tracking [15], to name a few. Similarly, several studies have attempted to automate this mechanical enucleation process by using different micromanipulation tools, such as micropipettes, magnetically driven microtools, and microfluidic chips [16]. Note that all the aforementioned approaches were applied to large-scale cells of ~80 μm. Single cell biopsy for small cells, such as many human cells (~20 μm), is more challenging because of their small size, irregular shape, and flexible membranes, all of which demand high-precision manipulation.

Cell injection is a process of inserting foreign materials into cells. Several cell injection systems that allow semi- [17] and fully automatic [13,18] injection processes have been proposed. However, cell biopsy differs from cell injection in several aspects. First, cell injection is accomplished at an arbitrary location within the cell, usually at the center [19]. In cell biopsy, the micropipette tip must be precisely inserted inside the cell at the position of the desired organelle. Precisely accessing the spatial position of the organelle within the cell, especially for small suspended cells, such as most human cells, in dimensions of ~20 μm, is challenging. Thus, a highly precise positioning control is needed for cell biopsy. Second, cell biopsy requires a number of extra operations, such as organelle aspiration control inside the micropipette and cell holding control during extraction. Therefore, cell organelle biopsy is more complex than cell injection.

In this chapter, an automated organelle biopsy system for small dimension cells is introduced. The main contributions of this work are described as follows. First, a robotic single cell organelle biopsy system is developed for small cells with dimensions of less than 20 μm. This system is capable of performing a complete automated single cell biopsy task, including cell positioning, organelle detection, aspiration, and organelle release, at a specific location. Second, to meet the high requirements for organelle positioning accuracy, a microfluidic single cell patterning device is designed, which can compress and pattern cells inside rectangular channels in a 1D array, hence, simplifying the automated process. Third, under a visual-based robust control algorithm, the system can be applied to biopsy a broad range of cells with high manipulation accuracy while addressing model uncertainties and external disturbances. To demonstrate that the proposed system can be applied to perform organelle biopsy of different categories of cells such as cancer and human cell lines, which differ in their cytoskeleton and filament structures that surround the organelles, both suspension cell (NB_4 cancer cell) and adherent cell (human dermal fibroblast cell (HDF)) were processed in our experiments. The extraction rate of mitochondria biopsy and cell viability for different size of micropipette is analyzed. The functionality of mitochondria and the viability of the biopsied cell after removal of mitochondria is also investigated.

Development of an automatic system and enabling technologies to enhance organelle biopsy can greatly benefit its clinical applications. For example, embryo biopsy at early stage of embryonic development provides an alternative to termination of pregnancy for parents who are at risk of transmitting genetic disease to their offspring [20]. Organelle biopsy can help improve therapeutic cloning or reproductive cloning [21,22]. The technique can help identify compartmentalization of mRNAs. This is important after occurrence of an injury, because an altered spatial distribution of mRNA can result in neurodegenerative diseases and intellectual disability [23].

The rest of this chapter is organized as follows. Section 2 introduces the cell surgery system setup and the microfluidic cell patterning device. Section 3 describes the organelle biopsy process and control strategy. Section 4 demonstrates the effectiveness of the automatic organelle biopsy approach through experimental studies. Finally, conclusions of this study are presented in Section 5.

2 Generic small cell biopsy system

2.1 System development

The proposed single cell biopsy system comprises three modules, for execution, control, and sensing. The executive module comprises an X-Y-Z positioning stage (Prior Scientific, ProScan) with a 10 nm resolution along the X-Y direction and a z-axis that controls the height of the objective lens. A micropipette (B-100-78-15, Sutter Instrument) for organelle extraction is mounted on an in-house assembled 3-DOF micromanipulator (Newport, 9067-XYZ-R-V), which is connected to a modified microinjector based on an original product of CellTramvario, Eppendorf. A microfluidic cell patterning device and a homemade pressure regulator for vacuum generation inside the cell patterning device are also included in the executive module. The control module includes two motion controllers for controlling the 3-DOF manipulator and the X-Y-Z positioning stage. The sensing

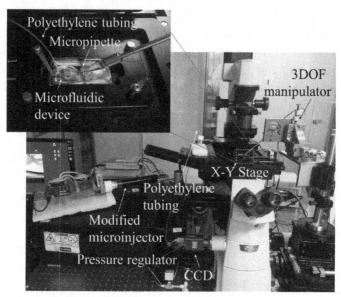

Fig. 6.1 Automated single cell biopsy system [24].

module includes a CCD camera (FOculus, FO124SC) and a microscope (Nikon TE2000) with a 60× objective. A GUI interface is developed with Visual C++. All mechanical setups are supported by an antivibration table. Fig. 6.1 illustrates the system setup.

2.2 Cell patterning

To simplify the cell manipulation process, cells are patterned in a 1D array. The PDMS-based microfluidic cell patterning device (MCPD) is constructed by placing many rectangular shaped channels in parallel. This design eliminates the need for a second manipulator (for cell holding) and simplifies the cell/organelle searching and extraction processes. Fig. 6.2 presents the schematic design of the microfluidic device. The MCPD consists of an aspirating layer and a cell patterning layer. The arrangement of these two layers produces a rectangular hole to pattern and hold each cell firmly. A small cylindrical gap called the "store" is added near the front of the channel to hold and store the biopsied sample, as shown in Fig. 6.2. The store was created to keep the extracted organelles at the defined locations for easy relocation for further analysis, and to prevent them from being lost by drifting away. Moreover, the PDMS in front of the store helps to clear off cell debris from the micropipette tip when it penetrates into the PDMS to eject the

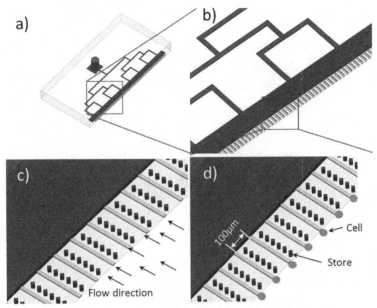

Fig. 6.2 Schematic design of the cell patterning microfluidic device. (A) Full view of the microfluidic device, (B) close-up view of the inset, (C) flow direction, and (D) cell trapping [24].

nuclei into the store. Two nearby channels are separated by a distance of 100 μm to ensure that only one channel and a store can be visualized within the field of view, thereby simplifying the image processing for detecting the position of the organelle for a cell present in the field of view.

2.3 Cell compression for organelle positioning

Figs. 6.3 and 6.4 illustrate the cell assembly, which uses hydrodynamic drag force inside channels for organelle positioning. Depending on the size of the organelle to be biopsied, either the cell can be trapped at the opening of the

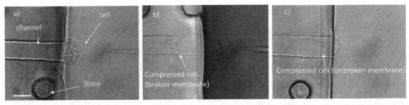

Fig. 6.3 Cell (NB_4) patterning by a microfluidic device. (A) Cell is trapped in front of channel, (B) cell compressed and hold inside channel, and (C) cell condition after compressing it [24].

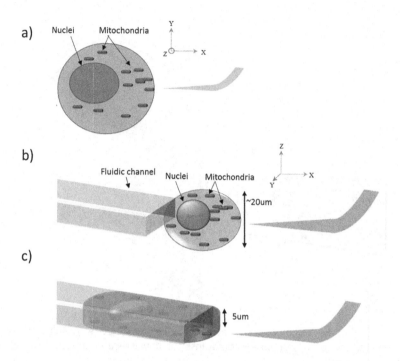

Fig. 6.4 Confining the position of the mitochondria along the z-axis by compressing the cell. (A) 2D camera image view (*top view*), (B) 3D representation of the cell before compression, and (C) 3D representation of the cell after compression inside the fluidic channel [24].

PDMS channel or compressed inside the PDMS channel. For a large organ-elle such as the nucleus (7–9 μm) of NB$_4$ cells, its position within a cell of ∼20 μm can be easily accessed. Therefore, the automated biopsy of a large organelle can be achieved after holding the cell at the opening of the chan-nel. However, for a small organelle, the cell must be compressed to confine the 3D position for the following reasons.

First, unlike the nucleus, small organelles such as mitochondria (∼2 μm) are distributed in a 3D space within the cell. The 2D images (see Fig. 6.4A) obtained through florescence microscopy are combinations of stacked images that are obtained at distinct depths along the z-axis. To perform mitochondria biopsy in an automatic way, it is challenging to precisely position the 1–2 μm micropipette tip to coincide with the position of the mitochondria along the z-axis (see Fig. 6.4B), which requires the use of complex image processing algorithms to obtain the accurate position of small organelles along the z-axis.

Second, compressing the cell helps to accurately determine the negative pressure, applied by the motor-controlled injector, needed to aspirate the mitochondria into the micropipette. If the micropipette tip is inserted into the cell where the mitochondrion is positioned relatively far from the micropipette tip along the z-axis (which is common for cells in their suspended state), then a large volume of cytoplasm must also be aspirated along with the mitochondria, which can cause significant damage to the cell. Without restricting the position of the small organelle along the z-axis, the negative pressure applied through the computer-controlled injector to aspirate the organelle may vary for each cell, thereby making the biopsy results inconsistent, or requiring a human operator to control the amount of aspiration content.

Considering the adherent cells that contain organelles in a confined position within the height of the cell, a careful compression of suspended cells can restrict the position of small organelles without damaging cell. Therefore, by compressing the cell, the distance between the micropipette tip and the small organelle along the z-axis is restricted by the height of the channel, which is 5 µm only, as shown in Fig. 6.4C.

Furthermore, the beveled micropipette, along with the compressed mode of the cell, which tightens the cell membrane, facilitates the perforation of the cell membrane. For the above reasons, instead of using complex image processing algorithms to determine the 3D position of the mitochondria as well as complex control of the motion along the z-axis and of the biopsy injector, the organelle biopsy process was simplified by aspirating approximately 95% of the cell into the soft channel of the PDMS microfluidic device (see Fig. 6.4C). Experiments were conducted to observe how the different heights and widths of rectangular channels affected cell trapping and its vitality. It was found that a channel with a 5 µm height and 15 µm width can hold and compress a cell of 20 µm diameter without rupturing its cell membrane (Fig. 6.3C).

3 Cell biopsy process

3.1 Procedures of automatic organelle extraction and release

The procedure to conduct automatic organelle extraction is outlined as follows. The system detects the position of the florescence-labeled cell organelle, such as the nucleus or mitochondria, by using an image processing algorithm. The stage then automatically moves along the y-axis to align the organelle with the micropipette tip, as shown in Fig. 6.5A and B. After

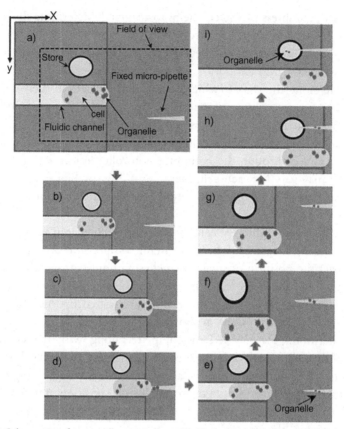

Fig. 6.5 Schematic of organelle extraction. (A–E) Steps of the organelle extraction process and (F–I) steps of releasing the organelle into the store [24].

alignment, the stage approaches the micropipette tip along the x-axis to ensure that the intended organelle coincides with the micropipette tip (Fig. 6.5C). Then, the microinjector aspirates the organelle into the micropipette tip (Fig. 6.5D).

The controller then moves the stage away from the micropipette to extract the micropipette from the cell (Fig. 6.5E). The aspirated organelle can be placed in the store (see Fig. 6.5F–I) automatically for further analysis. To release the organelle inside the store, the stage moves 15 μm along the y-axis to bring the store within the microscope field of view (Fig. 6.5F). The center of the store is detected with an image processing technique, and the stage moves the cell patterning device along the y-axis to align the micropipette with the center of the store (Fig. 6.5G). The stage then moves further along the x-axis to perforate the micropipette tip into PDMS, such that the center of the store coincides with the micropipette tip (Fig. 6.5H). Finally, the microinjector ejects the organelle into the store, automatically (Fig. 6.5I).

3.2 Motion control

The dynamic equation of a 3-DOF motorized stage (see Fig. 6.6) can be determined by using the following Lagrange's equation of motion:

$$M\ddot{q} + B\dot{q} + G(q) = \tau \tag{6.1}$$

where $M \in \mathfrak{R}^{3\times3}$ denotes the inertial matrix of the system, $q = \left[q_x, q_y, q_z\right]^T \in \mathfrak{R}^{3\times1}$ is the stage position coordinate, $B \in \mathfrak{R}^{3\times3}$ denotes the damping friction coefficient, $G = [0, 0, mg] \in \mathfrak{R}^{3\times1}$ models the gravity

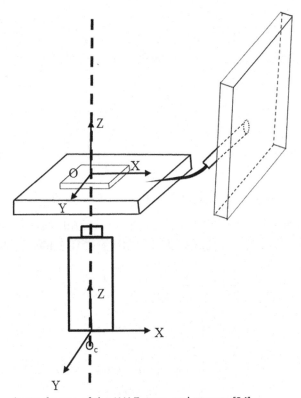

Fig. 6.6 Coordinate frames of the *X-Y-Z* stage and camera [24].

force, and $\tau = [\tau_x, \tau_y, \tau_z]^T$ is the input torque. Both M and B are diagonal and positive definite.

Given the model uncertainty and external disturbances (e.g., mechanical vibration), the dynamic model of the motorized stage can be generalized as follows:

$$(M_0 + \Delta M)\ddot{q} + (B_0 + \Delta B)\dot{q} + (G_0 + \Delta G) + \tau_d = \tau \qquad (6.2)$$

where M_0, B_0, and G_0 denote the nominal values of the system parameters, ΔM, ΔB, and ΔG represent the system uncertain values, and τ_d denotes the external disturbance torque.

A sliding variable is defined as follows:

$$S_r = \dot{q} - \dot{q}_r = \dot{q} - (\dot{q}_d - \alpha q_e) = \dot{q}_e + \alpha q_e \qquad (6.3)$$

where $q_e = q - q_d$ is the position error, and α is the positive constant.

The dynamic model of the motorized stage in Eq. (6.1) can be expressed in terms of S_r as follows:

$$M\dot{S}_r + BS_r + G(q) = \tau + M(-\ddot{q}_d + \alpha\dot{q}_e) + B(-\dot{q}_d + \alpha q_e) \qquad (6.4)$$

Then, a robust sliding controller is designed as follows to achieve trajectory tracking:

$$\tau = \tau_0 + \tau_s \qquad (6.5)$$

where

$$\tau_0 = -M_0(-\ddot{q}_d + \alpha\dot{q}_e) - B_0(-\dot{q}_d + \alpha q_e) + G_0 \qquad (6.6)$$

$$\tau_s = -K\tanh(S_r) \qquad (6.7)$$

$$\tanh(S_r) = [\tanh(S_{r1}), \tanh(S_{r2}), \tanh(S_{r3})]^T \qquad (6.8)$$

where $K = \text{diag}\{k_1, k_2, k_3\}$ is the positive control gain matrix, $S_r = [S_{r1}, S_{r2}, (S_{r3})]^T \in \Re^{3\times1}$, and $\tanh(\cdot)$ is the standard hyperbolic tangent function.

Substituting Eq. (6.5) together with Eqs. (6.6) and (6.7) into Eq. (6.2) yields the following closed-loop dynamics:

$$M_0\dot{S}_r + B_0 S_r + K\tanh(S_r) + \Delta f = 0 \qquad (6.9)$$

where $\Delta f = \Delta M\ddot{q} + \Delta B\dot{q} + \Delta G + \tau_d$.

The stability of the closed-loop dynamics in Eq. (6.9) can be analyzed by defining a Lyapunov function candidate as follows:

$$V = \frac{1}{2}S_r^T M S_r \qquad (6.10)$$

By differentiating V with respect to time and utilizing the closed-loop Eq. (6.9),

$$
\begin{aligned}
\dot{V} = S_r^T M \dot{S}_r &= -S_r^T K S_r - S_r^T (K \tanh(S_r) + \Delta f) \\
&= -S_r^T K S_r - \sum_{i=1}^{3} S_{ri}[k_i \tanh(S_{ri}) + \Delta f_i]
\end{aligned}
\tag{6.11}
$$

When $S_{ri} > 0$, given $k_i > |\Delta f_i|$ for $i = 1, 2, 3$,

$$
\sum_{i=1}^{3} S_{ri}[k_i \tanh(S_{ri}) + \Delta f_i] > 0
\tag{6.12}
$$

Note that k_i is easily chosen to be larger than $|\Delta f_i|$, based on the fact that $\dot{q}$ and $\ddot{q}$ are not so large in cell manipulation process and, hence, Δf_i is bounded.

When $S_{ri} < 0$, given $k_i > |\Delta f_i|$,

$$
\sum_{i=1}^{3} S_{ri}[k_i \tanh(S_{ri}) + \Delta f_i] > 0
\tag{6.13}
$$

Therefore,

$$
\dot{V} \leq 0
\tag{6.14}
$$

By using the LaSalle invariance principle, it is concluded that the system states are eventually driven into the sliding surface $S_r = 0$, and thus, the tracking errors q_e and $\dot{q}_e$ converge to zero.

3.3 Biological tests

Biological tests were further conducted on the extracted organelles to characterize their biological functionality. The mitochondria membrane potential has been widely used as an indicator of mitochondria heath [25,26]. A significant reduction in membrane potential of mitochondria is an indicator of mitochondria death. By using the proposed organelle biopsy system, it must be verified that the membrane potential of the extracted mitochondria can remain the same as that before the mitochondria biopsy, therefore, demonstrating that the extracted mitochondria has survived. The invasiveness of biopsy process on cell by calcium imaging before and after the mitochondria biopsy was reported in [6]. It can be demonstrated that the membrane potential of the extracted mitochondria and calcium flux of the remaining cell after mitochondria biopsy remained almost unchanged, as reported in detail in the next section.

4 Experiments

4.1 Material preparation

NB_4 and HDF cells were used in the organelle biopsy experiments. NB_4 cells were maintained in a petri dish containing RPMI 1640 with 2 mM L-glutamine inside a humidified incubator with an atmosphere of 37°C and 5% CO_2. HDF cells were kept in Dulbecco's modified Eagle's medium (DMEM, Gibco), 100 U/mL penicillin supplemented with 10% fetal bovine serum (FBS, Gibco), and 100 U/mL streptomycin inside a humidified incubator with an atmosphere of 37°C and 5% CO_2. The cells were then stained with MitoTracker Red CMXRos (M7512, Thermo Fisher Scientific) and Hoechst (33342, Thermo Fisher Scientific) for the florescence labeling of mitochondria and nucleus, respectively. The HDF cells were treated with trypsin (Sigma) for detachment from the petri dish, while the NB_4 cells were directly used in the suspended state.

A glass micropipette (OD 1 mm and ID 0.78 mm) was pulled using a micropipette puller (P-2000, Sutter Instrument), and a gradual taper with a final tip size of 0.5 μm was obtained. To ensure the perforation of the flexible cell membrane and minimize damage to the cell (for the mitochondria biopsy), the micropipette tip was beveled using a BV-10 microelectrode beveller (Sutter Instrument) at 30 degree with a final tip size of about 1–3 μm for mitochondria biopsy and 5–8 μm for nucleus biopsy. The micropipette was bent 4–5 mm from the tip to ensure that the tip can access the cell when the latter is docked in the cell patterning device.

Images (640 × 480) were captured using a CCD camera under 60× magnification. The physical size of each pixel was approximately 0.12 μm. To evaluate motion performance, the position tracking errors of the X-Y-Z stage were measured during the experiments.

4.2 Organelle extraction

To biopsy mitochondria and nucleus of single cells, successful extraction of these organelles from cells is the first important action. Fig. 6.7 shows images of the automatic extraction of mitochondria from HDF cells. These images have been converted into grayscale for better visualization. To perform mitochondria extraction in a minimally invasive way, a beveled and nearly cylindrical micropipette with a tip size of about 2–3 μm was utilized (Fig. 6.7A). The mitochondria position (Fig. 6.7B) was automatically identified using the proposed image processing algorithm and was aligned with the micropipette tip along the x-axis, as shown in Fig. 6.7C. The cell patterning

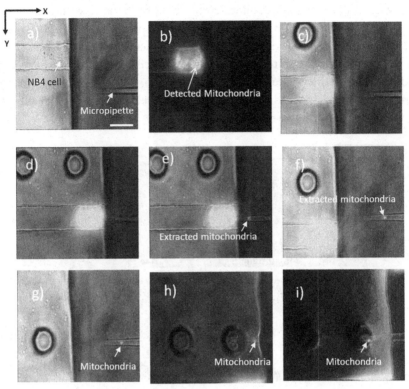

Fig. 6.7 Automatic extraction of the mitochondria. Image is presented in gray scale for better visualization. (A) Micropipette and cell before biopsy. Scale bar is 10 μm. (B) Mitochondria detection. The small square pointed with an arrowhead shows the detected position of the mitochondria. (C) Moving the stage to the aligned micropipette with the detected mitochondria. (D) Moving the stage toward the micropipette. (E) Aspirated mitochondria. (F) Moving the stage back to extract the micropipette out of the cell and moving the stage 15 μm along the y-axis to bring the store within the field of view. (G) Moving the stage along the y-axis to align the micropipette with the center of the store. (H) The micropipette reaching the center of the store. (I) Releasing the mitochondria [24].

device then moved toward the micropipette tip until the micropipette tip coincided with the detected position of the mitochondria after perforating the cell membrane (Fig. 6.7D). The injector aspirated the mitochondria into the micropipette (Fig. 6.7E), and the cell patterning device moved toward its original position (along the x-axis) to extract the micropipette from the cell. To ensure that the store is completely present within the field of view, the stage moved 15 μm along the y-axis (Fig. 6.7F) and was positioned along the y-axis to align the center of the store with the micropipette tip. The stage moved along the x-axis to collocate the center of the store with the micro-pipette tip. The injector ejected the aspirated mitochondria (Fig. 6.7F) by applying positive pressure at the back of the micropipette.

Extraction of the mitochondria from NB_4 cells was also conducted by following the same sequence as explained above, with the results presented in Fig. 6.8. The aspirated mitochondria can be released outside the cell pat-terning device as shown in Fig. 6.8F.

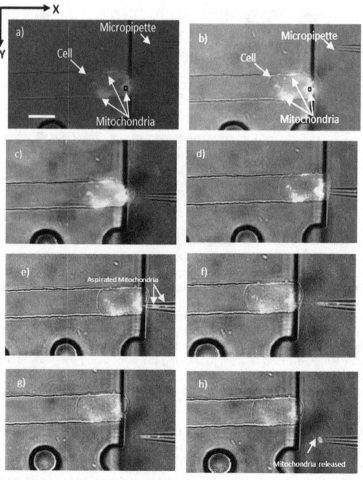

Fig. 6.8 Automated extraction of the mitochondria from an NB_4 cell. (A) Detected position of the mitochondria shown with small black rectangle. (B) Gray scale image of (A). Moving the stage toward the micropipette. (C) Micropipette alignment with mitochondria. Aspirated of mitochondria. (D) Moving the stage back toward its home position. (E) Moving the stage along the y-axis. (F) Ejecting the mitochondria. Scale bar is 15 μm [24].

Fig. 6.9A and B shows the position error obtained during the mitochondria extraction experiment. The initial positions of the detected mitochondria and microneedle tip were (17.40 μm, 32.48 μm) and (33.29 μm, 15.49 μm), respectively. The stage moved along the y-axis to align the mitochondria with the micropipette tip and reached the (17.40 μm, 15.48 μm) position within 5 s, during which the position error along the y-axis converged to zero. The stage then moved along the x-axis to coincide with the micropipette tip. The stage eventually reached the (33.27 μm, 15.48 μm) position with a final position accuracy of 0.02 and 0.01 μm along the x and y-axes, respectively.

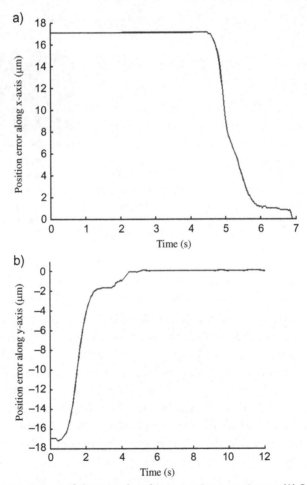

Fig. 6.9 Position errors of the mitochondria extraction experiment. (A) Position error along x-axis and (B) position error along y-axis [24].

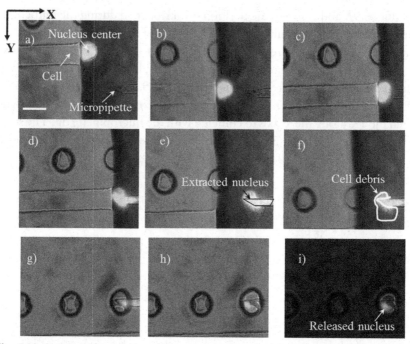

Fig. 6.10 Automated extraction of the nucleus from an NB$_4$ cell. (A) Detection of the nucleus center. Scale bar is 15 μm. (B) The nucleus aligned with the tip of the micropipette. (C) Moving the stage toward the micropipette. (D) Aspirating the nucleus. (E) Moving the stage toward its original position and 15 μm along the y-axis. (F) Alignment the micropipette with the center of the store. (G) Penetrating the micropipette tip into the store. (H, I) Releasing the nucleus [24].

Fig. 6.10 shows the automatic extraction of the nucleus from an NB$_4$ cell. Given the large organelle to cell size ratio, cell compression is not necessary for nucleus biopsy and the cell can be trapped at the channel inlet. The center of the nucleus was automatically identified using the image processing algorithm and was aligned with the micropipette tip along the x-axis, as shown in Fig. 6.10B. The cell patterning device moved toward the micropipette tip for perforation until the micropipette tip coincided with the center of the nucleus after perforating the cell membrane (Fig. 6.10C). The injector aspirated part of the nucleus into the micropipette (Fig. 6.10D). Afterward, the cell patterning device moved toward its original position along the x-axis, while the stage moved 15 μm along the y-axis (Fig. 6.10E) to bring the store within the field of view. An image processing algorithm was used to detect the position of the organelle store space, and the cell patterning device was controlled to move along the y-axis to align the center of the store with the micropipette tip (Fig. 6.10F). The stage then moved along x-axis until the store center met the micropipette tip (Fig. 6.10G). Cell debris was cleared off from the micropipette tip during perforation of micropipette tip into the MCPD and only the biopsied nucleus remained in the micropipette tip (Fig. 6.10G). The injector ejected the biopsied nucleus from the tip into the organelle store space (Fig. 6.10H and I) by applying positive pressure.

The micropipette can easily be pulled out from the cell if the extracted organelle is small (e.g., mitochondria), but this task is challenging if the extracted organelle is large (e.g., nucleus). For aspirating large organelles, a greater negative pressure is applied to the micropipette during organelle aspiration, which may restrict the separation of the micropipette and the cell. For example, the nucleus, when occupying more than 60% of the cell volume, requires a relatively high negative pressure to be aspirated into the micropipette tip with a size ranging from 5 to 8 μm. If same amount of negative pressure is applied at the back of the fluidic channel, the cell will rupture during the extraction of the micropipette, while some of its parts (cell debris) remain attached to the micropipette. Given the large size of the micropipette for nucleus biopsy, the biopsy of the nucleus is very invasive and the remaining part of the cell is considered cell debris. Consequently, instead of controlling the pressure at the back of the channel or extracting the micropipette from the cell, the cell debris is removed by penetrating the micropipette into PDMS. This rinsing process can be achieved automatically during the movement of the micropipette toward the store (see Fig. 6.10F–G). The negative pressure required at the back of the micropipette during organelle aspiration depends on the amount of the cellular content to be biopsied and the tip size of the micropipette. These control parameters were fixed through calibration experiments.

Fig. 6.11 shows the experimental results of the nuclei extraction performed on the HDF cell in the same sequence as explained above for the nuclei biopsy of NB$_4$ cell.

4.3 Biological tests on extracted organelles and the remaining cells

Upon completion of organelle extraction, a number of biological tests on the extracted organelles and the remaining cells were conducted. The functionality test of the extracted mitochondria was performed by observing the mitochondria membrane potential before and after the mitochondria biopsy. Cells were stained with JC-1 Dye, Mitochondrial Membrane

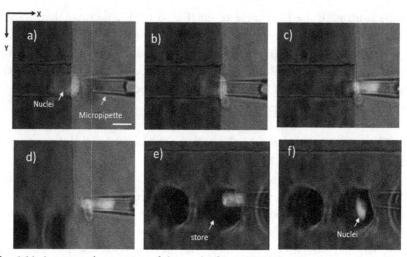

Fig. 6.11 Automated extraction of the nuclei from HDF cells. (A) Aligning the nucleus with the tip of the micropipette. Scale bar is 15 μm. (B) Moving the stage toward the micropipette. (C) Aspirating nucleus. (D) Moving the stage toward its original position and 15 μm along the y-axis. (E) Penetration into the PDMS. (F) Releasing the nucleus [24].

Potential Probe (T3168, Thermo Fisher Scientific), prior to the biopsy experiment. Depolarized mitochondria shows a change in the florescence emission from yellowish orange or reddish orange (depending on the filter type) to green. The testing results of our study are illustrated in Fig. 6.12. After extraction from NB_4 cells, the mitochondrial membrane potential remained unchanged, as shown in Fig. 6.13. Mitochondria stained with JC-1 were biopsied from 25+ NB_4 cells and in all these cases, mitochondrial membrane potential was observable, which indicated 100% mitochondria survival rate.

The survival rate of the remaining cell after mitochondria extraction was also examined. The cells were stained with Fluo-4 Calcium Imaging

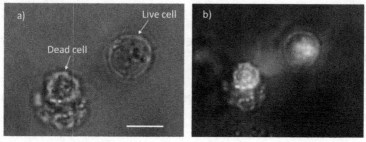

Fig. 6.12 JC-1 staining of NB_4 cells. (A) Bright field image and (B) fluorescence image (GFP filter) [24].

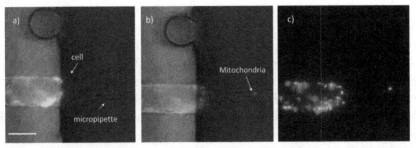

Fig. 6.13 Functionality test of mitochondria after mitochondria biopsy. (A) Florescence image before biopsy, (B) florescence image after biopsy, and (C) florescence image after biopsy (RFP filter) [24].

(F10489, Thermo Fisher Scientific) and calcium flux was detected before and after the extraction of mitochondria [5]. The size and shape of the micropipette tip, the amount of biopsied content from the cell, and depth of penetration of the micropipette tip are the major factors that affect disruption of the cell during cell biopsy. With smaller micropipettes and less cell penetration, the mitochondria biopsy can be achieved in a minimally invasive way. A larger sized micropipette tip as well as larger aspirated volume of the cell can increase the chance of successful mitochondria biopsy; however, it increases cell disruption as well. It was found that, with the proposed methodology in this chapter, the mitochondria extraction rate and cell viability was inversely proportional. With the small micropipette tip and less aspiration of cellular content, less damage to the cell is seen. However, with these smaller micropipettes, the chance of aspiration of mitochondria into the micropipette tip is also reduced.

Table 6.1 shows the extraction rate of mitochondria and nuclei of NB_4 and HDF cells achieved with our automated organelle biopsy system. The organelle extraction rate is calculated from the ratio of the number of cells that are successfully biopsied to the total number of cells used in

Table 6.1 Organelle extraction rate.

Cell	Organelle	No. of cells used for organelle extraction	No. of cells with successful organelle extraction	Organelle extraction rate (%)
NB_4	Mitochondria	125	78	62.4
NB_4	Nuclei	125	86	68
HDF	Mitochondria	50	29	58
HDF	Nuclei	50	36	72

experiments. The mitochondria and nuclei biopsy for the NB_4 cell could achieve an extraction rate of 62.4% and 68%, respectively. Similarly, the organelle extraction rate of HDF cells was 58% for the mitochondria and 72% for the nuclei. The mitochondria biopsy obtained a relatively low efficiency because, in some cases, the detected mitochondria may be located near the back portion of the cell and the micropipette may be unable to perforate the membrane of the nuclei. The extraction rate of nuclei was affected by the sliding of the micropipette tip over soft cell membrane due to the relatively large size of the micropipette tip. The cell survival rate is calculated from the ratio of the number of cells that are successfully biopsied to the number of cells surviving after the mitochondria biopsy. The negative pressure at the back of the micropipette was calibrated while monitoring the successful aspiration of the mitochondria into the micropipette. A larger negative pressure corresponds to a large volume change inside micropipette, which consequently aspirates more cellular content during biopsy, and thus, increases the mitochondria extraction rate.

5 Conclusions

A robotic surgery system to achieve automated single cell biopsy for small cells with dimensions of less than 20 μm in diameter is developed in this study. The cells are patterned with a microfluidic device, and a template-matching-based image processing algorithm is developed to automatically measure the position of the desired organelles inside the cell. A position control algorithm is used to drive the microfluidic device toward a micropipette, and organelle extraction is then performed in an automatic way. A sliding nonlinear PID controller is developed to enhance the manipulation accuracy and robustness. The experiments were performed successfully on small suspended (NB_4 ~18 μm) and adherent cells (HDF ~20 μm), in which the mitochondria and nucleus were biopsied using the proposed system and methodology. The experimental results have shown that the developed system can automatically perform organelle extraction with high precision and repeatability, thereby reducing the human fatigue and error involved in complex single cell biopsy procedures. The presented single cell organelle biopsy system can be further implemented to perform mitochondria transfer and injection, somatic cell nuclear transfer, cell injection, preimplantation genetic diagnosis, and cloning.

Due to heterogeneity among cells of different sizes, a pressure of 24.9 psi may be large enough for cells that are smaller in size compared to others, and therefore, some neighboring channels may become empty. Similarly, larger cells may not be properly aspirated into the channels, and thus, may not be compressed properly, which can reduce the efficiency of mitochondria extraction. Another limitation is the deep penetration of the needle tip, which may occur when the organelle is located near the center of the cell.

References

[1] J. Liu, et al., Voyage inside the cell: microsystems and nanoengineering for intracellular measurement and manipulation, Adv. Mat. Res. 1 (2015) 15020.

[2] H. Pertoft, T.C. Laurent, Isopycnic Separation of Cells and Cell Organelles by Centrifugation in Modified Colloidal Silica Gradients, 1977.

[3] C. Schubert, Single-cell analysis: the deepest differences, Nature 480 (7375) (2011) 133.

[4] P. Braude, et al., Preimplantation genetic diagnosis, Nat. Rev. Genet. 70 (12) (2002) D18.

[5] P. Actis, et al., Compartmental genomics in living cells revealed by single-cell nano-biopsy, ACS Nano 8 (1) (2013) 546.

[6] Hayakawa, et al., Fabrication of an on-chip nanorobot integrating functional nanomaterials for single-cell punctures, IEEE Trans. Robot. 30 (1) (2014) 59–67.

[7] R. Singhal, Z. Orynbayeva, R.V.K. Sundaram, J.J. Niu, S. Bhattacharyya, E.A. Vitol, M.G. Schrlau, E.S. Papazoglou, G. Friedman, Y. Gogotsi, Multifunctional carbon-nanotube cellular endoscopes, Nat. Nanotechnol. 6 (1) (2011) 57–64.

[8] K. Sakaki, et al., Localized, macromolecular transport for thin, adherent, single cells via an automated, single cell electroporation biomanipulator, IEEE Trans. Biomed. Eng. 60 (11) (2013) 3113–3123.

[9] M. Xie, et al., Out-of-plane rotation control of biological cells with a robot-tweezers manipulation system for orientation-based cell surgery, IEEE Trans. Biomed. Eng. (2018) 1.

[10] H. Ladjal, J.L. Hanus, A. Ferreira, Micro-to-nano biomechanical modeling for assisted biological cell injection, IEEE Trans. Biomed. Eng. 60 (9) (2013) 2461–2471.

[11] N. Ogawa, et al., Microrobotic visual control of motile cells using high-speed tracking system, IEEE Trans. Robot. 21 (4) (2005) 704–712.

[12] D.H. Kim, S. Yun, B. Kim, Mechanical force response of single living cells using a microrobotic system (Proceedings. ICRA '04. 2004 IEEE International Conference on Robotics and Automation, 2004), 2004.

[13] H.B. Huang, et al., Robotic cell injection system with position and force control: toward automatic batch biomanipulation, IEEE Trans. Robot. 25 (3) (2009) 727–737.

[14] W.H. Wang, X.Y. Liu, Y. Sun, Contact detection in microrobotic manipulation, Int. J. Robot. Res. 26 (8) (2007) 821–828.

[15] Z. Wang, W.T. Ang, Automatic dissection position selection for cleavage-stage embryo biopsy, IEEE Trans. Biomed. Eng. 63 (3) (2016) 563–570.

[16] M. Hagiwara, et al., High Speed Enucleation of Oocyte Using Magnetically Actuated Microrobot on a Chip, IEEE, 2012.

[17] Y. Sun, B.J. Nelson, Biological Cell Injection Using an Autonomous MicroRobotic System, Sage Publications, Inc., 2002.

[18] X. Liu, et al., Nanonewton force sensing and control in microrobotic cell manipulation, Int. J. Robot. Res. 28 (8) (2009) 1065–1076.

[19] Y.T. Chow, et al., A high-throughput automated microinjection system for human cells with small size, IEEE/ASME Trans. Mechatron. 21 (2) (2016) 838–850.

[20] J.R. Vermeesch, T. Voet, K. Devriendt, Prenatal and pre-implantation genetic diagnosis, Nat. Rev. Genet. 17 (10) (2016) 643.

[21] G.D. Greggains, et al., Therapeutic potential of somatic cell nuclear transfer for degenerative disease caused by mitochondrial DNA mutations, Sci. Rep. 4 (2014) 3844.

[22] H. Li, et al., Generation of biallelic knock-out sheep via gene-editing and somatic cell nuclear transfer, Sci. Rep. 6 (2016) 33675.

[23] E. Tóth, et al., Single-cell nanobiopsy reveals compartmentalization of mRNAs within neuronal cells, J. Biol. Chem. 293 (13) (2018) 4940–4951.

[24] A. Shakoor, M. Xie, et al., Achieving automated organelle biopsy on small single cells using a cell surgery robotic system, IEEE Trans. Biomed. Eng. 66 (8) (2019) 2210–2222.

[25] Jérme, et al., Cancer cell mitochondria are direct proapoptotic targets for the marine antitumor drug lamellarin D, *Cancer* Res. 66 (6) (2006) 3177–3187.

[26] E.M. Pasini, et al., A novel live-dead staining methodology to study malaria parasite viability, Malar. J. 12 (1) (2013) 190.

CHAPTER 7

3D force-feedback optical tweezers for experimental biology

Edison Gerena and Sinan Haliyo
Sorbonne Université, CNRS, Institut des Systèmes Intelligents et de Robotique, Paris, France

1 Robotic bio-manipulation

The individuality of cells has been recognized since they were first observed through optical microscopy. However, this fact has been underestimated by the bulk cell culture experiments, which interpret the phenomena through an "average" cell [1,2]. It is only recently that the heterogeneity of cells of the same type started to be considered relevant to biological phenomena [3,4], for instance, in targeted therapeutics [5] and drug resistance studies [6].

In this context, the ability to manipulate and characterize individual biological objects has become a major scientific challenge. The need to perform tasks such as single-cell deformation, stimulation, rotation, or transportation has called for new emerging techniques that allow working at the unicellular scale (Fig. 7.1).

Extending the interaction and manipulation capabilities at the submillimeter scale, microrobotics have become a key tool in micro/nanoscale science and technology for both industry and academic research, in a wide range of fields such as healthcare, biotechnology, and manufacturing [7].

Operating at the microscopic scale is highly demanding, mainly due to the specific physical effects governing the microworld and the size limitations. Specific methods for working in the microworld in terms of fabrication, actuation, and sensing have been proposed in the last two decades. The scaling of physical laws, robotic micromanipulation fundamentals, and bioengineering application have been extensively reviewed in recent years [8–12].

1.1 Contact methods for robotic bio-manipulation

Single-cell manipulation tasks are usually performed using 3-axis cartesian robots consisting of motor-driven micromanipulators with prismatic joints and equipped with different end effectors.

Robotics for Cell Manipulation and Characterization
https://doi.org/10.1016/B978-0-323-95213-2.00010-7
145

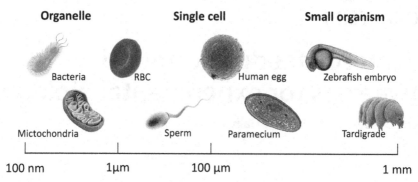

Fig. 7.1 Microrobotic aims to manipulate microobjects such as organelles (<1 μm), single cells (1–100 μm), and small organisms (<1 mm).

Those micromanipulators are not themselves micron-sized necessary, however, as the robots performs task in the microworld their end effector must be micron-sized. These robots must have a submicron positioning resolution and a range of motion of a few millimeters. Micromanipulators are generally driven with DC motors and controlled in closed loop with a typical accuracy of 0.2 μm. Piezoelectric manipulators may have a higher resolution, but, due to the resolution limit of optical microscopes (light diffraction limit ∼0.25 nm), they are slightly less used in biological manipulation. End effectors are tools mounted on a micromanipulator to interact with microobjects for tasks such as pick and place. The usual end effectors are micropipettes, microgrippers or atomic force microscopy (AFM) cantilevers.

1.2 External energy-fields for robotic bio-manipulation

There is an increasing demand to manipulate objects in confined environments such as microfluidic devices in order to decrease flow disturbances, contamination, or evaporation of the culture medium, rendering external actuators unusable. All these constraints call for the replacement of current techniques with noncontact manipulation methods. Remote actuation using different external energy fields, like magnetic, acoustic, or optical, has appeared as a very promising solution in applications where a high spatial maneuverability and precision are required. These contact-free solutions eliminate adhesive effects, which have a significant impact on handling tasks. However, these processes are often limited to a restricted range of materials.

Accordingly, a great effort has been made in the search for solutions for the remote actuation and power supply of untethered microrobots (i.e., untethered and mobile robots where the entire body is micrometer-sized) to serve as remote manipulators. These remote manipulators are able to handle different types of objects, in terms of shape and physical properties, and more advanced capabilities can be included by design, such as sensing or targeted drug delivery [8]. The new capabilities of these untethered micro-manipulators promise immense potential for a variety of in vivo and in vitro biomedical applications.

2 Optical micromanipulation

Among the variety of noncontact methods, optical trapping offers several advantages in the handling and the mechanical characterizations of small biological samples, ranging from hundreds of nanometers to tenths of millimeters, in a confined environment such as microfluidic devices.

Optical manipulation exploits light radiation pressure to noninvasively trap and position suspended microobjects and cells with a nanometer resolution; resulting in a contamination-free, contact-free, and label-free method for cell manipulation in their original culture medium. Its compatibility with other optical techniques, especially microscopy, implies that it is highly appropriate for lab-on-chip systems and microfluidic devices. In addition, controlling multiple focal spots enables it to simultaneously trap and manipulate several objects in 3D [13].

Optical manipulation has become a popular tool for manipulating single biological samples, successfully demonstrated in a large range of in vivo [14] and in vitro [15] experiments such as the trapping of red blood cells in living animals [16], the immobilization of bacterial cells for nanoscopy [17], and cell rotation for tomographic imaging [18], among others.

In the next section, the principle of optical tweezers (OTs) and design considerations will be discussed. For further details, readers are directed to two excellent reviews, written by Neuman and Block [19], and Bowman and Padgett [20].

2.1 Principle and development of optical tweezers

Optical trapping uses the optical force generated by the energy and momentum exchange between light and particles to drive the mechanical motions of microobjects with nanometer resolution. When an individual light ray is refracted by the surface of an object, its path is altered, resulting in a light momentum change. This change of momentum generates a pressure upon the surface, called radiation pressure.

In his paper, "History of optical trapping and manipulation" [21], Ashkin relates his first experiment to look for particle motion by laser radiation pressure using transparent latex spheres and a Gaussian laser beam. The speeds corresponds to Ashkin's estimation, demonstrating that radiation pressure is pushing them. However, an additional unexpected force also appeared, which strongly pulled particles located in the fringes of the beam into the high-intensity region in the beam center.

From the ray optics theory and the intensity profile of the Gaussian beams, Ashkin deduced that the pressure force can be decomposed into two components. An optical force related to the gradient of light intensity, called gradient force, and a force component pointing in the direction of the incident light, called scattering force. Based on this, he conceived the first stable optical trap between two opposing diverging Gaussian beams [22].

Ashkin et al. reported in 1986 the use of a single strongly focused Gaussian beam to stably trap particles in three dimensions, which Ashkin originally called single-beam gradient force trap, today known as OTs [23].

OTs use the pressure of light radiation to trap an object by providing a steep potential well in all axes, generated by the balance between the gradient forces and the scattering force. An equilibrium position is created near the narrowest focal point, known as beam waist, as the gradient force becomes dominant above the scattering force (cf. Fig. 7.2). Dielectric particles around this position are attracted and trapped by the OTs. Then, by controlling the position of the beam waist, the motion of the trapped object is controlled.

This method was then extended, to manipulate atoms [24], cells [25], and viruses [26], by the same group. Since then, the applications of the OTs have not stopped increasing in diverse fields such as molecular biology, cell biology, materials science, and quantum physics. Ashkin was awarded the Nobel Prize in Physics in 2018 for this development and its application to biological systems.

2.2 Biological applications of optical manipulation

OTs appear frequently in experiments on biological samples. They are used to trap and manipulate red blood cells within subdermal capillaries in living mice, inducing an artificial clot [16]. They are used to impose local forces on cell contacts in the early Drosophila embryo and measure tension at cell junctions [27] and in the study of propulsion forces and motility efficiency of the unicellular parasites' trypanosomes [28]. Optical manipulation of

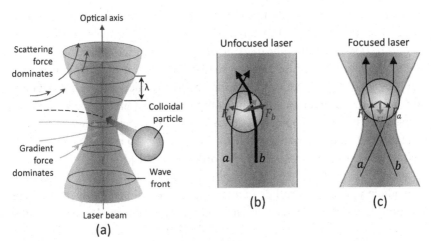

Fig. 7.2 An optical tweezers uses forces exerted by a strongly focused beam of light to trap small objects. By controlling the position of the focal point, the position of the trapped object can be controlled. (A) The variation in the momentum of photons produces a force in the direction of the intensity gradient (gradient force) and a force in the direction of the beam propagation (scattering force). (B) In an unfocused laser, the gradient is directed toward the laser center and the scattering force pushes the object in the laser direction. (C) In a focused laser, near the focal point, the intensity gradient is directed toward the laser focus. Thus, the gradient force is able to balance out the scattering force, and a particle can be trapped in three dimensions. ((A) Adapted from D.G. Grier, A revolution in optical manipulation, Nature 424 (6950) (2003) 810–816.)

injected nanoparticles (fluorescent, red) inside living zebrafish was shown [29]. Out-of-plane cell rotation through two OTs has been demonstrated [30]. Large deformation of human red blood cells causes them to be subjected to direct stretching by OTs [31]. OTs are used to measure the force during kinesin motor stepping and structured RNA molecule unfolding [32]. An OTs-based single-cell sorting device was proposed in Ref. [33]. Mouse embryonic stem cells have been patterned into precise complex cellular microenvironments using holographic OTs [34].

2.3 Optical manipulation setups

The essential elements to implement an optical trap are the laser and a high numerical aperture (NA) objective. A Gaussian laser beam is traditionally chosen as laser source, but other modes can also be used [35]. The wavelength depends on the size of the trapped objects and the environment. For the manipulation of micrometrical objects, laser wavelengths that fall into the visible or near-infrared spectral region are usually used. As a final consideration, if biological material is manipulated, a wavelength between 800 and 1064 nm is usually used in order to minimize photonic damage due to the laser absorption and the oxidation reaction of organic material. The power of the laser must be sufficient to produce a trap and depends on multiple parameters such as the components that are between the laser output and the microscope objective that may induce energy losses. In general terms when using micron-scale beads, for every 10 mW of power delivered, the maximum forces are in the order of 1 pN [36].

A microscope objective with high NA (typically 1.2–1.4 NA) is required to produce a gradient of sufficient intensity to overcome the dispersion force and produce a stable optical trap for microscopic objects. In this way, the microscope objective plays a fundamental role in the efficiency and stability of the traps. The working distance and the immersion medium of the objective (oil, water, or glycerol) must be chosen depending on the application specifications, such as the aqueous capture medium and the depth to which it will be manipulated. Spherical aberrations in the beam waste highly degrade the trap's performance [19]. The various optical manipulation setups differ mainly in how the force is measured and the way in which the position of the trap is controlled.

2.3.1 Force measurement in optical traps

In an optical trap, the tightly focused laser beam provides an equilibrium position near the beam waist. When the trapped objects are laterally or axially displaced away from the equilibrium position, the optical forces act to pull it back toward the equilibrium position. For small displacements, the force applied to the particle is a linear function, and the optical trap can be compared to a simple spring system [37], which follows Hooke's law. In this model, the restoring optical force is proportional to the distance from its equilibrium position, and the trap stiffness is proportional to the light intensity.

This particularity has led to the use of optical trapping for quantitative force measurements [15], such as the strength of intermolecular bonds [38], the stiffness of a cell membrane [39], or intracellular measurements for microrheology [40].

Force calibration of arbitrary objects with reasonable accuracy is still a challenge as the optical stiffness depends on several factors such as the material properties, the refractive index of both object and environment, the sizes and shapes of trapped objects, etc. For these reasons, silicon or polystyrene microspheres are often used as a force *probe* to indirectly manipulate and sense the force during micromanipulation tasks. By using artificial microbeads that are uniform in material and shape, the optical force applied on the microbead can be obtained. For small displacements, the optical force is described by a spring model [37]:

$$\mathbf{F}_{opt} = \mathbf{K} \times (\mathbf{P}_{laser} - \mathbf{P}_{probe})$$

where $\mathbf{P}_{laser}$ and $\mathbf{P}_{probe}$ represent the laser and the probe position, respectively; $\mathbf{F}_{opt}$ is the 3D optical force; $\mathbf{K}$:$[K_x, K_y, K_z]$ is the trap stiffness, where K_x, K_y, K_z represent the stiffness in x, y, z directions, respectively. Therefore, the optical force is obtained by measuring the particle's displacement from the equilibrium trap position.

Well-established methods exist for precise stiffness calibration of spherical objects, such as power spectrum analysis, equipartition theorem, and Stokes' drag [19,41,42]. Hence, force sensing directly stems from the capability to detect the position of the laser and one of the trapped objects. In quantitative measurements with OTs, the optical trap is often fixed, and only the trapped object position information is needed. The determination of this position can be done by image-based position detection where the images of video cameras are used to extract the information, or by interferometric methods where beam displacements after the laser passes through the sample are measured via quadrant photodiode quadrant photodiode (QDP). In Section 3.1, further discussion in the methods for 3D high-speed force detection in OTs is presented.

2.3.2 Position control in optical traps

Displacement in OT systems can be achieved by moving the platform or moving the laser beam, referred to as passive actuation or active actuation, respectively. The incorporation of a motorized stage allows dynamic control of the sample chamber, while the manipulated object remains fixed on the trap. This type of actuation technique, referred to as passive actuation, greatly facilitates the trap calibration, and does not interfere with the stability of the trap.

The motion of the trapped object can also be done by moving the laser beam directly, referred to as active actuation, through active optical components that change the laser direction from the light path. In addition, it is possible to multiplex the trapping laser beam in order to simultaneously trap several objects using a unique laser source.

By rapidly deflecting the laser beam between several trap positions, it is possible to trap several objects, in such a way that the objects do not have time to spread between two laser actuations. Laser-deflection can be done by actuated mirrors, acousto-optic deflectors (AODs), or electro-optic deflectors (EODs). These methods are referred to as time-shared methods and are commonly limited to two dimensions.

Using active diffractive optical elements [13] such as spatial light modulators (SLM) or a digital micromirror device (DMD), several traps can be dynamically controlled, creating a number of diffraction spots at different 3D positions. This method has the ability to move traps in a three-dimensional space and is referred to as holographic OTs.

3 3D real-time force sensing in optical manipulation

Optical instruments, from the microscope of van Leeuwenhoek to super-resolved fluorescence microscopy, have always been precursors to great advances, especially in life sciences. OTs offer a similar potential, by augmenting visual feedback by mechanical interactions, making it possible to probe and characterize biological samples in a liquid environment. In an optical trap, the restoring optical force is related to the distance from the equilibrium position. This particularity has led to the use of OTs for quantitative force measurements, such as the strength of intermolecular bonds [38] or the stiffness of a cell membrane [39].

In most of these experiments, an inert bio-compatible polystyrene bead attached to the molecule or the cell is used to indirectly handle the samples. This indirect approach is applied for two main reasons: because precise force calibration for nonspherical objects is still a complex challenge [43] and to avoid direct laser exposure in biological samples that can negatively influence their behavior or cause photo-damages. In the case of a spherical object, the effect of the trap is akin to a linear stiffness around the focal point of the beam. Indeed, the optical trap acts as a spring of stiffness proportional to the light intensity [19] and the force acting on the object can be obtained by measuring its motion [37].

Current robotics research on optical manipulation are focused on the implementation of control techniques based on position. In existing techniques, direct trapping and manipulation of objects are performed [44–46]. Some recent works introduce controls and planning approaches for indirect manipulation of cells using silica beads arranged into gripper formations [47,48]. However, to our best knowledge, none of the current robotics techniques exploit the inner capability of force sensing of optical trapping, possibly due to the high dynamics effects in the microworld (e.g., quasiinstantaneous acceleration) and the limitations in the current detection methods.

Considering the trapped probe as the end-effector, the external forces acting on the probe can be used in the feedback path to close the control loop. Those forces can be provided by tracking the motion of the probe under the optical microscope. Considering the force range and the trap stiffness, piconewton resolution is reachable. This kind of performance would make OTs a formidable apparatus for micromanipulation in general and for biology and biochemistry in particular.

3.1 Background and related work

As mentioned earlier, the precise force detection is directly related to the capability to track the position of the trapped object. Nevertheless, high dynamics effects at the microscale and the limitations of the optical microscope render most classical methods useless. Hence, the lack of robust 3D tracking reduces most applications to simple planar tasks. Also, the latency and low bandwidth hurts the system stability and its real-time capabilities [49]. Improving the automation requires making improvements to the force sensing performance. Moreover, making this feedback available with low latency at high bandwidth will open the road to complex applications requiring closed-loop control.

Several methods to improve the tracking have been developed. Most commonly, a QDP is used to sample the position of the target at tens of kHz with nanometric precision [50]. Nonetheless, this method is vulnerable to occlusions and disturbances and works reliably only on isolated objects. Hence, it is not suitable for micromanipulation tasks.

An alternative is image processing through video cameras integrated into the microscope. Visual tracking algorithms using CMOS cameras offer straightforward implementation, but their bandwidth is limited by the amount of data that should be transmitted and processed. Real-time force information is hardly available. State of the art of real-time visual tracking algorithms on commercial CMOS cameras can rarely exceed 60 Hz [51]. Limiting the imaging to a smaller region of interest (ROI) can accelerate processing, at the detriment of resolution and precision [52]. To obtain more than 1 kHz sampling rate, the ROI have to be decreased to 60×60 pixels, which compromises the working scale and resolution. By combining a high-speed CMOS with tracking implemented on graphics processing units (GPUs), 3D tracking at several kHz is reported [53]. This approach is, however, fairly complex as it requires special knowledge and hardware.

As an alternative to classical CMOS cameras, event-based cameras were proposed [54]. These bio-inspired sensors are frame-free and eliminate data redundancy by design. They have been shown to allow 2D tracking at a speed in the order of tens of kilohertz [54,55]. Because of the particularity of the image-data provided, well-known processing techniques and algorithms cannot be used. Further investigations are especially needed for real-time 3D robust tracking.

In this section, a 3D motion tracking technique using an event-based algorithm taking advantage of an event-based sensor is presented. It provides piconewton resolution and its bandwidth reaches 10 kHz. Its capabilities are demonstrated in a teleoperated 3D manipulation scenario with a haptic user interface. This kind of control scheme is very demanding and requires a feedback loop at 1 kHz for stability. Reliable and reproducible 3D exploration of biological surfaces for nonexpert users has been demonstrated for the first time (Tables 7.1 and 7.2).

3.2 Asynchronous time-based image sensor

Asynchronous time-based image sensor (ATIS) is the latest generation prototype of silicon retinas [56]. In this sensor, every pixel combines a brightness

Table 7.1 Comparison of tracking techniques under optical microscopy.

Method	Acquisition rate	Processing rate	Advantage	Limitation
QDP	MHz	MHz	Fast, high-precision	Small working space, only isolated objects
H-CMOS	kHz	<60 Hz	High precision	Not for real-time use
H-CMOS + GPU	kHz	kHz	Fast, high precision	Not for long time continuous use
H-CMOS + ROI	kHz	1 kHz	Fast, high precision	Small working space
EV-C	>100 kHz	>10 kHz	Fast, low consumption	No 3D applications

EV-C, event-based camera; *GPU*, graphics processing unit; *H-CMOS*, high-speed CMOS camera; *QDP*, quadrant photo-diode; *ROI*, region of interest.

Table 7.2 Summary of the force sensor parameters.

Parameters	Value		
OTs stiffness (pN/µm)[a]	$x = 12.3$	$y = 12.6$	$z = 1.5$
Force detection range (pN)[a]	$x = 36.9$	$y = 37.8$	$z = 4.5$
Force resolution (pN)[a]	$x = 0.3$	$y = 0.3$	$z = 0.25$
Haptic loop	1 kHz hard real-time		

[a]Depends on the laser power. Values for 300 mW laser power.

change detector and an additional exposure measurement circuit, which provide illuminance polarity. The most distinctive difference between the conventional CCD/CMOS camera and the silicon retina is the image capture mechanism. For the digital camera, a brighter image at a given pixel produces a larger intensity value at that pixel. The image sequence is collected frame by frame with a fixed sampling interval. While for silicon retina, for each pixel, as long as the outside stimuli or the input light intensity surpasses a threshold, an "event" is triggered independently and asynchronously without any global clock (see Fig. 7.3).

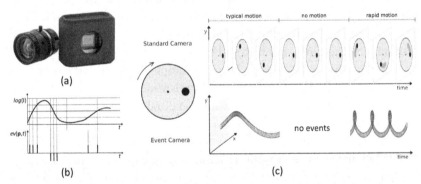

Fig. 7.3 Event-based camera principle. (A) Picture of a commercial event-based camera (Gen3 VGA-CD, Prophesee, https://www.prophesee.ai/). (B) Events generation principle of event camera single pixel. Events with +1 or −1 polarity are emitted when the change in log intensity exceeds a predefined threshold. (C) Data generation comparison between standard cameras and event-based cameras. In contrast to standard video frames, shown in the *upper graph*, a stream of events from an event camera, plotted in the *lower graph*, offers no redundant data output, only informative pixels or no events at all. *Gray dots* represent negative or positive events. *(Adapted from H. Kim, S. Leutenegger, A.J. Davison, Real-time 3D reconstruction and 6-DoF tracking with an event camera, in: European Conference on Computer Vision, Springer, 2016, pp. 349–364.)*

Each event contains the information of position, polarity, and time stamp. Thus, by conveying the time-stamped events instead of the whole frames, ATIS shows the advantages of having low latency of 15 µs for each coming event and being free of redundancy. Also ATIS achieves a larger dynamic range of 143 dB compared to conventional cameras, which means it is more capable of capturing light in different environments without under- or over-saturation. The temporal resolution is 100 kfps under 100 lux scene illumination, and the pixel bandwidth may be decreased by low illumination conditions in microscopy.

The absence of events when there are no changes in scene luminance implies the suppression of data redundancy, which is a common problem in high-speed CMOS tracking. ATIS has the capacity to be adaptable to high-speed on-line tracking due to its efficient encoding of movements. Thereby, the system captures only the dynamic information and the amount of data to be processed is therefore considerably reduced, which makes it possible to process at speeds up to 10 kfps [57]. More details of this technology are discussed in Refs. [56,58].

The earliest application of silicon retina is the vehicle detection in traffic monitoring [59]. Delbruck introduced several event-based vision methods on background noise filter, orientation feature extraction, and cluster tracking [60]. The event-based optical flow is presented by Benosman's group, which is shown to significantly reduce running time compared to the conventional method [61]. Event-based blob tracking algorithms are proposed by Lagorce, which allow the tracking of multiple shapes with the processing rate of several hundred kilohertz on a standard desktop PC [62]. By combining the epipolar geometric with the temporal constraints, real-time 3D reconstruction was proposed [63].

Although many applications of event-based vision are presented in macroscale, its usage in microscale is an almost unexploited area. Regnier's group proposed for the first time to use an event-based camera for tracking of objects under an optical microscope [55] and the same group has proposed the 2D particle tracking under OTs [54]. In the next section, we demonstrate the first 3D high-speed tracking system of microparticles under OTs with an event-based camera.

3.3 System description

An OTs set-up has been built. The optical scheme of the system is shown in Fig. 7.4. The unique feature of this set-up is the inclusion of an ATIS. It is

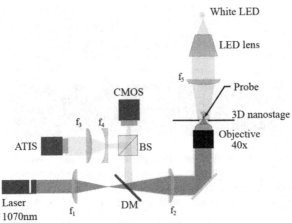

Fig. 7.4 Custom-designed optical system. The source is a 1070 nm laser with a maximum output of 10 W. An oil immersion objective (Olympus UPlanFLN 40x, NA 1.3) produces a fixed optical trap. Two microstages (x-y) and a 3D nanostage provide, respectively, coarse and fine positioning of samples. The illumination (LED, 3W) is reflected by a long pass dichroic mirror (900 nm cut-off) then is divided by an unpolarized beam splitter (R9:T1) into the silicon retina camera (ATIS, 240 × 304 pixels) and the CMOS camera (Basler, 659 × 494 pixels), where 90% of the light is led to ATIS for better brightness. $f_1 = 30$, $f_2 = 125$, $f_3 = 100$, $f_4 = -50$, $f_5 = 45$.

based on an inverted microscope where the same objective is used for both imaging and producing the optical trap. The laser source is a 1070 nm wavelength with a maximum output of 10 W. This laser beam is expanded first and directed into the oil immersion objective (Olympus UPlanFLN 40x, NA 1.3) to produce a fixed optical trap. Two microstages are used for x-y coarse positioning and a 3D nanostage is applied for x-y-z fine positioning of a petri dish. The illumination (LED, 3W) passes through a lens to create the collimated beam. After passing through the sample, all the outcome visible light is reflected by a long pass dichroic mirror to the cameras. Then the beam is divided by an unpolarized beam splitter cube (9:1) into the event-based camera (ATIS, 240 × 304 fully autonomous pixels) and the CMOS camera (Basler, 659 × 494 pixels), where 90% of the light is led to ATIS for better brightness. The CMOS camera provides the environment information and visual perception to operators, while ATIS is used to calculate the relative position between probe and OTs laser spot.

In a scene with a stable light environment, only the dynamic information stimulated by the moving object is recorded. In the case of the presented system, the trap center has a fixed position in the ATIS image. With an appropriate threshold, image variations are mostly generated on the contour

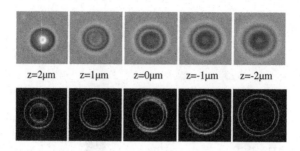

Fig. 7.5 Image of microbead (3 μm polystyrene) under the system. *First column*: The CMOS image at different z-displacements relative to the focus plane (*z* = 0). *Second column*: The corresponding ATIS image with 33 ms accumulation time. *Different colors* indicate positive or negative polarity of the events. The displacements are in micrometers.

of the trapped bead, as shown in Fig. 7.5. This information is exploited to infer the 3D motion of the probe.

3.4 3D tracking

Fig. 7.5 shows the conventional and event-based images of the probe. In ATIS image, accumulated events can be grouped in two concentric circles. Their center position is related to the planar motion of the probe while the radius is linked to the depth position. An event-based ring tracking algorithm is developed. It is used to recognize both circles, then to selectively extract center position and radius parameters from the inner one. Indeed, when the probe is in contact with other samples, the inner one is much more robust and stable. However, it is necessary to track both circles to be able to set them apart. The principle of ring tracking is to minimize the distance of the events' spatial coordinates and the ring model [64], as presented in Fig. 7.6.

Denoting $\mathbf{E}(\mathbf{p}, t)$ as an event occurs at time t with spatial location $\mathbf{p} = (x, y)$ in ATIS coordinate. $U(t)$ is defined as the set of useful events' locations at time t:

$$U(t) = \{\mathbf{E}(\mathbf{p}, t) | \mathbf{p} \in ROI(t)\} \qquad (7.1)$$

where $ROI(t)$ is the ROI of one circle model at time t [65]. Since the fitting methods are vulnerable to noise, the ROI is used as solution to filter the outliers.

Suppose that the unknown circle model's parameters at time t is $\mathbf{C_t}(\mathbf{P_t}, R_t)$, where $\mathbf{P_t} = (X_t, Y_t)$ is the circle center's position and R_t is the radius. Then, a fast noniterative algebraic fit [66] minimizes the cost function:

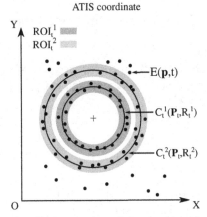

Fig. 7.6 Working principle of the event-based ring tracking algorithm.

$$\min_{\mathbf{P_t}\in\mathbb{R}^2 R_t\in\mathbb{R}} \sum_{\mathbf{p_k}\in U(t)} \| d(\mathbf{p_k}, \mathbf{P_t})^2 - R_t^2 \|^2 \qquad (7.2)$$

where $\mathbf{p_k}$ is the kth event's location in $U(t)$. $d(\mathbf{p_k}, \mathbf{P_t})$ is Euclidean distance between $\mathbf{p_k}$ and the circle center $\mathbf{P_t}$. Introducing parameters $A = -2X_t$, $B = -2Y_t$, and $C = X_t^2 + Y_t^2 - R_t^2$, Eq. (7.2) can be written as a linear least square problem:

$$\min_{\mathbf{P_t}\in\mathbb{R}^2 R_t\in\mathbb{R}} \sum_{\mathbf{p_k}\in U(t)} \| Ax_k + By_k + C + x_k^2 + y_k^2 \|^2 \qquad (7.3)$$

Solving A, B, and C gives the parameters of circle $(\mathbf{P_t}, R_t)$. This single circle tracking method is used to parameterize the two concentric circles of the ring (inner $\mathbf{C_t^1}$ and outer $\mathbf{C_t^2}$). All incoming events occurring in a considered time interval are assigned to the closest circle model. Events at the intersection of two ROIs are discarded to reduce the ambiguity. This condition makes sure that the inner and outer circle will not merge into one. The ROIs are updated accordingly (Algorithm 7.1).

4 Evaluation of tracking

4.1 Range and resolution

A 3 μm silica microsphere fixed to the sample holder is used to evaluate the tracking. Fig. 7.7 gives the real displacement, as reported by the nanostage versus tracked motion in pixels for all directions as well as the calculated pixel/micrometer transform.

ALGORITHM 7.1 Event-based robust ring fitting.
Require : Events $E(p, t)$
  1: **for** every *step* **do**
  2:    Update the content of $U^1(t)$ and $U^2(t)$ according to Eq. (7.1)
  3:    Estimate C_t^1 and C_t^2 parameters according to Eq. (7.2).
  4:    Update output : $[X_t^1, Y_t^1, R_t^1]$
  5:    Update ROI_{t+1}^1 and ROI_{t+1}^2
  6: **end for**

The detectable motion range is 6 and 7 μm for x and y, respectively, with less than 3% standard deviation (SD). On the z-axis, the relative radii variance is shown in Fig. 7.7C. The linear detection range is about ±2 μm around the focus plane with 5% SD. This 3D detection range (6 × 7 × 4μm) is sufficient since the linearity of the trap stiffness is valid around one diameter of the trapped bead [37], here 3 × 3 × 3μm^3.

With 204 × 304 pixels, the theoretical resolution is 23.8 nm/pixel in x and y and 166.6 nm/pixel in z. Practically, the bead center position and radius were estimated with subpixel accuracy using the circle tracking algorithm. This sensitivity varies in different illumination conditions and working environment.

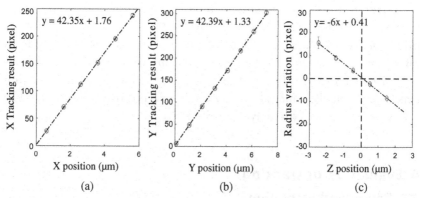

(a)   (b)   (c)

Fig. 7.7 The 3D detection range of the system. (A, B) The inner center position x and y (in pixels) with the microsphere's position (in μm). (C) The relative inner radii change (in pixels) with the microsphere's axial position (in μm). At focus plane, the inner radius is 64 pixels. The linear regression coefficients show the conversion between pixels in ATIS coordinates and micrometers in the world coordinates.

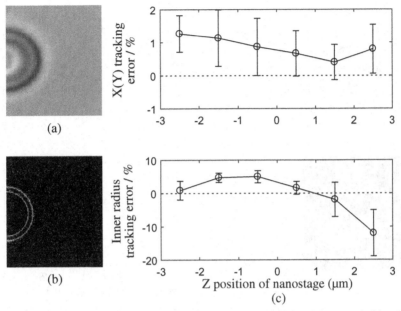

Fig. 7.8 Robustness under 30% partial occlusion situation. (A) Image recorded by the CMOS camera. (B) Corresponding image recorded by ATIS with accumulation time of 33 ms. (C) 3D tracking error under 30% occlusion at different depths. The tracking error is determined as the difference between the detection and the ground truth, divided by the detection range in corresponding axis.

4.2 Robustness

Robustness here is defined as the ability to extract the 3D position of the target from noisy data, or even a small subset of data. It will be tested under the two most commonly encountered situations partial occlusion, that is, only part of the target image is captured, and obstacles disturbances, that is, to track the target among many obstacles.

In the occlusion test, the microsphere's image is partially out of the view of ATIS. The tracking errors for 30% occlusion of the inner circle are shown in Fig. 7.8C. As can be seen, for less than 30% occlusion, the tracking error and the SD are less than 5% for both lateral and axial detection.

In the obstacles' disturbance test, the target bead (in the center) is surrounded by two other similar objects as shown in Fig. 7.9A. They are fixed on petri dish and animated with a sinusoidal movement. The tracking result is shown in Fig. 7.9C. The lateral tracking errors are less than ±5% with 2% SD in within ±2.5 μm. The axial errors are less than ±20% around the focus plane with less than 10% SD. As the image plane moves far away from focus plane, the radius error increases up to 30%.

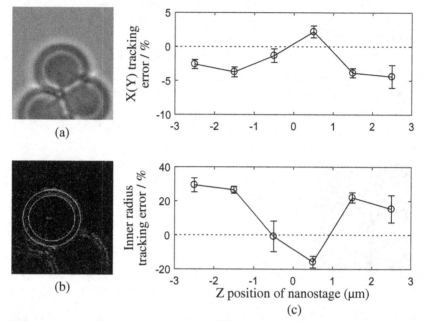

Fig. 7.9 Robustness in obstacles disturbance situation. (A) Image recorded by the CMOS camera. (B) Corresponding image recorded by ATIS with accumulation time of 33 ms. (C) The mean and SD of the 3D tracking error at different depths.

4.3 Computational load

The algorithm is implemented in C++ on a hard real-time framework. The testing relies on a 2.9 GHz dual core desktop PC, with a total CPU load about 50% of its power and a memory consumption of about 4 MB. The average running time for each iteration is less than 60 μs, with less than 2 μs deviation. Then the system is successfully pushed to 10 kHz real-time sampling rate.

To sum up, a cutting-edge solution for obtaining 3D robust high-speed force during the micromanipulation tasks has been developed. It is the first time to our knowledge to allow this performance in an OTs system. This development paves the way for achieving 3D stable closed-loop OTs manipulation systems. It also brings great potential to various applications as a powerful 3D high-speed force sensor.

4.4 3D haptic feedback optical tweezers

The proposed force sensor methods discussed in this section could be used in fully automated systems or teleoperated systems. In order to show its benefits and validate the proposed force sensing, a teleoperation scenario with haptic feedback is implemented. This is a demanding application from the control point of view as the stability and transparency of haptic feedback coupling requires a sampling of 1 kHz [49,67]. Previous works presented haptic feedback on OTs but were limited to 2D [68,69]. The 3D tracking allows for a real spatial coupling in this case.

4.5 Haptic coupling

Haptic feedback requires a bilateral control scheme, as presented in Fig. 7.10. In this scheme, the user handles the haptic interface (Omega.3, ForceDimension). Its position is used to control the motion of the sample through the nanostage. The force measured on the probe is feed back to the user with an appropriate gain.

Motion of the master device is scaled down by 6×10^{-4} to drive the trap position relatively to sample holder. Actually, the trap position is fixed, and the mobile part is the nanostage holding the samples. The measured forces are magnified by 1×10^{12} and sent back to the user. A single PC (Intel Xeon core, 2.93 GHz) operating under a real-time cokernel Linux and RTOS APIs Xenomai is used to control the system. The control-loop runs with a force refresh rate of 1 kHz.

4.6 Calibration

The force on the trap is calculated using the optical force model [37]:

$$\mathbf{F}_{opt} = \mathbf{K} \times \left(\mathbf{P}_{laser} - \mathbf{P}_{probe}\right) \tag{7.4}$$

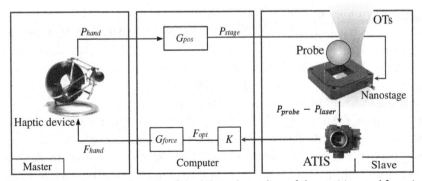

Fig. 7.10 The schematic diagram of the bilateral coupling of the position and force in the haptic OTs system.

where $\mathbf{P}_{laser} - \mathbf{P}_{probe}$ represents the displacement between the laser and the probe position as obtained from the tracking method. $\mathbf{K}$ is the stiffness of the trap.

Practically, to obtain the laser position, the positions of trapped probe before touching anything are recorded for a period of time and the average position is considered as the position of OTs, which is around (120, 150) in the ATIS coordinate. The stiffness is calculated experimentally using the equipartition method [19]. For an object in a harmonic potential:

$$\frac{1}{2}k_B T = \frac{1}{2}\mathbf{K}\langle d\rangle^2 \tag{7.5}$$

where k_B is Boltzmann's constant, T is absolute temperature, and $\langle d\rangle^2$ is the displacement variance (Brownian motion) of the particle from its trapped equilibrium position. Considering 300 mW laser power it is estimated in x-axis, y-axis, and z-axis as $K_x = 12.3$ pN/μm, $K_y = 12.6$ pN/μm, and $K_z = 1.5$ pN/μm, respectively, under room temperature of 25.5°C. The axial stiffness is less than the lateral stiffness, which is normal according to [70]. This indicates that during manipulation, the loss of the trapped object is more likely caused by the reacting force in the axial direction. The same laser power will be used for the following experiments, under the same condition, that is, temperature, laser power, medium, etc. Stiffness adjustments may be required depending on the specific applications, then the calibration process will be reconducted for obtaining the OTs' stiffness.

5 3D haptic experiments on biologic samples

Biologic samples are chosen to illustrate the use of the system in a real-world scenario. Red blood cells (RBC) are easy to acquire and have an 3D irregular dumbbell-shaped profile, and hence, are well suited for this illustration. They are fixed in 4% formaldehyde for biological stability. Probes are 3 μm polystyrene beads incubated in PBS and ethylenediaminetetraacetic acid (EDTA) solution to prevent the surface sticking. First, we squeeze the cell from the top to verify the axial force sensation, after that, we contour the shape of an RBC from the sides in a 3D haptic exploration, and then, we will explore two transparent RBCs attached together by following along a ∞ shaped path. Finally, preliminary user experiments will be implemented (Fig. 7.11).

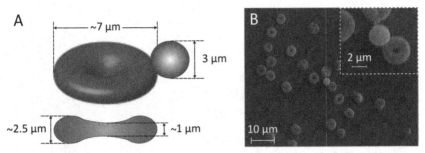

Fig. 7.11 (A) Schematic illustration with dimensions of RBCs and probe. (B) Scanning electron microscope (SEM) image of in vitro RBCs.

5.1 *Z*-axis haptic feedback

The first experiment will validate the force feedback on *z*-axis. The RBC sample is fixed on the bottom of a petri dish. The user first traps a bead to serve as a probe, then moves the sample holder as to position the probe above a fixed RBC. The planar motion of the nanostage is then artificially blocked and the user controls only the *z* motion until the probe touches the cell and he/she feels an obvious counterforce. The experimental results are shown in Fig. 7.12. Small fluctuations are caused by the 3D Brownian motion of the probe, which are largely present at the considered scale. The contact point is shown in region II in (A), (B), and (C). At contact, a sudden reaction force in the *z*-axis of about 0.3 pN is detected, and the Brownian motion is attenuated compared to region I. Then the probe is pushed deeper into the cell until reaching 1.6 pN in the *z*-direction, as shown in III. At this time, the user feels about 1.6 N force in the *z*-axis. The cell is contacted twice during the presented experiment. Similar results are obtained during the two passes, which also proves the repeatability of the axial force detection.

Notice that this experiment aims to validate the axial haptic feedback during biological manipulations. In addition, the stiffness of the RBCs can be roughly obtained from the result. By using the Hertz model, which considers the cell as a homogeneous smooth semisphere, the elastic modulus of the RBC is in the order of magnitude of 30 kPa [71]. This result is similar to literature, with the measure of the elastic modulus of adherent living cells [72]. This is an approximated result; for a proper mechanical characterization, the trap stiffness should be calibrated in the neighborhood of the cell in advance [73]. The system would also make it possible to automatize the experiment and to repeat it accurately over a large number of trials.

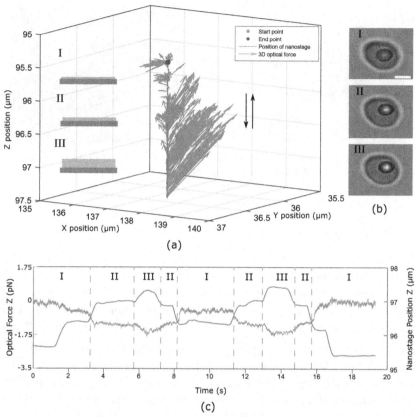

Fig. 7.12 Pressing on a red blood cell from above. (A) The 3D force applied on the probe during the experiment. The x-, y-, and z-coordinates are the position of nanostage in world coordinates. *Inset picture I*: the optically trapped probe is not in contact with the cell; II: by increasing the position of the nanostage, the probe is coming into contact with the RBC; III: the probe pushes deeper into the RBC. (B) Pictures of RBC and the probe corresponding to the three stages I, II, and III. (C) The optical force and the position of nanostage in z-direction during the cell pressing process. The scale bars are 3 μm.

5.2 3D haptic exploration

This experiment is dedicated to touching the 3D contour of cells and exploring their shapes. The difficulty of these tasks comes from the uncertain contour of biological objects. Since the visual information may be blurry or lost at some parts, the haptic feedback will help users to maintain contact and decrease the possibility of losing the trap.

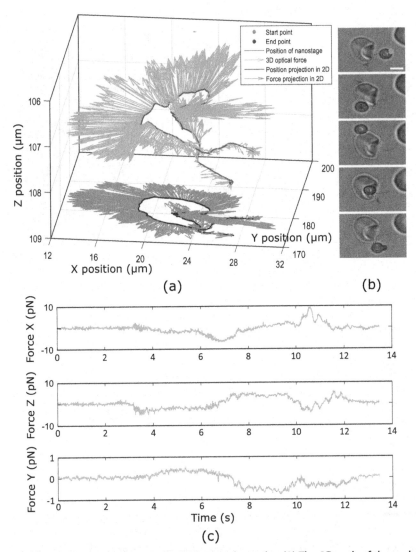

(a)

(b)

(c)

Fig. 7.13 Touching the contour of a RBC using the probe. (A) The 3D path of the probe and the 3D contact forces during the RBC contour exploration. (B) Pictures of this process under microscopy. (C) 3D optical force during this process. The scale bars are 3 μm.

First, haptic exploration of an isolated RBC is conducted (see Fig. 7.13). Then two flat and transparent RBCs, damaged after hemoglobin leakage, stuck together forming a ∞ shape are explored (Fig. 7.14). The 3D contact force is successfully perceived and maintained by operators. Haptic feedback

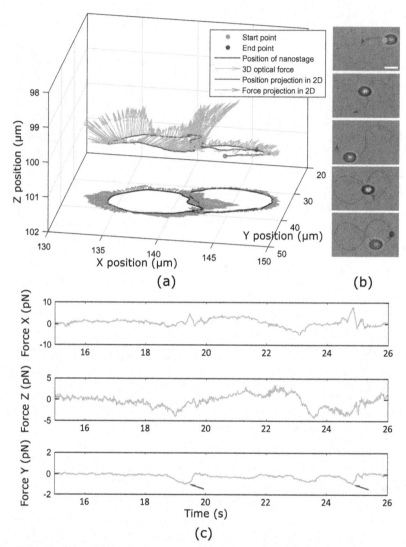

Fig. 7.14 Haptic exploration of two connected transparent RBCs. (A) The 3D path of probe and 3D contact forces during the contour exploration. (B) Corresponding pictures under microscope during this process. (C) 3D optical force. When the probe passes through the connected part of two cells, a burst of axial force of 1 pN is detected (marked with *arrows*). The scale bars are 3 μm.

makes it possible to maintain contact even when the probe is occluded and users were able to achieve a surface exploration with consistent force feedback.

From the figures, it is observed that the 3D piconewton contact forces are maintained throughout the exploration. Furthermore, the 3D force direction indicates a counterforce from the cell's surface. For the two-cell exploration, when the probe passes through the connected part of two cells, a burst of axial force of 1 pN is detected, as shown in Fig. 7.14A and C, indicated by the red arrow. Since the connected part of the two cells is higher in the z-direction than the regular path of the probe, when the probe crosses over this part, a force in the z-axis is detected, demonstrating that there are obstacles below the probe. Biological objects are often transparent, so their manipulation is even more difficult. This experiment is a good illustration of scenarios where haptic force will be useful, as the cells are really hard to see. The haptic feedback helps users to maintain contact even if they cannot see the object.

References

[1] J.M. Levsky, R.H. Singer, Gene expression and the myth of the average cell, Trends Cell Biol. 13 (1) (2003) 4–6.
[2] S. Huang, Non-genetic heterogeneity of cells in development: more than just noise, Development 136 (23) (2009) 3853–3862.
[3] R.N. Zare, S. Kim, Microfluidic platforms for single-cell analysis, Annu. Rev. Biomed. Eng. 12 (2010) 187–201.
[4] S.M. Weiz, M. Medina-Sánchez, O.G. Schmidt, Microsystems for single-cell analysis, Adv. Biosyst. 2 (2) (2018) 1700193.
[5] R. Fisher, L. Pusztai, C. Swanton, Cancer heterogeneity: implications for targeted therapeutics, Br. J. Cancer 108 (3) (2013) 479–485.
[6] N.C. Turner, J.S. Reis-Filho, Genetic heterogeneity and cancer drug resistance, Lancet Oncol. 13 (4) (2012) e178–e185.
[7] J. Li, B.E.-F de Ávila, W. Gao, L. Zhang, J. Wang, Micro/nanorobots for biomedicine: delivery, surgery, sensing, and detoxification, Sci. Robot. 2 (4) (2017) eaam6431.
[8] H. Ceylan, J. Giltinan, K. Kozielski, M. Sitti, Mobile microrobots for bioengineering applications, Lab Chip 17 (10) (2017) 1705–1724.
[9] N. Chaillet, S. Régnier, Microrobotics for Micromanipulation, John Wiley & Sons, 2013.
[10] E. Diller, M. Sitti, Micro-scale mobile robotics, Found. Trends Robot. 2 (3) (2013) 143–259.
[11] X.-Z. Chen, B. Jang, D. Ahmed, C. Hu, C. De Marco, M. Hoop, F. Mushtaq, B.J. Nelson, S. Pané, Small-scale machines driven by external power sources, Adv. Mater. 30 (15) (2018) 1705061.
[12] Z. Zhang, X. Wang, J. Liu, C. Dai, Y. Sun, Robotic micromanipulation: fundamentals and applications, Annu. Rev. J. Control. Robot. Autom. Syst. 2 (2019) 181–203.
[13] J.E. Curtis, B.A. Koss, D.G. Grier, Dynamic holographic optical tweezers, Opt. Commun. 207 (1–6) (2002) 169–175.
[14] S.P. Gross, Application of optical traps in vivo, Methods Enzymol. 361 (2003) 162–174.
[15] H. Zhang, K.-K. Liu, Optical tweezers for single cells, J. R. Soc. Interface 5 (24) (2008) 671–690.
[16] M.-C. Zhong, X.-B. Wei, J.-H. Zhou, Z.-Q. Wang, Y.-M. Li, Trapping red blood cells in living animals using optical tweezers, Nat. Commun. 4 (2013) 1768.
[17] R. Diekmann, D.L. Wolfson, C. Spahn, M. Heilemann, M. Schüttpelz, T. Huser, Nanoscopy of bacterial cells immobilized by holographic optical tweezers, Nat. Commun. 7 (2016) 13711.

[18] Y.-C. Lin, H.-C. Chen, H.-Y. Tu, C.-Y. Liu, C.-J. Cheng, Optically driven full-angle sample rotation for tomographic imaging in digital holographic microscopy, Opt. Lett. 42 (7) (2017) 1321–1324.

[19] K.C. Neuman, S.M. Block, Optical trapping, Rev. Sci. Instrum. 75 (9) (2004) 2787–2809.

[20] R.W. Bowman, M.J. Padgett, Optical trapping and binding, Rep. Prog. Phys. 76 (2) (2013) 026401.

[21] A. Ashkin, History of optical trapping and manipulation of small-neutral particle, atoms, and molecules, IEEE J. Sel. Top. Quantum Electron. 6 (6) (2000) 841–856.

[22] A. Ashkin, Acceleration and trapping of particles by radiation pressure, Phys. Rev. Lett. 24 (4) (1970) 156.

[23] A. Ashkin, J.M. Dziedzic, J.E. Bjorkholm, S. Chu, Observation of a single-beam gradient force optical trap for dielectric particles, Opt. Lett. 11 (5) (1986) 288–290.

[24] S. Chu, J.E. Bjorkholm, A. Ashkin, A. Cable, Experimental observation of optically trapped atoms, Phys. Rev. Lett. 57 (3) (1986) 314.

[25] A. Ashkin, J.M. Dziedzic, T. Yamane, Optical trapping and manipulation of single cells using infrared laser beams, Nature 330 (6150) (1987) 769–771.

[26] A. Ashkin, J.M. Dziedzic, Optical trapping and manipulation of viruses and bacteria, Science 235 (4795) (1987) 1517–1520.

[27] K. Bambardekar, R. Clément, O. Blanc, C. Chardès, P.-F. Lenne, Direct laser manipulation reveals the mechanics of cell contacts in vivo, Proc. Natl. Acad. Sci. 112 (5) (2015) 1416–1421.

[28] E. Stellamanns, S. Uppaluri, A. Hochstetter, N. Heddergott, M. Engstler, T. Pfohl, Optical trapping reveals propulsion forces, power generation and motility efficiency of the unicellular parasites Trypanosoma brucei brucei, Sci. Rep. 4 (2014) 6515.

[29] P.L. Johansen, F. Fenaroli, L. Evensen, G. Griffiths, G. Koster, Optical micromanipulation of nanoparticles and cells inside living zebrafish, Nat. Commun. 7 (1) (2016) 1–8.

[30] M. Xie, A. Shakoor, Y. Shen, J.K. Mills, D. Sun, Out-of-plane rotation control of biological cells with a robot-tweezers manipulation system for orientation-based cell surgery, IEEE Trans. Biomed. Eng. (2018) 199–207.

[31] C.T. Lim, M. Dao, S. Suresh, C.H. Sow, K.T. Chew, Large deformation of living cells using laser traps, Acta Mater. 52 (7) (2004) 1837–1845.

[32] F.M. Fazal, S.M. Block, Optical tweezers study life under tension, Nat. Photonics 5 (6) (2011) 318–321.

[33] X. Wang, S. Chen, M. Kong, Z. Wang, K.D. Costa, R.A. Li, D. Sun, Enhanced cell sorting and manipulation with combined optical tweezer and microfluidic chip technologies, Lab Chip 11 (21) (2011) 3656–3662.

[34] G.R. Kirkham, E. Britchford, T. Upton, J. Ware, G.M. Gibson, Y. Devaud, M. Ehrbar, M. Padgett, S. Allen, L.D. Buttery, Precision assembly of complex cellular microenvironments using holographic optical tweezers, Sci. Rep. 5 (2015) 8577.

[35] M. Padgett, R. Bowman, Tweezers with a twist, Nat. Photonics 5 (6) (2011) 343–348.

[36] K. Svoboda, S.M. Block, Biological applications of optical forces, Annu. Rev. Biophys. Biomol. Struct. 23 (1) (1994) 247–285.

[37] A. Ashkin, Forces of a single-beam gradient laser trap on a dielectric sphere in the ray optics regime, Biophys. J. 61 (2) (1992) 569.

[38] A.L. Stout, Detection and characterization of individual intermolecular bonds using optical tweezers, Biophys. J. 80 (6) (2001) 2976–2986.

[39] D. Raucher, M.P. Sheetz, Characteristics of a membrane reservoir buffering membrane tension, Biophys. J. 77 (4) (1999) 1992–2002.

[40] K. Mandal, A. Asnacios, B. Goud, J.-B. Manneville, Mapping intracellular mechanics on micropatterned substrates, Proc. Natl. Acad. Sci. USA 113 (46) (2016) E7159–E7168.

[41] K. Berg-Sørensen, H. Flyvbjerg, Power spectrum analysis for optical tweezers, Rev. Sci. Instrum. 75 (3) (2004) 594–612.

[42] P.H. Jones, O.M. Maragò, G. Volpe, Optical Tweezers: Principles and Applications, Cambridge University Press, 2015.

[43] F. Català, F. Marsà, M. Montes-Usategui, A. Farré, E. Martín-Badosa, Extending calibration-free force measurements to optically-trapped rod-shaped samples, Sci. Rep. 7 (2017) 42960.

[44] D. Sun, H. Chen, Moving groups of microparticles into array with a robot-tweezers manipulation system, IEEE Trans. Robot. 28 (5) (2012) 1069–1080.

[45] A.G. Banerjee, S. Chowdhury, W. Losert, S.K. Gupta, Real-time path planning for coordinated transport of multiple particles using optical tweezers, IEEE Trans. Autom. Sci. Eng. 9 (4) (2012) 669–678.

[46] D. Sun, X. Li, C.C. Cheah, S. Hu, Dynamic trapping and manipulation of biological cells with optical tweezers, Automatica 49 (6) (2013) 1614–1625.

[47] S. Chowdhury, A. Thakur, P. Svec, C. Wang, W. Losert, S.K. Gupta, Automated manipulation of biological cells using gripper formations controlled by optical tweezers, IEEE Trans. Autom. Sci. Eng. 11 (2) (2014) 338–347.

[48] C.C. Cheah, Q.M. Ta, R. Haghighi, Grasping and manipulation of a micro-particle using multiple optical traps, Automatica 68 (2016) 216–227.

[49] A. Bolopion, B. Cagneau, D.S. Haliyo, S. Régnier, Analysis of stability and transparency for nanoscale force feedback in bilateral coupling, J. Micro-Nano Mechatron. 4 (4) (2008) 145.

[50] D. Ruh, B. Tränkle, A. Rohrbach, Fast parallel interferometric 3D tracking of numerous optically trapped particles and their hydrodynamic interaction, Opt. Express 19 (22) (2011) 21627–21642.

[51] A. Handa, R.A. Newcombe, A. Angeli, A.J. Davison, Real-time camera tracking: when is high frame-rate best? in: Computer Vision-ECCV 2012, Springer, 2012, pp. 222–235.

[52] R. Bowman, D. Preece, G. Gibson, M. Padgett, Stereoscopic particle tracking for 3D touch, vision and closed-loop control in optical tweezers, J. Opt. 13 (4) (2011) 044003.

[53] A. Huhle, D. Klaue, H. Brutzer, P. Daldrop, S. Joo, O. Otto, U.F. Keyser, R. Seidel, Camera-based three-dimensional real-time particle tracking at kHz rates and Ångström accuracy, Nat. Commun. 6 (2015) 5885.

[54] Z. Ni, C. Pacoret, R. Benosman, S. Régnier, 2D high speed force feedback teleoperation of optical tweezers, in: 2013 IEEE International Conference on Robotics and Automation (ICRA), IEEE, 2013, pp. 1700–1705.

[55] Z. Ni, A. Bolopion, J. Agnus, R. Benosman, S. Régnier, Asynchronous event-based visual shape tracking for stable haptic feedback in microrobotics, IEEE Trans. Robot. 28 (5) (2012) 1081–1089.

[56] C. Posch, D. Matolin, R. Wohlgenannt, A QVGA 143 dB dynamic range frame-free PWM image sensor with lossless pixel-level video compression and time-domain CDS, IEEE J. Solid State Circuits 46 (1) (2011) 259–275.

[57] X. Lagorce, C. Meyer, S.-H. Ieng, D. Filliat, R. Benosman, Asynchronous event-based multikernel algorithm for high-speed visual features tracking, IEEE Trans. Neural Netw. Learn. Syst. 26 (8) (2014) 1710–1720.

[58] C. Posch, D. Matolin, Sensitivity and uniformity of a 0.18 μm CMOS temporal contrast pixel array, in: 2011 IEEE International Symposium of Circuits and Systems (ISCAS), IEEE, 2011, pp. 1572–1575.

[59] M. Litzenberger, C. Posch, D. Bauer, A.N. Belbachir, P. Schon, B. Kohn, H. Garn, Embedded vision system for real-time object tracking using an asynchronous transient vision sensor, in: 2006 IEEE 12th Digital Signal Processing Workshop & 4th IEEE Signal Processing Education Workshop, IEEE, 2006, pp. 173–178.

[60] T. Delbruck, Frame-free dynamic digital vision, in: Proceedings of International Symposium on Secure-Life Electronics, Advanced Electronics for Quality Life and Society, 2008, pp. 21–26.

[61] R. Benosman, S.-H. Ieng, C. Clercq, C. Bartolozzi, M. Srinivasan, Asynchronous frameless event-based optical flow, Neural Netw. 27 (2012) 32–37.

[62] X. Lagorce, C. Meyer, S.-H. Ieng, D. Filliat, R. Benosman, Asynchronous event-based multikernel algorithm for high-speed visual features tracking, IEEE Trans. Neural Netw. Learn. Syst. 26 (8) (2015) 1710–1720.

[63] L.A. Camuñas-Mesa, T. Serrano-Gotarredona, B. Linares-Barranco, S. Ieng, R. Benosman, Event-driven stereo vision with orientation filters, in: 2014 IEEE International Symposium on Circuits and Systems (ISCAS), IEEE, 2014, pp. 257–260.

[64] W. Gander, G.H. Golub, R. Strebel, Least-squares fitting of circles and ellipses, BIT Numer. Math. 34 (4) (1994) 558–578.

[65] P.J. Huber, Robust Statistics, Springer, 2011.

[66] I. Kåsa, A circle fitting procedure and its error analysis, IEEE Trans. Instrum. Meas. 1001 (1) (1976) 8–14.

[67] R.T. Verrillo, Effect of contactor area on the vibrotactile threshold, J. Acoust. Soc. Am. 35 (12) (1963) 1962–1966.

[68] Z. Ni, A. Bolopion, J. Agnus, R. Benosman, S. Régnier, Asynchronous event-based visual shape tracking for stable haptic feedback in microrobotics, IEEE Trans. Robot. 28 (5) (2012) 1081–1089.

[69] K. Onda, F. Arai, Multi-beam bilateral teleoperation of holographic optical tweezers, Opt. Express 20 (4) (2012) 3633–3641, https://doi.org/10.1364/OE.20.003633.

[70] A. Rohrbach, Stiffness of optical traps: quantitative agreement between experiment and electromagnetic theory, Phys. Rev. Lett. 95 (16) (2005) 168102.

[71] N. Guz, M. Dokukin, V. Kalaparthi, I. Sokolov, If cell mechanics can be described by elastic modulus: study of different models and probes used in indentation experiments, Biophys. J. 107 (3) (2014) 564–575.

[72] E. Planus, R. Fodil, M. Balland, D. Isabey, Assessment of mechanical properties of adherent living cells by bead micromanipulation: comparison of magnetic twisting cytometry vs optical tweezers, J. Biomech. Eng. 124 (4) (2002) 408–421.

[73] S. Nawaz, P. Sánchez, K. Bodensiek, S. Li, M. Simons, I.A.T. Schaap, Cell viscoelasticity measured with AFM and optical trapping at sub-micrometer deformations, PLoS One 7 (9) (2012) e45297.

CHAPTER 8

Magnetically driven robots for clinical treatment

Xingzhou Du[a,b], Yuezhen Liu[a], and Jiangfan Yu[a,b]
[a]School of Science and Engineering, The Chinese University of Hong Kong, Shenzhen, China
[b]Shenzhen Institute of Artificial Intelligence and Robotics for Society (AIRS), Shenzhen, China

1 Introduction

In the science fiction film "Fantastic Voyage," a crew of surgeons are shrunk to micrometer scale and injected inside the human body to cure diseases, which represents people's dream of minimally invasive and noninvasive surgeries. Surgical procedures with less trauma lead to reduced risk, relieved pain for the patient, and decreased difficulty of recovery [1,2]. In decades of development, microrobots have begun to shed light on realizing these targets [3,4]. Micrometer or nanometer scale microrobots have the potential to actively conduct diagnostic and therapeutic tasks in hard-to-reach regions inside body in a minimally invasive manner.

Owing to the small size, typical devices on a traditional robot, such as sensors, electrical motors, and control circuits, can hardly be integrated to microrobots. The actuation process relies on external fields. For example, electrostatic fields can be implemented to manipulate small objects through electrostatic force [5,6]. Chemical reactions with the surrounding environments can drive microrobots [7–9]. Acoustic fields and light fields can also provide energy for the locomotion of microrobots [10–15]. Considering biomedical applications, magnetic fields show high potential because of their compatibility with biological samples and wide control bandwidth [3,16,17]. They are also programmable and have rapid response [18].

Since the early 1900s, magnetic fields have been employed to manipulate magnetic objects inside the body, for instance, to extract bullets or bomb fragments in wounded soldiers [19,20], or to steer surgical tools remotely [21,22]. Commercialized magnetic actuation systems have also been designed for clinical deployment of magnetic catheters for cardiovascular interventions [23,24] and magnetic capsule endoscopes for active inspection of the gastrointestinal tract [25,26]. Such clinical practice has shown the

Robotics for Cell Manipulation and Characterization
https://doi.org/10.1016/B978-0-323-95213-2.00016-8
173

safety and performance of magnetic fields in conducting minimally invasive surgery. At small scales, magnetic fields can drive various types of microrobots with different functions, such as helical robots with screw-like motions [27], robots with 5 degrees of freedom (DoF) movements [28], microrobots rolling on surfaces [29], soft robots with multimodel locomotion [30], and microrobotic swarms that consists of millions of agents [31–33]. The control of magnetic fields has also endowed magnetic microrobots with capabilities of conducting precise operations, such as targeted delivery in hard-to-reach regions to reduce required drug dose and side effects [34–36], the selective killing of cancer cells [37,38], conducting cellular-scale manipulations [39], and thrombolysis in tiny vessels [40,41].

In this chapter, the actuation techniques for magnetic microrobots and their clinical applications are summarized. We review different types of actuation magnetic fields and microrobots, the strategies to deal with physiological environments, the methods to monitor the status and track the positions of microrobots using medical imaging devices, and the clinical tasks that can be accomplished by magnetic microrobots. Moreover, a conclusion and future development directions are provided, such as enhanced safety with less trauma, smarter materials with promising biocompatibility and functionality, and automated surgical procedures with higher accuracy.

2 Actuating microrobots using magnetic fields

Magnetic forces and magnetic torques are two methods used to actuate magnetic microrobots. The exerted magnetic force $\mathbf{F}$ and torque $\boldsymbol{\tau}$ on a microrobot in a magnetic field are defined as the following equations, respectively:

$$\mathbf{F} = (\mathbf{m} \cdot \nabla)\mathbf{B} \tag{8.1}$$

$$\boldsymbol{\tau} = \mathbf{m} \times \mathbf{B} \tag{8.2}$$

in which $\mathbf{m}$ is the magnetic moment of the magnetized microrobot and $\mathbf{B}$ is the magnetic flux density at the position of the robot.

Magnetic-force actuation relies on the magnetic field gradient, and the robot moves toward the position with a higher magnetic flux density. Magnetic-torque actuation relies on the alignment between the magnetization of magnetic objects and external magnetic fields. External magnetic fields $\mathbf{B}$ change directions and magnitudes continuously with time t, which is described as $\mathbf{B}(t)$. Each type of magnetic field $\mathbf{B}(t)$ for magnetic-torque actuation can be regarded as the rotation of a basic magnetic field pattern $\mathbf{B}_{basic}(t)$ to the demanded direction angle φ and pitch angle θ. The process can be expressed as:

$$\mathbf{B} = \mathbf{T}_\varphi \mathbf{T}_\theta \mathbf{B}_{\text{basic}} \tag{8.3}$$

in which $\mathbf{T}_\theta$ is the rotation matrix of the pitch angle and $\mathbf{T}_\varphi$ is the rotation matrix of the direction angle.

The two matrices are defined as:

$$\mathbf{T}_\theta = \begin{bmatrix} \cos\theta & 0 & \sin\theta \\ 0 & 1 & 0 \\ -\sin\theta & 0 & \cos\theta \end{bmatrix} \tag{8.4}$$

$$\mathbf{T}_\varphi = \begin{bmatrix} \cos\varphi & -\sin\varphi & 0 \\ \sin\varphi & \cos\varphi & 0 \\ 0 & 0 & 1 \end{bmatrix} \tag{8.5}$$

To drive the robot, the commonly used forms of magnetic fields are discussed as follows.

2.1 Gradient magnetic fields

Magnetic forces will be exerted on the microrobot in gradient fields. The magnitude and the direction of the force is determined by Eq. (8.1), which is related to multiple factors, such as the magnetization of the robot, the direction and speed in which the field changes, and the shape and position of the external magnet or electromagnetic coils [28,42–44]. As a magnetized object has the tendency of moving along the field gradient, the gradient fields do not need to change with time t to propel the continuous movements of microrobots. The mathematical models to generate demanded magnetic force through different magnetic actuation systems in terms of robotic control have been summarized [45]. Li et al. designed a magnetic scaffold for cell delivery in vivo, which was actuated by magnetic gradient [34]. Ongaro et al. designed an electromagnetic coil system with nine coils to actuate two magnetic microspheres independently through magnetic fields [44].

If magnetic force is combined with magnetic torque, 5-DoF control of miniature robots in an open space can be accomplished. Kummer et al. developed the Octomag system, which had eight electromagnetic coils to complete 3-DoF position control and 2-DoF orientation control of a microrobot [28]. Visual serving of the robot for retinal surgery has also been explored [46]. Furthermore, 5-DoF control of a microrobot could also be accomplished by using a permanent magnet-based system. Ryan et al. completed the 5-DoF actuation task employing a system with eight rotatable magnets [43]. 5-DoF actuation provides higher dexterity for microrobots so that they are capable of conducting precise operations in which the positions and orientations all matter to the results.

As for actuation using gradient fields, there are few requirements on the shapes and designs of microrobots, thus, the structure of the robots can be simple, and many types of microrobots can be driven using field gradient. The disadvantage lies in the reduced actuation efficiency with the decrease of the robot's size. According to the scaling law, the exerted magnetic forces will reduce faster compared with magnetic torques when the microrobots get smaller [47]. As a result, stronger magnetic gradients will be demanded for actuating small-sized microrobots in low Reynolds number environments or against dynamic flow.

2.2 Rotating magnetic fields

Rotating magnetic fields are also widely implemented to actuate microrobots. This involves applying a rotating magnetic flux density to induce torque on the robot, resulting in a rotational motion. The basic field pattern is shown as:

$$\mathbf{B}_{\text{basic}} = [b \, \cos(2\pi ft), b \, \sin(2\pi ft), 0]^{\text{T}} \tag{8.6}$$

in which b is the magnitude of the magnetic flux density, f is the rotating frequency, and t is the time.

The rotating field is a periodic time-varying field that has a plane of rotation. The angle between the z axis and the normal line of the rotating plane is the pitch angle θ, and the angle between the x axis and the projection of the normal line on the x-y plane is the direction angle φ (Fig. 8.1B). When under the influence of these rotating fields, the microrobots can exploit two main types of moving methods: corkscrew-like motion or surface rolling.

To achieve a corkscrew-like motion, the helical structure is an essential feature of the shape of the robot. The direction of movement of the robot is along the normal line of the rotating field and is adjusted by changing the values of pitch angle and direction angle. In the literature, magnetic microrobots in the shapes of helices or with helical structures have been designed for corkscrew-like motion, such as artificial bacterial flagella (Fig. 8.1B1) [27,56–58] and biohybrid microrobots based on natural spirulina [59,60]. Helical microrobots have shown promising performance in low Reynolds number environments, in which the dimension of the robot is small or the viscosity of the fluid is high [47]. Owing to the helical robots' capabilities of three-dimensional locomotion and reliable mathematical model, closed-loop control methods have been widely studied. The robots were capable of

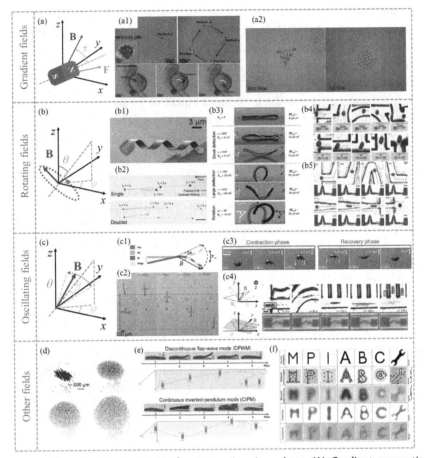

Fig. 8.1 The actuation strategies for magnetic microrobots. (A) Gradient magnetic fields: (A1) Actuation of a microrobot in vivo using magnetic force under the gradient field [34]. (A2) 5-DoF actuation of a microrobot using magnetic torque and magnetic force, aiming at retinal surgery [28]. (B) Rotating magnetic fields: (B1) Actuation of a helical robot using rotating field [27]. (B2) Rolling motion of microrobots driven by rotating fields [48]. (B3) Multimodel locomotion of a soft robot actuated by rotating fields with different field strengths [30]. (B4) A vortex-like microrobotic swarm that was formed and actuated under rotating fields [31]. (B5) A microrobotic swarm with pattern changing capability under elliptical fields [49]. (C) Oscillating magnetic fields: (C1) Undulation motion of a microrobot under oscillating fields [50]. (C2) Locomotion of a soft MagnetoSperm with a soft tail [51]. (C3) Jellyfish-like motion of a robot actuated by an oscillating field in a vertical direction [52]. (C4) Formation, actuation, and pattern changing of a microrobotic swarm using a customized magnetic field [32]. (D) Spreading of a swarm pattern [53]. (E) Crawling locomotion of a soft robot [54]. (F) Patterned magnetic field to form swarm patterns [55].

moving along an optimal trajectory, crossing obstructions, and reaching the demanded targets [61–63]. Helical propulsion can also be implemented on soft structures. Huang et al. designed a soft microrobot that can conduct oar-like motion in low viscosity and corkscrew motion in high-viscosity environments [64].

As for surface rolling, the pitch angle equals to 90 degree, indicating that the rotating plane is perpendicular to the substrate. The rolling direction is determined by the direction angle. The locomotion of the microrobot relies on the interaction between the robot and the substrate, or the different flow induced on the upper and lower regions of the microrobots owing to the boundary effects. An example showing the surface rolling of a microrobot against flow is shown in Fig. 8.1B2 [48]. Furthermore, by adjusting the strength of the fields when actuating a soft robot with heterogeneous magnetization, multimodel locomotion was achieved, including crawling and surface rolling (Fig. 8.1B3) [30].

Rotating fields can also be implemented to trigger the magnetic and fluidic interactions between microscale agents to form microrobotic swarms. A vortex-like paramagnetic nanoparticle swarm was developed by Yu et al., which consisted of millions of Fe_3O_4 nanoparticles and was formed by a rotating magnetic field parallel to the substrate (Fig. 8.1B4) [31]. The swarm was able to conduct controlled locomotion with a high delivery rate by adding a pitch angle to the rotating field, accomplish merging and splitting of swarms, and complete reversible spreading and shrinking of swarm pattern [31,65]. Xie et al. implemented rolling hematite colloids to generate microswarms with multiple modes. The chain, vortex, and ribbon-like swarms, which could be formed for multiple purposes, were actuated by different rotating magnetic fields [66]. Automated formation and active locomotion have also been achieved [67]. Furthermore, by adjusting the shape of rotating fields to the elliptical, Yu et al. developed a microrobotic swarm of elliptical shape with actively adjustable aspect ratio. The controllable elongation and shrinking of the swarm pattern made it possible to pass through narrow structures and take sharp turns inside mazes while keeping the swarm pattern stable (Fig. 8.1B5) [49]. As well as nanoparticles, microrobotic swarms can be formed by helical swimmers, actuated by rotating magnetic fields, and with weak interactions existing among the agents [17,68–70].

2.3 Oscillating magnetic fields

Oscillating field is another type of magnetic field for actuating microrobots through magnetic torque. The basic field pattern can be expressed as:

$$\mathbf{B}_{\text{basic}} = \left[0, \quad b\sin\left(\frac{\beta}{2}\right)\sin(2\pi ft), \quad b\sqrt{1 - \left(\sin\left(\frac{\beta}{2}\right)\sin(2\pi ft)\right)^2} \right]^T \quad (8.7)$$

where b is the magnitude, f is the frequency, and β is the oscillating angle, i.e., the angle between the center line of oscillation and the outermost direction of $\mathbf{B}$ during oscillation (Fig. 8.1C).

Furthermore, if the oscillating plane demands tilting, the fields pattern can be written as:

$$\mathbf{B}_{\text{basic}} = \begin{bmatrix} -b\sin\left(\frac{\beta}{2}\right)\sin\sigma\sin(2\pi ft) \\[2mm] b\sin\left(\frac{\beta}{2}\right)\cos\sigma\sin(2\pi ft) \\[2mm] b\sqrt{1 - \left(\sin\left(\frac{\beta}{2}\right)\sin(2\pi ft)\right)^2} \end{bmatrix} \quad (8.8)$$

in which σ is the angle between the oscillating plane of $\mathbf{B}_{\text{basic}}$ with the y-z plane.

Anticlockwise rotation of the oscillating plane around the z axis leads to the positive values of σ. $\sigma = 0$ indicates that $\mathbf{B}_{\text{basic}}$ is oscillating on the y-z plane, and $\sigma = \pi/2$ means that the oscillating plane of $\mathbf{B}_{\text{basic}}$ is on x-z plane. Oscillating fields are commonly used for microrobots with soft tails or multiple links. The fields will lead to an undulation movement of the robot, which will induce a translational wave in the surrounding flow to propel the movement of the robot [47]. This swimming method has been applied to different microrobots. A microrobot consists of multiple sections has been actuated by oscillating fields to move against flow (Fig. 8.1C1) [50]. A soft MagnetoSperm with soft tail also implemented undulation movements, mimicking the swimming methods of a natural sperm (Fig. 8.1C2) [51]. A soft robot with nonuniform magnetization that was actuated using oscillating fields was capable of swimming upstream against flow speeds of 2–3 mm/s [71]. Furthermore, oscillating fields along the z axis propelled the jellyfish-like motion of a soft robot, making it capable of conducting multiple functions using the flow around the body, such as object selective transportation and locally mixing fluids (Fig. 8.1C3) [52].

Overall, oscillating magnetic fields are feasible to actuate the locomotion of soft magnetic microrobots, making the robots capable of conducting demanded tasks without a rigid body. On the other hand, the process of generating oscillating fields increases the requirement for magnetic actuation systems. Electromagnetic coil-based systems with multiple coils have advantages in providing programmed oscillating fields with satisfying frequency, as the fields can be produced by simply changing the currents in the coils without the demands of mechanical movements of the system [72].

Customized oscillating fields can also trigger swarm behavior. The ribbon-like paramagnetic nanoparticle swarm proposed by Yu et al. implemented a customized oscillating magnetic field that consisted of a static magnetic field and a perpendicular alternating magnetic field (Fig. 8.1C4). Millions of Fe_3O_4 nanoparticles could be actuated and form dynamically stable collectives [32]. By adjusting the ratio between the strength of the static field and magnitude of the alternating field, the swarm was able to conduct reversible elongation and shrinking of the swarm pattern, merging and splitting of swarms, and controllable locomotion through the channels toward the intended targets (Fig. 8.1C4). Furthermore, independent pattern elongation and shrinking of two swarms under the same oscillating field were achieved by changing the agents in the swarms [33].

2.4 Other forms of magnetic fields

Apart from the abovementioned magnetic fields, some magnetic microrobots are actuated by customized magnetic fields to complete required functions. A customized magnetic field that consisted of a vertical rotating field with a steering rotating plane was implemented to conduct spreading and fragmentation of nanoparticle chains (Fig. 8.1D) [53]. This magnetic field made it possible to conduct reversible swelling or disassembly of the pattern of the paramagnetic nanoparticle swarm [65]. Disassembled nanoparticle swarms will have enhanced efficiency of surface reactions in specific applications, such as targeted delivery or hyperthermia [73]. Shrinking of the swarm pattern or assembly of the swarm for improved delivery rate in locomotion can be completed by using rotating fields in the horizontal plane [31].

Another example using customized magnetic fields was in the actuation of soft robots. The customized magnetic fields which combined time-varying magnetic flux density and magnetic gradient have made a multilegged soft robot crawling on the surface (Fig. 8.1E) [54]. The magnetic fields generated by the mechanical movement of a permanent magnet endowed the soft robot with fast locomotion speed, high carrying capacity, and high flexibility to avoid obstacles. Furthermore, an array of electromagnetic coils or permanent magnets was able to generate patterned magnetic fields for actuating multiple microrobots. A patterned external magnetic field was capable of producing collective behaviors of ferromagnetic particles at the air–water interface. Pattern formation and cargo delivery were accomplished through the swarm (Fig. 8.1F) [55]. An array of electromagnetic coils was implemented for actuating ferromagnetic droplet robots. Splitting, merging, locomotion, and targeted cargo delivery have been accomplished by droplet robots under patterned magnetic fields [74]. Closed-loop independent control of multiple microrobots has been accomplished with patterned magnetic fields as well. By controlling the currents in the coil array that was printed on a circuit board, local magnetic fields were generated and the movements of multiple microrobots were actuated individually [75].

3 Clinical applications for microrobots

3.1 Dealing with physiological environments

The physiological environment inside the body is one of the major challenges for magnetic microrobots to conduct clinical tasks. Biological fluid, e.g., mucus, complicated structures, and dynamic environments will affect the operations of microrobots. Finding strategies to tackle the challenges has been an attractive but challenging topic for researchers.

Active locomotion inside body fluids with different viscoelastic properties is the first task that microrobots should deal with. For example, mucus contains mucins, proteins, lipids, and electrolytes [76]. A network of mucin fibers in the mucus will prohibit the movements of microrobots. In this case, a helical microrobot can be a promising choice [47]. Modifications have been made to improve the biocompatibility and locomotion efficiencies for helical micropropellers in different biological media. Ferrite coating on the microrobot has improved the chemical stability of the structure over time in whole blood (Fig. 8.2A) [77]. Surface-immobilized urease, which is a type of enzyme, has been coated on helical robots to liquefy the surrounding mucus for easier penetration, and the microrobots were capable of moving through purified porcine mucus (Fig. 8.2B) [78]. A liquid layer on the surface and the small dimension of helical microrobots enabled them to penetrate the vitreous humor of the eye, which also contains meshed structures [69].

Different properties of diverse surface coatings can be implemented for controlling multiple microrobots inside biofluids as well. By adjusting the

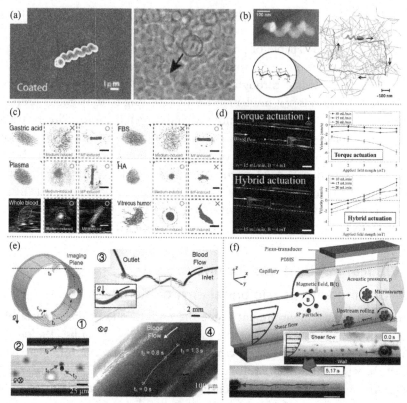

Fig. 8.2 Designs of magnetic microrobots to deal with physiological environments. (A) Conformal ferrite coatings on helical microrobots for enhanced cytocompatibility and chemical stability in whole blood [77]. (B) The urease-functionalization of nanorobots for improved penetration capability in mucus [78]. (C) Swarm behavior of nanoparticles in different biofluids [79]. (D) Magnetic torque and force-hybrid propulsion of a helical robot upstream within the blood flow [80]. (E) Rolling locomotion of a Janus microrobot climbing the wall of a tube and moving against blood flow in a 3D microchannel [81]. (F) Upstream mobility of microswarm robots propelled by magnetic and acoustic fields [82].

hydrophilic or hydrophobic properties of surface coating, controlling individual magnetic helical microswimmers inside a swarm has been conducted inside bovine serum albumin solution [70]. Besides helical propulsion, gradient fields with exerted magnetic force and magnetic torque on the robots are also capable of actuating microrobots in biological fluids with meshed structures. 3-DoF position and 2-DoF orientation control of a microrobot can be conducted in viscous fluids [28,43], for example, in vitreous humor during vitreoretinal surgery [83]. The collective behaviors of the paramagnetic nanoparticle swarms were also effective in biological fluids (Fig. 8.2C). Formation, pattern changing, and locomotion of swarms have been accomplished in various biological fluidic environments with different viscosities, ionic strengths, and mesh-like structures [79].

In addition to the viscoelastic nature of biological fluids, dynamic blood flow is another factor that will influence the performance of microrobots in endovascular applications. To propel magnetic microrobots within the blood flow, optimizing the magnetic actuation procedure is a solution. A feasible method is to combine the time-varying magnetic fields with magnetic gradients. By using a horizontally rotating magnetic field with a gradient pointing to the center, a Fe_3O_4 nanoparticle swarm was assembled and actuated in the dynamic blood flow [84]. The field was produced by a rotating spherical magnet with its dipole moment parallel to the rotating plane. Ex vivo experiments demonstrated that the swarm could be navigated along and against the blood flow in porcine coronary artery with real-time position feedback through ultrasound Doppler signal. The flow rate in the artery ranged from 8 to 12 mL/min.

The combination of rotating fields and magnetic gradients can also be produced by systems with multiple mobile electromagnetic coils. By optimizing the positions and orientations of the coils, magnetic gradients can be produced in the required directions during rotating-field actuation. A helical miniature robot was actuated to move against a continuous blood flow with flow rate up to 20 mL/min and a pulsatile flow with flow rate of 24 mL/min through this hybrid propulsion (Fig. 8.2D) [80]. Locomotion along the wall of the vessels is a promising strategy for microrobots as well, as the flow stream will significantly decrease near the boundary [82,84]. A Janus microrobot that had a silica core and nickel/gold (Ni/Au) half caps was actuated by rotating magnetic fields to produce rolling locomotion on the vessel wall [81]. The microrobot was propelled against the blood flow with a physiologically relevant pressure up to 2.5 dyn/cm^2. It also showed capabilities of moving on three dimensional surfaces and in microfluidic channels (Fig. 8.2E). In the subsequent work, anisotropic magnetic microrollers that consisted of two Janus microrollers demonstrated better capabilities of moving on vessel-like 3D surfaces and crossing barriers in environments with obstructions [48].

Magnetic stem cell spheroid microrobots (MSCSMs) also implemented a rolling locomotion strategy [85]. The microrobots consisted of stem cells and polydopamine (PDA)-coated magnetic iron particles, which were fabricated through a coculture process on a nonadhesive surface. After actuation in the required location, the robots could be anchored at that position by gradient magnetic fields against flow with speeds of up to 108 mm/s. Furthermore, combining acoustic fields with magnetic actuation will endow the robots with upstream mobility. Acousto-magnetic microswarm robots were assembled inside an acoustic pressure node and actuated by a vertically rotating magnetic field [82,86]. Upstream rolling was conducted against a stream with velocity up to 1.2 mm/s (Fig. 8.2F) [82].

3.2 Tracking using medical imaging devices

In clinical operations, controllable navigation and operation of microrobots is essential to ensure the safety, precision, and performance of the therapeutic procedure [87]. Localization and tracking in vivo are crucial to achieve targeted navigation inside the body, acquire feedback for closed-loop control, and monitor the status of the operation. In clinical scenarios, various imaging techniques have been employed for diagnosis and imaging-guided surgery, such as magnetic resonance imaging (MRI), ionizing radiation-based techniques, optical-based techniques, and ultrasound (US) imaging [88]. Efforts have been made in developing the tracking of microrobots using these medical imaging methods to enable clinical applications.

MRI is one of the most common imaging techniques for diagnosis in clinics. Three dimensional images showing the internal structures of the human body can be acquired with spatial resolution at the submillimeter scale [88,89]. The imaging procedure is based on the relaxation of the atom spin. After applying pulses of electromagnetic radio waves to the atoms, the nuclear spin will tend to realign with the high-intensity static magnetic fields, and energy excitation will be produced during this relaxation process. As for magnetic microrobots, MRI can be implemented for tracking the positions in vivo. For instance, a swarm of biohybrid helical microrobots in the stomachs of mice have been tracked with MRI. The movements of the microrobots can be clearly seen in the imaging (Fig. 8.3A) [59]. The MRI device can also be used for actuating microrobots using its strong magnetic fields. Time-sharing methods can allow both imaging and actuation to be conducted inside the same MRI machine. Various microrobots, such as ferromagnetic spheres, magnetotactic bacteria, and tethered magnetic devices can be well actuated [93–96]. However, owing to the strong magnetic fields in the imaging process, it is difficult to integrate MRI with other magnetic actuation systems that can generate programmable magnetic fields, and the formats of magnetic fields generated by MRI are limited,

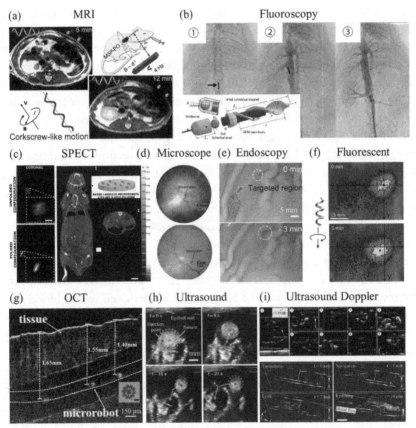

Fig. 8.3 Imaging and tracking methods for magnetic microrobots in vivo. (A) Tracking of a swarm of biohybrid helical microrobots using MRI [59]. (B) Tracking and monitoring of a guide-wired helical microrobot under X-ray fluoroscopy for thrombolysis [90]. (C) Imaging of hydrogel-based magnetic microrobots using SPECT for targeted drug release [91]. (D) Tracking of a microrobot using an optical microscope for retinal surgery [46]. (E) Releasing and monitoring of magnetic stem cell spheroid microrobots using medical endoscopy [85]. (F) Magnetic actuation of a swarm of artificial bacteria flagella inside a mouse using fluorescent imaging [17]. (G) Tracking the movement of a magnetic microrobot in vessels using OCT [92]. (H) Tracking of microrobotic swarms in bovine eye by ultrasound imaging [79]. (I) Tracking of helical miniature robot through the ultrasound Doppler signal and monitoring the thrombolysis using power Doppler mode [80].

which raises higher requirements for the design of actuation strategies for microrobots.

As for ionizing radiation–based techniques, high-frequency electromagnetic waves with wavelengths ranging from 100 to 10 nm are employed as the source. The ionizing radiation is capable of penetrating the human body.

The internal bones and tissues of the patient can be imaged owing to their different radiation absorption rates. Typical ionizing radiation-based techniques include X-ray fluoroscopy (X-ray), X-ray computed tomography (X-ray CT), positron emission tomography (PET), and single photon emission computed tomography (SPECT). The X-ray-based methods have high penetration capability, submillimeter-scale spatial resolution, and millisecond-scale temporal resolution [97–100]. The γ-ray-based PET and SPECT also have high penetration depth, but the millimeter-scale special resolution and the temporal resolution of seconds to minutes are not as high as the X-ray-based methods [99,101–104]. The disadvantage of the ionizing radiation-based methods lies in the excessive exposure to strong energy, which will be harmful to the human body [88].

Ionizing radiation-based techniques for tracking magnetic microrobots have been widely studied. A guide-wired helical microrobot (GHM) for thrombolysis applications consisted of a guidewire and a magnetic helical tip. Fluoroscopy was implemented to track its position in vivo and evaluate the results of thrombolysis. The robot could be actuated to the blood clot through vessels, drill through the clot mechanically, and be retracted from the body (Fig. 8.3B) [90]. SPECT imaging was implemented for tracking hydrogel-based magnetic soft microrobots for a drug delivery task in vivo. Radioactive compounds were integrated into the soft thermo-responsive materials of the microrobot, making it possible to see the robots deep inside the body of mice (Fig. 8.3C) [91].

For optics-based techniques, the imaging procedures use optical sources, such as LEDs or lasers. The sources have no ionizing radiation and, therefore, guarantee the safety of patients. They can also have high spatial and temporal resolutions, which are related to the quality of optical components and processing units [88,89,105]. However, the limited tissue-penetration capability can be a drawback in tracking microrobots inside the body [88,89,106,107]. Microscopes and endoscopes are devices that directly use visible light for imaging. For example, a microscope was implemented in real-time navigation of a cylindrical miniature robot for retinal surgery, where the magnetic fields were provided by an OctoMag system (Fig. 8.3D) [46]. In the gastrointestinal tract, microrobots were delivered by endoscopy across long distances to improve the delivery rate, and local operations in relatively small region were conducted by the microrobots themselves (Fig. 8.3E) [85].

As for the fluorescence-based methods, imaging of magnetically actuated microrobots has also been demonstrated. The movements of large numbers of microrobots with markers can be tracked by fluorescence inside body (Fig. 8.3F) [17]. Gathering of the microrobots in vivo can be seen using fluorescent imaging as well [59]. Optical coherence tomography (OCT), which is another reflection-based method, can be used for imaging of microrobots in tissue with low depth. For example, in vivo tracking of a magnetic microrobot inside the portal vein of mice using OCT has been conducted (Fig. 8.3G) [92]. The OCT has also been implemented to quantify the spatial distribution of a swarm of helical microrobots near the retina of the porcine eye [69].

Ultrasound imaging relies on the gradient of acoustic impedance to examine the soft tissue. By analyzing the echo produced by different tissues with different reflection properties, a cross-sectional image of the internal body structure can be acquired [88]. The deep-tissue imaging ability that can reach 20 cm, the high temporal resolution of 10–20 ms, and promising spatial resolution of 0.1–2 mm make it feasible for real-time tracking of microrobots with no ionizing radiation [88,89,108,109]. B-mode ultrasound imaging is a commonly used mode to show the internal tissues through gray-scale images. With B-mode ultrasound imaging feedback, closed-loop control of untethered microrobots, such as microparticles and soft grippers, has been explored [110,111]. Guiding a helical magnetic microrobot to perform mechanical thrombolysis tasks using ultrasound imaging has been conducted on imaging phantoms [112]. Real-time imaging of the paramagnetic nanoparticle swarm using B-mode ultrasound imaging has also been accomplished [79,113]. A vortex-like swarm was formed and actuated inside a bovine eye, and the status was monitored clearly (Fig. 8.3H) [79]. Comparable to the vortex-like swarm, a ribbon-like swarm could also be tracked under ultrasound imaging [41,79].

Navigation of collective cell microrobots within the blood stream has also been conducted utilizing ultrasound feedback [114]. In addition to B-mode, the ultrasound Doppler signal is also widely used for tracking microrobots. The rotating motion of a helical robot will induce a flow in the direction perpendicular to the blood flow in the vessels. The generated Doppler signal surrounding the robot on the cross-sectional plane of the vessel has been employed for tracking the robot's movements for endovascular navigation. In addition, the results of thrombolysis were monitored by power Doppler mode (Fig. 8.3I) [80]. A microrobotic swarm will also induce a Doppler signal. Real-time tracking of the rotating field-induced nanoparticle swarm during navigation has been demonstrated in porcine coronary artery [84].

3.3 Microrobots for clinical operations

Their small dimensions and active locomotion capabilities are the major merits of microrobots, so that precise targeted operations can be conducted inside hard-to-reach regions of a patient's body with minimal invasiveness, decreased time for recovery, reduced side effects, and enhanced therapeutic performance. Thus, on the way toward clinical applications, endowing the microrobots with the required medical functions is essential to exploit their advantages.

By using magnetic microrobots, targeted delivery can be conducted actively and precisely, which will increase drug dose locally, and thus, reduce the overall required dose. Therefore, the side effects to the patients will be reduced. Strategies to design a microrobot for targeted delivery have been widely explored, and researchers have provided multiple solutions. Stem cells have shown promising capability in regenerative medicine [115–117]. 3D printed microscaffolds with nickel and titanium coatings are a feasible type of magnetic microrobot for transporting stem cells. The spherical and cubic microrobots could be actuated by gradient fields to deliver cells in mice (Fig. 8.4A) [34], and helical shaped scaffolds were propelled by rotating fields for delivery in fluidic environments [122,123].

Porous magnetic micro scaffolds fabricated by a gelatin leaching process and magnetized by Fe_3O_4 nanoparticles have also been used for active transporting and fixation of stem cells for cartilage repair. In vivo experiments have shown promising results in recovery (Fig. 8.4B) [35]. Besides stem cells, microrobots have also been implemented to deliver drugs to target sites. Biohybrid microrobots, consisting of natural neutrophil with endocytic drug-loaded magnetic nanogels, were capable of actively penetrating the blood-brain barrier (BBB) through magnetic actuation and chemotactic motion. The purpose was to treat malignant glioma with improved efficiency of drug delivery compared with traditional methods (Fig. 8.4C) [36]. Magnetoaerotactic bacteria with drug-loaded liposomes have shown magnetotactic and aerotactic migration behavior. They have been

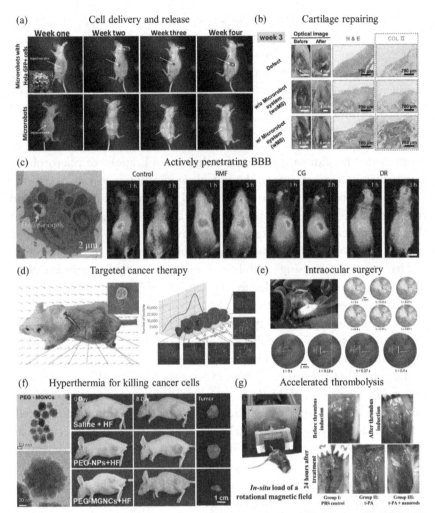

Fig. 8.4 Magnetic microrobots for in vivo surgical applications. (A) Magnetic microrobots in the form of scaffold for actively delivering cells inside body [34]. (B) Stem cell-loaded microrobots with active locomotion and fixation capabilities for cartilage repair [35]. (C) Biohybrid neutrobots that consisted of drug-loaded magnetic nanogels in natural neutrophil with capability of penetrating the blood-brain barrier (BBB) [36]. (D) Magneto-aerotactic bacteria with drug-loaded liposomes for targeted therapy in tumor hypoxic regions [118]. (E) Intraocular surgery conducted by a magnetic microrobot inside the eye of a rabbit [119]. (F) Locally killing cancer cells using magnetic hyperthermia [120]. (G) Promoting the spreading of tissue plasminogen activator (tPA) using rotating magnetic nanomotors to accelerate the thrombolysis [121].

developed for reaching tumor hypoxic regions targeted therapy (Fig. 8.4D) [118]. Furthermore, drug-loading smart materials on magnetic microrobots have also been designed for selectively killing cancer cells [37], such as sperm-hybrid microrobots with a drug-loaded sperm cell and a releasing mechanism [124], nanowires with drug releasing capability triggered by magnetic stimulation [125], and swarm of nanoparticles with hydrogel coating [38].

Minimally invasive surgeries have also obtained extensive attention from researchers who aim to promote the clinical applications of microrobots. Intraocular operation is one of the suitable application scenarios, as untethered microrobots are less invasive compared with traditional rigid tools and needles, and the remote actuation method can limit the maximum force on the tissue, which will reduce the risk of permanent damage [126]. Magnetic microrobots have been designed for intraocular surgery and retinal injection and have shown promising capabilities, e.g., helical robots with surface coating [69], microrobots assembled from electroplated pieces [28,46], and paramagnetic nanoparticle swarms [79]. In vivo demonstration of a cylindrical magnetic microrobot in the eye of a rabbit has been conducted, showing that the microrobot was capable of being tracked and actuated during an intraocular surgical process (Fig. 8.4E) [119].

Localized hyperthermia is another therapeutic process that fits the capabilities of microrobots. Polyethylene glycol (PEG)–coated Fe_3O_4 multigranule nanoclusters have been introduced for locally killing cancer cells through magnetic hyperthermia. The nanoclusters were able to accumulate at the tumor site in mice and successfully suppressed the growth of the tumor in vivo (Fig. 8.4F) [120]. A vortex-like microrobotic swarm formed by Fe_3O_4 nanoparticles could be actuated and gathered at the target region. Local hyperthermia was conducted with increased concentration to kill cancer cells only in the required area [127]. Furthermore, microrobots are capable of conducting thrombolysis tasks in narrow and complicated vessels in endovascular interventions. Mechanical abrasion using helical robots was capable of removing the blood clots in vessels [90,112,128]. However, the debris that could be generated from clots during mechanical abrasion might induce new blockages in smaller branches [41].

Promoting the spreading of tissue plasminogen activator (tPA) using induced flow around microrobots or microrobotic swarms is another feasible method. The thrombolysis rate has been improved up to 4.8 times by using helical swimmers and up to 3.13 times by using ribbon-like microrobotic swarms, compared with natural diffusion of tPA [41,80]. Magnetic nanomotors that composed of polystyrene beads and nickel nanorods also accelerated the thrombolysis process. The lysis rate in low-concentration tPA with nanomotors has been improved up to twofold compared with that in high-concentration tPA without nanomotors (Fig. 8.4G) [121]. Using tPA-loaded microrobots is also a promising strategy to perform accurate thrombolysis with decreased dose. A porous magnetic microbubble platform has been developed for localized thrombolysis. It was capable of maintaining the activity of tPA, accumulating at blood clot, and actively releasing tPA triggered by ultrasound. In vivo experiments showed that the residual thrombus had decreased by 67.5% compared with simply injection of tPA [129]. A biomimetic magnetic microrobot that consisted of aligned magnetic nanoparticle chains inside a microgel shell has been designed. The microrobot was able to deliver tPA using magnetic actuation and release tPA to remove blood clots triggered by magnetic hyperthermia [40].

4 Conclusion

Microrobots can be designed to conduct minimally invasive surgical tasks inside the human body. The small size and active locomotion capability of the magnetic microrobots enable them to play an important role in different clinical operations, such as targeted delivery, precise surgery, local hyperthermia, and thrombolysis. In this chapter, magnetic microrobots for clinical applications have been summarized, and the actuation strategies for microrobots introduced, including gradient fields, rotating fields, oscillating fields, and other customized magnetic fields. The strategies for microrobots to deal with physiological environments, the imaging and tracking methods inside body, and the medical functions are discussed as well.

Although progress has been made toward clinical applications, further efforts are still needed. Safety is always an important factor that is of concern to the users of microrobots. To improve safety, smarter materials can be a solution. Microrobots should be fabricated using materials with promising biocompatibility, and their toxicity to biological samples needs to be avoided. Microrobots should also be degradable or retrievable after the required procedures, ensuring that the remaining structures of microrobots will not cause additional trauma to the patients. Smart materials can endow microrobots with more practical functions as well. For example, responsive hydrogel can be triggered in specific environments for drug release [130], and stem cell-loaded membranes can promote the recovery of ulcers [131,132].

Improved controllability in the magnetic actuation process is also beneficial for higher safety and higher success rates of operation. From the aspect of hardware design, this can be achieved through system integration of magnetic actuation systems and medical imaging devices. The status of the robot can be monitored through the feedback from the imaging and localization devices during operation. From the aspect of software, more precise models and advanced control algorithms will be helpful [45]. Furthermore, the abilities of the robots in dealing with complicated dynamic environments and completing required tasks should be enhanced. Decision making process combined with robust control algorithms will enhance the performance of microrobots to precisely execute the commands of the surgeons and complete demanded tasks [63,133,134]. Finally, an automated surgical procedure using magnetic microrobots is one of the ultimate purposes for their development, so that routine surgery can become automated to relieve the burden on surgeons while the crucial steps can be operated by experienced clinicians to ensure the therapeutic outcome [135,136].

In summary, the development of magnetic microrobots has provided a promising tool to conduct precise minimally invasive and noninvasive surgeries. It still needs further efforts from experts working in physics, chemistry, smart materials, and control science. We envision that microrobots will appear more and more often in the operating room to fight diseases.

References

[1] M.J. Mack, Minimally invasive and robotic surgery, JAMA 285 (5) (2001) 568–572.
[2] M. Sitti, Voyage of the microrobots, Nature 458 (7242) (2009) 1121–1122.
[3] B.J. Nelson, I.K. Kaliakatsos, J.J. Abbott, Microrobots for minimally invasive medicine, Annu. Rev. Biomed. Eng. 12 (2010) 55–85.
[4] M. Sitti, H. Ceylan, W. Hu, J. Giltinan, M. Turan, S. Yim, E. Diller, Biomedical applications of untethered mobile milli/microrobots, Proc. IEEE 103 (2) (2015) 205–224.
[5] F. Ma, S. Wang, D.T. Wu, N. Wu, Electric-field-induced assembly and propulsion of chiral colloidal clusters, Proc. Natl. Acad. Sci. 112 (20) (2015) 6307–6312.
[6] J. Yan, M. Han, J. Zhang, C. Xu, E. Luijten, S. Granick, Reconfiguring active particles by electrostatic imbalance, Nat. Mater. 15 (10) (2016) 1095–1099.
[7] B.E.-F. de Ávila, P. Angsantikul, J. Li, M.A. Lopez-Ramirez, D.E. Ramrez-Herrera, S. Thamphiwatana, C. Chen, J. Delezuk, R. Samakapiruk, V. Ramez, et al., Micromotor-enabled active drug delivery for in vivo treatment of stomach infection, Nat. Commun. 8 (1) (2017) 272.
[8] J. Li, P. Angsantikul, W. Liu, B. Esteban-Fernández de Ávila, S. Thamphiwatana, M. Xu, E. Sandraz, X. Wang, J. Delezuk, W. Gao, et al., Micromotors spontaneously neutralize gastric acid for pH-responsive payload release, Angew. Chem. Int. Ed. 56 (8) (2017) 2156–2161.
[9] D. Kagan, S. Balasubramanian, J. Wang, Chemically triggered swarming of gold microparticles, Angew. Chem. Int. Ed. 50 (2) (2011) 503–506.

[10] V. Garcia-Gradilla, J. Orozco, S. Sattayasamitsathit, F. Soto, F. Kuralay, A. Pourazary, A. Katzenberg, W. Gao, Y. Shen, J. Wang, Functionalized ultrasound-propelled magnetically guided nanomotors: toward practical biomedical applications, ACS Nano 7 (10) (2013) 9232–9240.

[11] V. Garcia-Gradilla, S. Sattayasamitsathit, F. Soto, F. Kuralay, C. Yardmc, D. Wiitala, M. Galarnyk, J. Wang, Ultrasound-propelled nanoporous gold wire for efficient drug loading and release, Small 10 (20) (2014) 4154–4159.

[12] B. Esteban-Fernández de Ávila, C. Angell, F. Soto, M.A. Lopez-Ramirez, D.F. Báez, S. Xie, J. Wang, Y. Chen, Acoustically propelled nanomotors for intracellular siRNA delivery, ACS Nano 10 (5) (2016) 4997–5005.

[13] H. Eskandarloo, A. Kierulf, A. Abbaspourrad, Light-harvesting synthetic nano-and micromotors: a review, Nanoscale 9 (34) (2017) 12218–12230.

[14] Z. Li, N.V. Myung, Y. Yin, Light-powered soft steam engines for self-adaptive oscillation and biomimetic swimming, Sci. Robot. 6 (61) (2021) eabi4523.

[15] S. Palagi, A.G. Mark, S.Y. Reigh, K. Melde, T. Qiu, H. Zeng, C. Parmeggiani, D. Martella, A. Sanchez-Castillo, N. Kapernaum, et al., Structured light enables biomimetic swimming and versatile locomotion of photoresponsive soft microrobots, Nat. Mater. 15 (6) (2016) 647–653.

[16] J.F. Schenck, Safety of strong, static magnetic fields, J. Magn. Reson. Imaging 12 (1) (2000) 2–19.

[17] A. Servant, F. Qiu, M. Mazza, K. Kostarelos, B.J. Nelson, Controlled in vivo swimming of a swarm of bacteria-like microrobotic flagella, Adv. Mater. 27 (19) (2015) 2981–2988.

[18] H. Zhou, C.C. Mayorga-Martinez, S. Pané, L. Zhang, M. Pumera, Magnetically driven micro and nanorobots, Chem. Rev. 121 (8) (2021) 4999–5041.

[19] C. Jackson, Foreign bodies in the trachea, bronchi and oesophagus—the aid of oesophagoscopy, bronchoscopy, and magnetism in their extraction, Laryngoscope 15 (4) (1905) 257–281.

[20] C. Jackson, A bullet in the lung: bronchoscopic removal with the aid of magnetic fixation, JAMA 90 (16) (1928) 1272–1273.

[21] H. Tillander, Magnetic guidance of a catheter with articulated steel tip, Acta Radiol. 35 (1) (1951) 62–64.

[22] G. Gillies, R. Ritter, W. Broaddus, M. Grady, M. Howard III, R. McNeil, Magnetic manipulation instrumentation for medical physics research, Rev. Sci. Instrum. 65 (3) (1994) 533–562.

[23] M.P. Armacost, J. Adair, T. Munger, R.R. Viswanathan, F.M. Creighton, D.T. Curd, R. Sehra, Accurate and reproducible target navigation with the stereotaxis niobe magnetic navigation system, J. Cardiovasc. Electrophysiol. 18 (2007) S26–S31.

[24] J. Edelmann, A.J. Petruska, B.J. Nelson, Magnetic control of continuum devices, Int. J. Robot. Res. 36 (1) (2017) 68–85, https://doi.org/10.1177/02783649 16683443.

[25] H. Keller, A. Juloski, H. Kawano, M. Bechtold, A. Kimura, H. Takizawa, R. Kuth, Method for navigation and control of a magnetically guided capsule endoscope in the human stomach, in: 2012 4th IEEE RAS & EMBS International Conference on Biomedical Robotics and Biomechatronics (BioRob), IEEE, 2012, pp. 859–865.

[26] Ankon Technology Co., Navicam Stomach Capsule Endoscopy System, 2014. https://www.ankoninc.com.cn/official-web/productions.

[27] L. Zhang, J.J. Abbott, L. Dong, B.E. Kratochvil, D. Bell, B.J. Nelson, Artificial bacterial flagella: fabrication and magnetic control, Appl. Phys. Lett. 94 (6) (2009), 064107.

[28] M.P. Kummer, J.J. Abbott, B.E. Kratochvil, R. Borer, A. Sengul, B.J. Nelson, Octomag: an electromagnetic system for 5-dof wireless micromanipulation, IEEE Trans. Robot. 26 (6) (2010) 1006–1017.

[29] F. Mushtaq, A. Asani, M. Hoop, X.-Z. Chen, D. Ahmed, B.J. Nelson, S. Pané, Highly efficient coaxial tio2-ptpd tubular nanomachines for photocatalytic water purification with multiple locomotion strategies, Adv. Funct. Mater. 26 (38) (2016) 6995–7002.

[30] W. Hu, G.Z. Lum, M. Mastrangeli, M. Sitti, Small-scale soft-bodied robot with multimodal locomotion, Nature 554 (7690) (2018) 81–85.

[31] J. Yu, L. Yang, L. Zhang, Pattern generation and motion control of a vortex-like paramagnetic nanoparticle swarm, Int. J. Robot. Res. 37 (8) (2018) 912–930.

[32] J. Yu, B. Wang, X. Du, Q. Wang, L. Zhang, Ultra-extensible ribbon-like magnetic microswarm, Nat. Commun. 9 (1) (2018) 3260.

[33] X. Du, J. Yu, D. Jin, P.W.Y. Chiu, L. Zhang, Independent pattern formation of nanorod and nanoparticle swarms under an oscillating field, ACS Nano 15 (3) (2021) 4429–4439.

[34] J. Li, X. Li, T. Luo, R. Wang, C. Liu, S. Chen, D. Li, J. Yue, S.-h. Cheng, D. Sun, Development of a magnetic microrobot for carrying and delivering targeted cells, Sci. Robot. 3 (19) (2018) eaat8829.

[35] G. Go, S.-G. Jeong, A. Yoo, J. Han, B. Kang, S. Kim, K.T. Nguyen, Z. Jin, C.-S. Kim, Y.R. Seo, et al., Human adipose-derived mesenchymal stem cell-based medical microrobot system for knee cartilage regeneration in vivo, Sci. Robot. 5 (38) (2020).

[36] H. Zhang, Z. Li, C. Gao, X. Fan, Y. Pang, T. Li, Z. Wu, H. Xie, Q. He, Dual-responsive biohybrid neutrobots for active target delivery, Sci. Robot. 6 (2021) 9519eaaz.

[37] C.K. Schmidt, M. Medina-Sánchez, R.J. Edmondson, O.G. Schmidt, Engineering microrobots for targeted cancer therapies from a medical perspective, Nat. Commun. 11 (1) (2020) 1–18.

[38] D. Jin, K. Yuan, X. Du, Q. Wang, S. Wang, L. Zhang, Domino reaction encoded heterogeneous colloidal microswarm with on-demand morphological adaptability, Adv. Mater. 33 (37) (2021) 2100070.

[39] M. Medina-Sánchez, L. Schwarz, A.K. Meyer, F. Hebenstreit, O.G. Schmidt, Cellular cargo delivery: toward assisted fertilization by sperm-carrying micromotors, Nano Lett. 16 (1) (2016) 555–561.

[40] M. Xie, W. Zhang, C. Fan, C. Wu, Q. Feng, J. Wu, Y. Li, R. Gao, Z. Li, Q. Wang, et al., Bioinspired soft microrobots with precise magneto-collective control for microvascular thrombolysis, Adv. Mater. (2020), 2000366.

[41] Q. Wang, D. Jin, B. Wang, N. Xia, H. Ko, B.Y.M. Ip, T.W.H. Leung, S.C.H. Yu, L. Zhang, Reconfigurable magnetic microswarm for accelerating tpa-mediated thrombolysis under ultrasound imaging, IEEE/ASME Trans. Mechatron. 27 (4) (2022) 2267–2277.

[42] A.W. Mahoney, J.J. Abbott, Five-degree-of-freedom manipulation of an untethered magnetic device in fluid using a single permanent magnet with application in stomach capsule endoscopy, Int. J. Robot. Res. 35 (1–3) (2016) 129–147.

[43] P. Ryan, E. Diller, Magnetic actuation for full dexterity microrobotic control using rotating permanent magnets, IEEE Trans. Robot. 33 (6) (2017) 1398–1409.

[44] F. Ongaro, S. Pane, S. Scheggi, S. Misra, Design of an electromagnetic setup for independent three-dimensional control of pairs of identical and nonidentical microrobots, IEEE Trans. Robot. 35 (1) (2018) 174–183.

[45] J.J. Abbott, E. Diller, A.J. Petruska, Magnetic methods in robotics, Annu. Rev. Contr, Robot. Auton. Syst. 3 (2020) 57–90.

[46] C. Bergeles, B.E. Kratochvil, B.J. Nelson, Visually serving magnetic intraocular microdevices, IEEE Trans. Robot. 28 (4) (2012) 798–809.

[47] J.J. Abbott, K.E. Peyer, M.C. Lagomarsino, L. Zhang, L. Dong, I.K. Kaliakatsos, B.J. Nelson, How should microrobots swim? Int. J. Robot. Res. 28 (11–12) (2009) 1434–1447.

[48] U. Bozuyuk, Y. Alapan, A. Aghakhani, M. Yunusa, M. Sitti, Shape anisotropy-governed locomotion of surface microrollers on vessel-like microtopographies against physiological flows, Proc. Natl. Acad. Sci. 118 (13) (2021).

[49] J. Yu, L. Yang, X. Du, H. Chen, T. Xu, L. Zhang, Adaptive pattern and motion control of magnetic microrobotic swarms, IEEE Trans. Robot. 38 (3) (2022) 1552–1570.

[50] B. Jang, E. Gutman, N. Stucki, B.F. Seitz, P.D. Wendel-Garcá, T. Newton, J. Pokki, O. Ergeneman, S. Pané, Y. Or, et al., Undulatory locomotion of magnetic multilink nanoswimmers, Nano Lett. 15 (7) (2015) 4829–4833.

[51] I.S. Khalil, H.C. Dijkslag, L. Abelmann, S. Misra, Magnetosperm: a microrobot that navigates using weak magnetic fields, Appl. Phys. Lett. 104 (22) (2014), 223701.

[52] Z. Ren, W. Hu, X. Dong, M. Sitti, Multi-functional soft-bodied jellyfish-like swimming, Nat. Commun. 10 (1) (2019) 1–12.

[53] J. Yu, T. Xu, Z. Lu, C.I. Vong, L. Zhang, On-demand disassembly of paramagnetic nanoparticle chains for microrobotic cargo delivery, IEEE Trans. Robot. 33 (5) (2017) 1213–1225.

[54] H. Lu, M. Zhang, Y. Yang, Q. Huang, T. Fukuda, Z. Wang, Y. Shen, A bioinspired multilegged soft millirobot that functions in both dry and wet conditions, Nat. Commun. 9 (1) (2018) 1–7.

[55] X. Dong, M. Sitti, Controlling two-dimensional collective formation and cooperative behavior of magnetic microrobot swarms, Int. J. Robot. Res. 39 (5) (2020) 617–638.

[56] S. Tottori, L. Zhang, F. Qiu, K.K. Krawczyk, A. Franco-Obregón, B.J. Nelson, Magnetic helical micromachines: fabrication, controlled swimming, and cargo transport, Adv. Mater. 24 (6) (2012) 811–816.

[57] F. Qiu, S. Fujita, R. Mhanna, L. Zhang, B.R. Simona, B.J. Nelson, Magnetic helical microswimmers functionalized with lipoplexes for targeted gene delivery, Adv. Funct. Mater. 25 (11) (2015) 1666–1671.

[58] D. Schamel, A.G. Mark, J.G. Gibbs, C. Miksch, K.I. Morozov, A.M. Leshansky, P. Fischer, Nanopropellers and their actuation in complex viscoelastic media, ACS Nano 8 (9) (2014) 8794–8801.

[59] X. Yan, Q. Zhou, M. Vincent, Y. Deng, J. Yu, J. Xu, T. Xu, T. Tang, L. Bian, Y.-X.J. Wang, K. Kostarelos, L. Zhang, Multifunctional biohybrid magnetite microrobots for imaging-guided therapy, Sci. Robot. 2 (12) (2017) eaaq1155.

[60] X. Wang, J. Cai, L. Sun, S. Zhang, D. Gong, X. Li, S. Yue, L. Feng, D. Zhang, Facile fabrication of magnetic microrobots based on spirulina templates for targeted delivery and synergistic chemo-photothermal therapy, ACS Appl. Mater. Interfaces 11 (5) (2019) 4745–4756.

[61] A. Oulmas, N. Andreff, S. Régnier, 3D closed-loop swimming at low Reynolds numbers, Int. J. Robot. Res. 37 (11) (2018) 1359–1375.

[62] J. Liu, X. Wu, C. Huang, L. Manamanchaiyaporn, W. Shang, X. Yan, T. Xu, 3-D autonomous manipulation system of helical microswimmers with online compensation update, IEEE Trans. Autom. Sci. Eng. 18 (3) (2021) 1380–1391.

[63] S. Xu, J. Liu, C. Yang, X. Wu, T. Xu, A learning-based stable servo control strategy using broad learning system applied for microrobotic control, IEEE Trans. Cybern. 52 (12) (2022) 13727–13737.

[64] H.-W. Huang, F.E. Uslu, P. Katsamba, E. Lauga, M.S. Sakar, B.J. Nelson, Adaptive locomotion of artificial microswimmers, Sci. Adv. 5 (1) (2019) eaau1532.

[65] J. Yu, L. Zhang, Reversible swelling and shrinking of paramagnetic nanoparticle swarms in biofluids with high ionic strength, IEEE/ASME Trans. Mechatron. 24 (1) (2018) 154–163.

[66] H. Xie, M. Sun, X. Fan, Z. Lin, W. Chen, L. Wang, L. Dong, Q. He, Reconfigurable magnetic microrobot swarm: multimode transformation, locomotion, and manipulation, Sci. Robot. 4 (28) (2019) eaav8006, https://doi.org/10.1126/scirobotics.aav8006.

[67] H. Xie, X. Fan, M. Sun, Z. Lin, Q. He, L. Sun, Programmable generation and motion control of a snakelike magnetic microrobot swarm, IEEE/ASME Trans. Mechatron. 24 (3) (2019) 902–912.

[68] S. Tottori, L. Zhang, K.E. Peyer, B.J. Nelson, Assembly, disassembly, and anomalous propulsion of microscopic helices, Nano Lett. 13 (9) (2013) 4263–4268.

[69] Z. Wu, J. Troll, H.-H. Jeong, Q. Wei, M. Stang, F. Ziemssen, Z. Wang, M. Dong, S. Schnichels, T. Qiu, P. Fischer, A swarm of slippery micropropellers penetrates the vitreous body of the eye, Sci. Adv. 4 (11) (2018) eaat4388.

[70] X. Wang, C. Hu, L. Schurz, C. De Marco, X. Chen, S. Pané, B.J. Nelson, Surface-chemistry-mediated control of individual magnetic helical microswimmers in a swarm, ACS Nano 12 (6) (2018) 6210–6217.

[71] L. Manamanchaiyaporn, T. Xu, X. Wu, Magnetic soft robot with the triangular head–tail morphology inspired by lateral undulation, IEEE/ASME Trans. Mechatron. 25 (6) (2020) 2688–2699.

[72] X. Du, M. Zhang, J. Yu, L. Yang, P.W.Y. Chiu, L. Zhang, Design and real-time optimization for a magnetic actuation system with enhanced flexibility, IEEE/ASME Trans. Mechatron. 26 (3) (2021) 1524–1535.

[73] Q. Wang, J. Yu, K. Yuan, L. Yang, D. Jin, L. Zhang, Disassembly and spreading of magnetic nanoparticle clusters on uneven surfaces, Appl. Mater. Today 18 (2020), 100489.

[74] X. Fan, X. Dong, A.C. Karacakol, H. Xie, M. Sitti, Reconfigurable multifunctional ferrofluid droplet robots, Proc. Natl. Acad. Sci. 117 (45) (2020) 27916–27926.

[75] B.V. Johnson, S. Chowdhury, D.J. Cappelleri, Local magnetic field design and characterization for independent closed-loop control of multiple mobile microrobots, IEEE/ASME Trans. Mechatron. 25 (2) (2020) 526–534.

[76] F. Araújo, C. Martins, C. Azevedo, B. Sarmento, Chemical modification of drug molecules as strategy to reduce interactions with mucus, Adv. Drug Deliv. Rev. 124 (2018) 98–106.

[77] P.L. Venugopalan, R. Sai, Y. Chandorkar, B. Basu, S. Shivashankar, A. Ghosh, Conformal cytocompatible ferrite coatings facilitate the realization of a nanovoyager in human blood, Nano Lett. 14 (4) (2014) 1968–1975.

[78] D. Walker, B.T. Käsdorf, H.-H. Jeong, O. Lieleg, P. Fischer, Enzymatically active biomimetic micropropellers for the penetration of mucin gels, Sci. Adv. 1 (11) (2015), e1500501.

[79] J. Yu, D. Jin, K.-F. Chan, Q. Wang, K. Yuan, L. Zhang, Active generation and magnetic actuation of microrobotic swarms in bio-fluids, Nat. Commun. 10 (5631) (2019) 1–12.

[80] Q. Wang, X. Du, D. Jin, L. Zhang, Real-time ultrasound doppler tracking and autonomous navigation of a miniature helical robot for accelerating thrombolysis in dynamic blood flow, ACS Nano 16 (1) (2022) 604–616.

[81] Y. Alapan, U. Bozuyuk, P. Erkoc, A.C. Karacakol, M. Sitti, Multifunctional surface microrollers for targeted cargo delivery in physiological blood flow, Sci. Robot. 5 (42) (2020), https://doi.org/10.1126/scirobotics.aba5726.

[82] D. Ahmed, A. Sukhov, D. Hauri, D. Rodrigue, G. Maranta, J. Harting, B.J. Nelson, Bioinspired acousto-magnetic microswarm robots with upstream motility, Nat. Mach. Intell. 3 (2) (2021) 116–124.

[83] G. Chatzipirpiridis, O. Ergeneman, J. Pokki, F. Ullrich, S. Fusco, J.A. Ortega, K.M. Sivaraman, B.J. Nelson, S. Pané, Electroforming of implantable tubular magnetic microrobots for wireless ophthalmologic applications, Adv. Healthc. Mater. 4 (2) (2015) 209–214.

[84] Q. Wang, K.F. Chan, K. Schweizer, X. Du, D. Jin, C.H.S. Yu, B.J. Nelson, L. Zhang, Ultrasound doppler-guided real-time navigation of a magnetic microswarm for active

endovascular delivery, Sci. Adv. 7 (2021) eabe5914, https://doi.org/10.1126/sciadv. abe5914.

[85] B. Wang, K.F. Chan, K. Yuan, Q. Wang, X. Xia, L. Yang, H. Ko, Y.-X.J. Wang, J.-J.Y. Sung, P.W.Y. Chiu, et al., Endoscopy-assisted magnetic navigation of biohybrid soft microrobots with rapid endoluminal delivery and imaging, Sci. Robot. 6 (52) (2021).

[86] D. Ahmed, T. Baasch, N. Blondel, N. Läubli, J. Dual, B.J. Nelson, Neutrophil-inspired propulsion in a combined acoustic and magnetic field, Nat. Commun. 8 (1) (2017) 1–8.

[87] M. Medina Sánchez, O. Schmidt, Medical microbots need better imaging and control, Nature 545 (7655) (2017) 406.

[88] A. Aziz, S. Pane, V. Iacovacci, N. Koukourakis, J. Czarske, A. Menciassi, M. Medina-Sánchez, O.G. Schmidt, Medical imaging of microrobots: toward in vivo applications, ACS Nano 14 (9) (2020) 10865–10893.

[89] Z. Wu, Y. Chen, D. Mukasa, O.S. Pak, W. Gao, Medical micro/nanorobots in complex media, Chem. Soc. Rev. 49 (22) (2020) 8088–8112.

[90] K.T. Nguyen, S.-J. Kim, H.-K. Min, M.C. Hoang, G. Go, B. Kang, J. Kim, E. Choi, A. Hong, J. Park, et al., Guide-wired helical microrobot for percutaneous revascularization in chronic total occlusion in-vivo validation, I.E.E.E. Trans. Biomed. Eng. 68 (8) (2021) 2490–2498.

[91] V. Iacovacci, A. Blanc, H. Huang, L. Ricotti, R. Schibli, A. Menciassi, M. Behe, S. Pané, B.J. Nelson, High-resolution spect imaging of stimuli-responsive soft microrobots, Small 15 (34) (2019) 1900709.

[92] D. Li, D. Dong, W. Lam, L. Xing, T. Wei, D. Sun, Automated in vivo navigation of magnetic-driven microrobots using oct imaging feedback, I.E.E.E. Trans. Biomed. Eng. 67 (8) (2020) 2349–2358.

[93] J.-B. Mathieu, G. Beaudoin, S. Martel, Method of propulsion of a ferromagnetic core in the cardiovascular system through magnetic gradients generated by an mri system, I.-E.E.E. Trans. Biomed. Eng. 53 (2) (2006) 292–299.

[94] S. Martel, J.-B. Mathieu, O. Felfoul, A. Chanu, E. Aboussouan, S. Tamaz, P. Pouponneau, Automatic navigation of an untethered device in the artery of a living animal using a conventional clinical magnetic resonance imaging system, Appl. Phys. Lett. 90 (11) (2007), 114105.

[95] S. Martel, M. Mohammadi, O. Felfoul, Z. Lu, P. Pouponneau, Flagellated magneto-tactic bacteria as controlled MRI-trackable propulsion and steering systems for medical nanorobots operating in the human microvasculature, Int. J. Robot. Res. 28 (4) (2009) 571–582.

[96] A. Azizi, C.C. Tremblay, K. Gagné, S. Martel, Using the fringe field of a clinical mri scanner enables robotic navigation of tethered instruments in deeper vascular regions, Sci. Robot. 4 (36) (2019).

[97] E.L. Nickoloff, Aapm/rsna physics tutorial for residents: physics of flat-panel fluoroscopy systems: survey of modern fluoroscopy imaging: flat-panel detectors versus image intensifiers and more, Radiographics 31 (2) (2011) 591–602.

[98] M. Mahesh, D.D. Cody, AAPM/RSNA physics tutorial for residents: physics of cardiac imaging with multiple-row detector CT, Radiographics 27 (5) (2007) 1495–1509.

[99] E. Lin, A. Alessio, What are the basic concepts of temporal, contrast, and spatial resolution in cardiac CT? J. Cardiovasc. Comput. Tomogr. 3 (6) (2009) 403–408.

[100] J. Wang, D. Fleischmann, Improving spatial resolution at ct: development, benefits, and pitfalls, Radiology 289 (1) (2018) 261–262.

[101] W.W. Moses, Fundamental limits of spatial resolution in pet, Nucl. Instrum. Methods Phys. Res., Sect. A 648 (2011) S236–S240.

[102] T. Pan, S.A. Einstein, S.C. Kappadath, K.S. Grogg, C. Lois Gomez, A.M. Alessio, W.-C. Hunter, G. El Fakhri, P.E. Kinahan, O.R. Mawlawi, Performance evaluation of the 5-ring GE discovery MI PET/CT system using the national electrical manufacturers association NU 2-2012 standard, Med. Phys. 46 (7) (2019) 3025–3033.

[103] S. Vandenberghe, P. Moskal, J.S. Karp, State of the art in total body pet, EJNMMI Phys. 7 (2020) 1–33.

[104] Siemens Medical Solutions USA, Inc., Symbia intevo: SPECT/CT That Improves Your Image, 2021. https://www.siemens-healthineers.com/en-us/molecular-imaging/xspect/symbia-intevo.

[105] Olympus Europa SE & CO. KG, Endoscope Overview 2021, 2021. https://mdc.olympus.eu/asset/084438885177/0694ed24d569c1855888028a9fcf522a.

[106] P.H. Tomlins, R.K. Wang, Theory, developments and applications of optical coherence tomography, J. Phys. D Appl. Phys. 38 (15) (2005) 2519.

[107] S. Song, J. Xu, R.K. Wang, Long-range and wide field of view optical coherence tomography for in vivo 3d imaging of large volume object based on akinetic programmable swept source, Biomed. Opt. Express 7 (11) (2016) 4734–4748.

[108] J.W. Hunt, M. Arditi, F.S. Foster, Ultrasound transducers for pulse-echo medical imaging, I.E.E.E. Trans. Biomed. Eng. 8 (1983) 453–481.

[109] H. Yerli, S.Y. Eksioglu, Extended field-of-view sonography: evaluation of the superficial lesions, Can. Assoc. Radiol. J. 60 (1) (2009) 35–39.

[110] I.S. Khalil, P. Ferreira, R. Eleutério, C.L. de Korte, S. Misra, Magnetic-based closed-loop control of paramagnetic microparticles using ultrasound feedback, in: 2014 IEEE International Conference on Robotics and Automation (ICRA), IEEE, 2014, pp. 3807–3812.

[111] S. Scheggi, K.K.T. Chandrasekar, C. Yoon, B. Sawaryn, G. van de Steeg, D.H. Gracias, S. Misra, Magnetic motion control and planning of untethered soft grippers using ultrasound image feedback, in: 2017 IEEE International Conference on Robotics and Automation (ICRA), IEEE, 2017, pp. 6156–6161.

[112] I.S. Khalil, D. Mahdy, A. El Sharkawy, R.R. Moustafa, A.F. Tabak, M.E. Mitwally, S. Hesham, N. Hamdi, A. Klingner, A. Mohamed, et al., Mechanical rubbing of blood clots using helical robots under ultrasound guidance, IEEE Robot. Autom. Lett. 3 (2) (2018) 1112–1119.

[113] Q. Wang, L. Yang, J. Yu, P.W. Chiu, Y.-P. Zheng, L. Zhang, Real-time magnetic navigation of a rotating colloidal microswarm under ultrasound guidance, I.E.E.E. Trans. Biomed. Eng. (2020), https://doi.org/10.1109/TBME.2020.2987045.

[114] Q. Wang, Y. Tian, X. Du, H. Ko, B.Y.M. Ip, T.W.H. Leung, S.C.H. Yu, L. Zhang, Magnetic navigation of collective cell microrobots in blood under ultrasound doppler imaging, IEEE/ASME Trans. Mechatron. 27 (5) (2022) 3174–3185.

[115] F.H. Gage, Mammalian neural stem cells, Science 287 (5457) (2000) 1433–1438.

[116] O. Lindvall, Z. Kokaia, A. Martinez-Serrano, Stem cell therapy for human neurodegenerative disorders—how to make it work, Nat. Med. 10 (7) (2004) S42–S50.

[117] C. Madeira, A. Santhagunam, J.B. Salgueiro, J.M. Cabral, Advanced cell therapies for articular cartilage regeneration, Trends Biotechnol. 33 (1) (2015) 35–42.

[118] O. Felfoul, M. Mohammadi, S. Taherkhani, D. De Lanauze, Y.Z. Xu, D. Loghin, S. Essa, S. Jancik, D. Houle, M. Lafleur, et al., Magneto-aerotactic bacteria deliver drug-containing nanoliposomes to tumour hypoxic regions, Nat. Nanotechnol. 11 (11) (2016) 941–947.

[119] F. Ullrich, C. Bergeles, J. Pokki, O. Ergeneman, S. Erni, G. Chatzipirpiridis, S. Pané, C. Framme, B.J. Nelson, Mobility experiments with microrobots for minimally invasive intraocular surgery, Invest. Ophthalmol. Vis. Sci. 54 (4) (2013) 2853–2863.

[120] S. Jeon, B.C. Park, S. Lim, H.Y. Yoon, Y.S. Jeon, B.-S. Kim, Y.K. Kim, K. Kim, Heat-generating iron oxide multigranule nanoclusters for enhancing hyperthermic efficacy in tumor treatment, ACS Appl. Mater. Interfaces 12 (30) (2020) 33483–33491.

[121] R. Cheng, W. Huang, L. Huang, B. Yang, L. Mao, K. Jin, Q. ZhuGe, Y. Zhao, Acceleration of tissue plasminogen activator-mediated thrombolysis by magnetically powered nanomotors, ACS Nano 8 (8) (2014) 7746–7754.

[122] S. Jeon, S. Kim, S. Ha, S. Lee, E. Kim, S.Y. Kim, S.H. Park, J.H. Jeon, S.W. Kim, C. Moon, et al., Magnetically actuated microrobots as a platform for stem cell transplantation, Sci. Robot. 4 (eaav4317) (2019).

[123] E. Kim, S. Jeon, H.-K. An, M. Kianpour, S.-W. Yu, J.-y. Kim, J.-C. Rah, H. Choi, A magnetically actuated microrobot for targeted neural cell delivery and selective connection of neural networks, Sci. Adv. 6 (39) (2020) eabb5696.

[124] H. Xu, M. Medina-Sánchez, V. Magdanz, L. Schwarz, F. Hebenstreit, O.G. Schmidt, Sperm-hybrid micromotor for targeted drug delivery, ACS Nano 12 (1) (2018) 327–337.

[125] X.-Z. Chen, M. Hoop, N. Shamsudhin, T. Huang, B. Özkale, Q. Li, E. Siringil, F. Mushtaq, L. Di Tizio, B.J. Nelson, et al., Hybrid magnetoelectric nanowires for nanorobotic applications: fabrication, magnetoelectric coupling, and magnetically assisted in vitro targeted drug delivery, Adv. Mater. 29 (8) (2017) 1605458.

[126] M.P. Kummer, 5-Dof Wireless Micromanipulation Using Soft-Magnetic Core Electromagnets, PhD thesis,, ETH Zurich, 2010.

[127] B. Wang, K.F. Chan, J. Yu, Q. Wang, L. Yang, P.W.Y. Chiu, L. Zhang, Reconfigurable swarms of ferromagnetic colloids for enhanced local hyperthermia, Adv. Funct. Mater. (2018), 1705701.

[128] J. Leclerc, H. Zhao, D. Bao, A.T. Becker, In vitro design investigation of a rotating helical magnetic swimmer for combined 3-d navigation and blood clot removal, IEEE Trans. Robot. 36 (3) (2020) 975–982.

[129] S. Wang, X. Guo, X. Xiu, Y. Liu, L. Ren, H. Xiao, F. Yang, Y. Gao, C. Xu, L. Wang, Accelerating thrombolysis using a precision and clot-penetrating drug delivery strategy by nanoparticle-shelled microbubbles, Sci. Adv. 6 (31) (2020) eaaz8204.

[130] G.W. Ashley, J. Henise, R. Reid, D.V. Santi, Hydrogel drug delivery system with predictable and tunable drug release and degradation rates, Proc. Natl. Acad. Sci. 110 (6) (2013) 2318–2323.

[131] X. Xia, K.F. Chan, G.T.Y. Wong, P. Wang, L. Liu, B.P.M. Yeung, E.K.W. Ng, J.-Y.W. Lau, P.W.Y. Chiu, Mesenchymal stem cells promote healing of nonsteroidal anti-inflammatory drug-related peptic ulcer through paracrine actions in pigs, Sci. Transl. Med. 11 (516) (2019) eaat7455.

[132] X. Xu, X. Xia, K. Zhang, A. Rai, Z. Li, P. Zhao, K. Wei, L. Zou, B. Yang, W.-K. Wong, et al., Bioadhesive hydrogels demonstrating ph-independent and ultrafast gelation promote gastric ulcer healing in pigs, Sci. Transl. Med. 12 (558) (2020) eaba8014.

[133] T. Xu, C. Huang, Z. Lai, X. Wu, Independent control strategy of multiple magnetic flexible millirobots for position control and path following, IEEE Trans. Robot. (2022), https://doi.org/10.1109/TRO.2022.3157147.

[134] T. Xu, Z. Hao, C. Huang, J. Yu, L. Zhang, X. Wu, Multi-modal locomotion control of needle-like microrobots assembled by ferromagnetic nanoparticles, IEEE/ASME Trans. Mechatron. (2022), https://doi.org/10.1109/TMECH.2022.3155806.

[135] G.-Z. Yang, J. Bellingham, P.E. Dupont, P. Fischer, L. Floridi, R. Full, N. Jacobstein, V. Kumar, M. McNutt, R. Merrifield, B.J. Nelson, B. Scassellati, M. Taddeo, R. Taylor, M. Veloso, Z.L. Wang, R. Wood, The grand challenges of science robotics, Sci. Robot. 3 (14) (2018) eaar7650.

[136] P.E. Dupont, B.J. Nelson, M. Goldfarb, B. Hannaford, A. Menciassi, M.K. O'Malley, N. Simaan, P. Valdastri, G.-Z. Yang, A decade retrospective of medical robotics research from 2010 to 2020, Sci. Robot. 6 (60) (2021) eabi8017.

PART II

Robotic cell characterization

PART III

Robotic cell
characterization

CHAPTER 9

Robotic cell electrophysiological characterization for drug discovery

Riley E. Perszyk[a,b], Mighten C. Yip[a], Andrew Jenkins[c], Stephen F. Traynelis[b], and Craig R. Forest[a]

[a]George W Woodruff School of Mechanical Engineering, Georgia Institute of Technology, Atlanta, GA, United States
[b]Department of Pharmacology and Chemical Biology, Emory University School of Medicine, Atlanta, GA, United States
[c]Department of Pharmaceutical Sciences, University of Saint Joseph, West Hartford, CT, United States

1 Introduction

Patch-clamp electrophysiology is a key technique in the pharmacology, physiology, and neuroscience fields [1,2]. Patch-clamp recording is the most accurate method for electrophysiological characterization of cells. Specifically, it enables measurement of ionotropic receptors, is capable of fully resolving the time-course of postsynaptic or postjunctional currents, and can measure the ion flux and rapid transitions of even single ion channels [1,3–5]. However, the precision of this technique comes hand in hand with extensive effort and time commitment.

1.1 The current state-of-the-art robotics for electrophysiology characterization

Attempts have been made to accelerate the collection of electrophysiology data (or approximate electrical activity), i.e., fluorometric probes that sense particular ions or membrane voltage and machines that patch dissociated cells in parallel multiplexing to achieve high throughput [6–11]. However, these methods sacrifice the high precision of patch–clamp electrophysiology or the ability to patch all types of cells that a normal electrophysiology rig can record from in order to achieve higher throughput. For instance, fluorometric probes have limitations, i.e., their affinity for a particular ion or their sensitivity, that they must be tuned to a specific application, and that they may have a limited dynamic range. The lack of voltage clamping in imaging experiments also confounds voltage fluctuation associated with

Robotics for Cell Manipulation and Characterization
https://doi.org/10.1016/B978-0-323-95213-2.00013-2

the measured response. In these experiments, the voltage cannot be controlled, thus, fully resolving the activity of ionotropic receptors is typically not possible [12–15]. As for high-throughput planar patch-chip machines, the tested cells must be in suspension; adherent cells or cells embedded in intact tissues cannot be tested [10]. Planar patch-clamp is limited in the quality of the electrophysiological characterization as well, such as a lower signal-to-noise ratio, and poorer temporal resolution. Additionally, perfusion (e.g., for drug discovery) is typically less precise in volume, flow rate, and concentration, compared to what can be achieved on a traditional patch-clamp rig [10]. For high-throughput patch-clamp systems, the cost of both equipment and supplies are prohibitive for many studies.

In contrast to fluorometric probes and planar patch-clamp, whole-cell patch-clamp utilizes individual glass micropipette electrodes, with their tips placed delicately against a cell's membrane, to achieve the highest fidelity recordings of cell electrophysiology. While this technique traditionally requires human operator skill and has low throughput, recent advances in automation have enabled robotic cell electrophysiological characterization for drug discovery.

1.2 The patcherBot framework

A series of software and hardware developments have enabled a traditional intracellular electrophysiology rig to be capable of autonomous operation by integrating robotic vision, pipette pressure control, and electrode translation with a central computer controller [16,17]. Along with the discovery that patch-electrodes can be cleaned and reused, this resulting "patcherBot" can execute the required steps to perform fully autonomous patch-clamp electrophysiology. The patcherBot accelerates electrophysiology experimentation by reducing the experimental time of many tedious steps and decreases the amount of operator-rig interfacing time and the need for front loading most of the effort, e.g., calibrating the pipette and selecting cells to be patched. The patcherBot can be set up to run unattended for over 4 h (where it can patch over 30 cells sequentially), and operates at about a 70% success rate (the rate where whole-cell patch-clamp configuration is achieved per patching attempt) [16]. With these advances, the patcherBot can record spontaneous activity or voltage-dependent biological phenomena, which can be used to study a wide array of ionotropic receptors and can address questions concerning connectomics and neuronal intrinsic properties. However, despite its many capabilities, this version of the patcherBot cannot perform most of the assays needed to study ligand–gated ionotropic receptors or pharmacological studies on ionotropic receptors of all kinds.

A specialized version of the patcherBot, the patcherBot$_{Pharma}$, has also recently been created, which enables automated pharmacological electrophysiology experimentation (Fig. 9.1). The patcherBot$_{Pharma}$ can perform

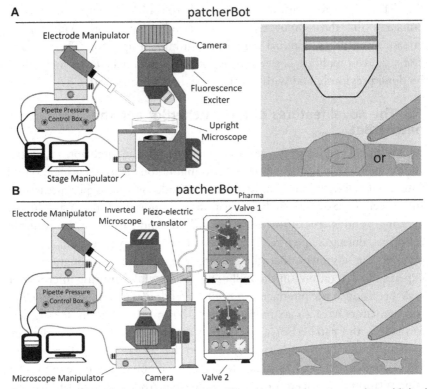

Fig. 9.1 Versions of the patcherBot. (A) Illustration of the previously published patcherBot [16], which is comprised of an upright microscope, a camera, a custom pressure control box, an electrode micromanipulator, and a motorized stage. In experiments, cells can be stimulated by photic stimulation or by electrode voltage or current control. (B) Illustration of the patcherBot$_{Pharma}$ [18], which is comprised of an inverted microscope, a camera, a custom pressure control box, an electrode micromanipulator, a motorized microscope manipulator, two motorized solution valves, and a solution exchange manifold controlled by a piezoelectric translator. In experiments, in addition to electrode voltage and current control, various solutions can be applied, and different solutions can be rapidly switched between to activate or modulate target ion-channels or receptors. *(Adapted with permission from R.E. Perszyk, M.C. Yip, O.L. McConnell, E.T. Wang, A. Jenkins, S.F. Traynelis, C.R. Forest. Automated intracellular pharmacological electrophysiology for ligand-gated ionotropic receptor and pharmacology screening. Mol. Pharmacol. 100 (1) (2021a) 73–82.)*

experiments on ligand-gated ionotropic receptors that require response to fast agonist exposure (ms exchange time) or concentration-response effects of pharmacological ligands. This version of the system is enabled by equipping the patcherBot with automated control of the bath solution and a solution manifold attached to solution valves and a piezoelectric translator. Additional modifications of the patcherBot$_{Pharma}$ software, to program the framework for various types of pharmacological experimentation, enhances the throughput rate via unattended operation, with further enhancement from minimal assistance of a human operator. The patcherBot$_{Pharma}$ can replicate conventional datasets substantially faster than could be done previously and with considerably less human effort.

1.3 The novel features of the patcherBot and the patcherBot$_{pharma}$

An additional engineering aspect that is required by the patchBot$_{pharma}$ is the ability to lift cells in the whole-cell configuration or by securing excised patches of cell membrane (inside-out or outside-out) for experimentation. Intact cell or excised patch translation is important for rapid solution exchange, and is crucial to ensure the viability and naivety of the available cell pool during sequential experimentation. The patcherBot$_{Pharma}$ can traverse the distances that the electrode must move throughout one experimental iteration (on the millimeter scale) while ensuring micrometer scale precision at the interface of a multibarrel flow pipe (as most piezoelectric translators have a maximum range $< 300\,\mu m$). In addition, for gigaseal formation, the patcherBot$_{Pharma}$ must place the electrode tip on the cell ($\sim 10\,mm$) and ideally on a specific spot on the cell (1–3 mm). Throughout one complete cycle of the patcherBot$_{Pharma}$ operation (patching, experiment manipulation, and electrode cleaning) the electrode can translate on the order of hundreds of milliseconds (Fig. 9.2).

A high level of accuracy and precision is required for efficient and productive experimentation with the patcherBot$_{Pharma}$; placement of the electrode at the solution manifold without manual, time-intensive error correction is required. Typically, this is done manually with visually placing the electrode tip (with cell or patch attached) at a predesignated location. Test pulses are generally conducted (to ensure proper placement), with the whole process taking at least 30 s for translation and testing. The ability to accurately return to the interface of the solution manifold is shown through four iterations of the patcherBot$_{Pharma}$ process (translating from the cells to the solution manifold to the clean and wash baths). The solution

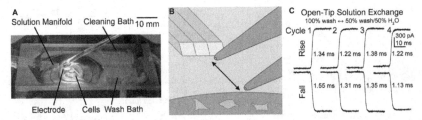

Fig. 9.2 Repeatability of electrode translations. (A) Image of the recording chamber and electrode cleaning baths. (B) Illustration depicting the distances (e.g., *X–Y* mm scale) the electrode translates during and experimental iteration. (C) Piezoelectric translator jump open-tip solution exchange times, across several experimental iterations (cell location to solution manifold interface to cleaning and wash baths). *(Adapted with permission from R.E. Perszyk, M.C. Yip, O.L. McConnell, E.T. Wang, A. Jenkins, S.F. Traynelis, C.R. Forest. Automated intracellular pharmacological electrophysiology for ligand-gated ionotropic receptor and pharmacology screening. Mol. Pharmacol. 100 (1) (2021a) 73–82.)*

exchange around the open-tip electrode was determined using the piezoelectric translation of the solution manifold.

Two of the standard methods used to study ligand-gated ionotropic receptors using rapid solution exchange manifolds are lifting cells in the whole-cell configuration and pulling patches (outside-out or inside-out). The patcherBot$_{pharma}$ can reliably perform both these techniques. The lifted cell method consists of a segmented (100 step) spiral translation trajectory while applying light suction to the patch pipette (-40 mbar), which enables a high success rate (Fig. 9.3A–C). The patch pulling method consists of a segmented (100 step) arc translation trajectory while the patch pipette is open to atmospheric pressure, which enables a high success rate (Fig. 9.3D–F). Characteristic low capacitance and high resistance of this patch-clamp conformation is routinely achieved (Fig. 9.3A and C).

1.4 patcherBot$_{pharma}$ implementation and performance

The patcherBot$_{Pharma}$ can perform rapid solution exchange experiments. This point was illustrated by experimental data from two synaptic ligand-gated ionotropic receptors, GABA$_A$R and NMDARs (Fig. 9.4). NMDAR responses were recorded from transfected HEK293 cells that were lifted off the bottom and from excised outside-out patches (Fig. 9.4A). GABA$_A$R responses were recorded from stably expressing cells testing long agonist applications as well as brief agonist applications (5 ms, Fig. 9.4B). In addition to rapid solution exchange protocols, the patcherBot$_{Pharma}$ is programmed with the ability to conduct many protocols (see supplemental figures in [18]). These

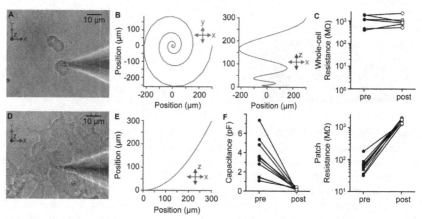

Fig. 9.3 Methods of removing cells or membrane patches from cells from coverslips. (A–C) Cell lifting method. (A) Image of a patched lone cell in the whole-cell configuration prior to lifting. (B) Lift trajectory (spiral path, 100 discrete segments) for isolated cells. (C) Resistance plot showing the electrode maintains a high resistance seal before and after the lifting process. (D–F) Patch pulling method. (D) Image of a patched cell in the whole-cell conformation prior to excising an outside-out patch. (E) Patch-pulling trajectory (arc path, 100 discrete segments) for removing outside-out patches from multicell colonies. (F) Capacitance and resistance plots illustrating that successful high-resistance, low-capacitance outside-out patches were obtained readily from adherent multicell colonies. The low resistance determinations prior to pulling a patch are due to direct electrical connections due (gap-junctions) between cells with direct physical contact to one another. *(Adapted with permission from R.E. Perszyk, M.C. Yip, O.L. McConnell, E.T. Wang, A. Jenkins, S.F. Traynelis, C.R. Forest. Automated intracellular pharmacological electrophysiology for ligand-gated ionotropic receptor and pharmacology screening. Mol. Pharmacol. 100 (1) (2021a) 73–82.)*

protocols include various rapid solution exchange paradigms as well as voltage-clamp and current-clamp protocols. These protocols can be employed to measure neuronal activity or study specific voltage-gated channels expressed in heterologous cells (adherent, lifted cell, or in excised patch conformations), replicating the capabilities of the original patcherBot but allowing for various pharmacological compounds to also be tested during various phases of experimentation.

The performance breakdown of the patcherBot$_{pharma}$ was determined using a series of rapid agonist application experiments on excised outside-out patches containing either GABA$_A$R or NMDARs (Table 9.1). Overall, the performance of the patcherBot$_{Pharma}$ for all experiments revealed that a giga-ohm resistance seal patch (gigaseal, or gigaseal patch) was readily obtained 81.2% of the time (108 of 133 attempts). Next, successful break

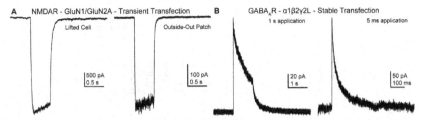

Fig. 9.4 Examples of fast-solution exchange data collected via the patcherBot$_{pharma}$. (A) NMDAR responses stimulated by 100 μM glutamate and 30 μM glycine. Recordings are from transiently transfected HEK cells lifted whole-cell (*left*) and an outside-out patch using a 4 MΩ electrode (*right*), at −60 mV in 0 mM Mg^{2+}. Note the more rapid rise of the response in the outside-out patch example due to inherent more rapid solution exchange as compared to the lifted cell. (B) Responses of stably transfected HEK cells expressing GABA$_A$R (α1β2γ2L) stimulated by long (*left*, 1 s application) or brief (*right*, 5 ms application) pulses of 1 mM GABA. (*Adapted with permission from R.E. Perszyk, M.C. Yip, O.L. McConnell, E.T. Wang, A. Jenkins, S.F. Traynelis, C.R. Forest. Automated intracellular pharmacological electrophysiology for ligand-gated ionotropic receptor and pharmacology screening. Mol. Pharmacol. 100 (1) (2021a) 73–82.)*

in to the whole-cell conformation occurred 96.3% of the time (104 of 108 gigaseals). The success rate of obtaining an outside-out patch was 76.0% (79 of 104 whole-cell conformations), thus allowing for the initiation of the experimental procedures. The completion of the experimental procedures of those that were initiated was found to be 74.7% (59 of 79 outside-out patches). Experiment failure typically arises from patch degradation (loss of the gigaseal) or from the lack of a receptor response to agonist. Overall success rate—percentage of the time that an experiment succeeded based on each patching attempt—of the patcherBot$_{Pharma}$ was 44.4% (59 of 133 attempts). There were two key factors that determined whether an attempt led to a successfully completed experiment or not: electrode placement and biological factors.

The transfection efficiency was a major contributing factor to overall experiment failure for the NMDARs experiments. There was a higher success rate in achieving a successful recording set from the stably expressing GABA$_A$R cells (31 successes out of 51 total attempts) than the transiently transfected NMDAR cells (28 successes out of 82 total attempts, Fisher's exact test, $P = .0039$). The use of a GFP plasmid included in the transient transfection protocol did not reliably return high enough expression levels of NMDARs (14 of the 42 pulled patches did not have a sufficient response). By contrast, the stable cell line had a sufficient GABA$_A$R response: 26 of the

Table 9.1 Success rates of individual steps in the patcherBot$_{Pharma}$ process.

	Attempts	GS	WC	O-O patch	Successful experiment	Zero current	Overall exp. success
All GABA$_A$R and NMDAR	133	108	104	79	59	19	
NMDAR (transient)	82	64	60	42	28	14	
GABA$_A$R (stable)	51	44	44	37	31	5	
(Computer vision)	15	9	9	6	6	0	
(Operator assisted)	36	35	35	31	25	5	
All GABA$_A$R and NMDAR		81.2%	96.3%	76.0%	74.7%	24.1%	44.4%
NMDAR (transient)		78.0%	93.8%	70.0%	66.7%	33.3%	34.1%
GABA$_A$R (stable)		86.3%	100.0%	84.1%	83.8%	13.5%	60.8%
(Computer vision)		60.0%	100.0%	66.7%	100.0%	0.0%	40.0%
(Operator assisted)		97.2%	100.0%	88.6%	80.6%	16.1%	69.4%

Success percentage for each individual step is calculated for each preceding step (i.e., O–O patch success is calculated based on the number of WC counts). An overall attempt was deemed a successful experiment if the full experimental procedures were performed, and the measured response was large enough to be measure for salient parameters. GS, gigaseal; WC, whole-cell; O-O, outside-out.

Adapted with permission from R.E. Perszyk, M.C. Yip, O.L. McConnell, E.T. Wang, A. Jenkins, S.F. Traynelis, C.R. Forest. Automated intracellular pharmacological electrophysiology for ligand-gated ionotropic receptor and pharmacology screening. Mol. Pharmacol. 100 (1) (2021a) 73–82.

31 outside-out patches. Optimizing expression is an important factor to the overall success of the patcherBot$_{pharma}$.

The other major source of patcherBot$_{Pharma}$ failure occurs at the gigaseal formation step and the patch excising step, both of which are due to slight errors (1–3 μm) in electrode placement. In this dataset, a subset of experiments was performed with an experimenter manually intervened by controlling the final placement of the electrode onto the cell. Overall, this is a trivial task, in that, in the patcherBot$_{Pharma}$ process, the system pauses just after the electrode is brought to a location directly above the next cell previously selected (100 μm above). In these operator-assisted experiments, the observed gigaseal yield was superior. Specifically, the operator-assisted trials resulted in 35 gigaseal successes from 36 attempts compared to 9 gigaseals successes from 15 attempts when computer-controlled placement was used (Fisher's exact test, $P = .0016$). Furthermore, the operator-assisted trials saw improvements in patch excising step, resulting in 31 successes from 35 attempts compared to six successes from nine attempts for the computer-controlled placement trials (Fisher's exact test, $P = .1383$). The operator-assisted trials had improved overall yield as a result with a 69.4% success rate (25 good experiments of 36 attempts), compared to the success rate of the fully automated trials (~40% success rate, 6 good experiments of 15 attempts, Fisher's exact test, $P = .0645$).

On top of differences in the success rate of these alternative methods, the current fully automated implementation of the patcherBot$_{Pharma}$ relies on cross-correlation machine vision methods. This method is comparatively slow, compared to manual manipulation, and has modest precision (~5–10 μm). For the improved success seen in the operator-assisted trials, the electrode tip placement must be improved by an order of magnitude (<1 μm). The fully automated method of the patcherBot$_{Pharma}$ takes, on average, 267 ± 35 s (mean $\pm$ standard deviation, SD) to using the autocorrelation method to correct the manipulators, place the electrode on the target cell, achieve a gigaseal, and break in to the whole-cell conformation, compared to only 74 ± 10 s (mean $\pm$ SD) for the operator-assisted patcherBot$_{Pharma}$. With this method, the robotic system moves the stage manipulator to the next cell and translates the electrode to a location (100 μm) above the cell, then the operator lands the electrode on the cell, forms a gigaseal, and the robotic system performs its automated break-in routine. As it stands, the patcherBot$_{Pharma}$ can operated in a fully autonomous mode; however, the operator-assisted mode has improved speed and overall performance.

With such operational proficiency, with brief operator assistance, experimental protocols can be designed to optimize the performance of the patcherBot$_{Pharma}$. Sets of experiments can be combined so that the time in-between data collection can be minimized; this was demonstrated by a representative run of the patcherBot$_{Pharma}$ (Fig. 9.5). In this experimental protocol, four phases of recording were collected by the patcherBot$_{Pharma}$; during each phase, five sweeps were collected, where each sweep had a 10 s duration and agonist was applied for 0.5 s. After the full protocol was collected the position of the electrode placement was validated by checking the open-tip exchange time (the patch can easily be cleared with high positive pressure using the pressure control box). In total, the data collection time and electrode placement validation totaled 11.2 min. If quality check deemed that the patcherBot$_{Pharma}$ failed at any of the process checkpoints, the patcherBot$_{Pharma}$ terminated the current iteration, cleaned the electrode, and initiated the next cycle—it typically takes only 1.4 min to start the next attempt.

In this representative run, 15 cells were selected to be patched, yielding 12 successful recordings over 3.8 h, as highlighted in Fig. 9.5. The operator only spent 15.5 min interacting with the patcherBot$_{Pharma}$ (7.1% of the experimental run time) after the initial 10.3 min to calibration the system and select cells. The patcherBot$_{Pharma}$ spent a total of 2.6 h recording data,

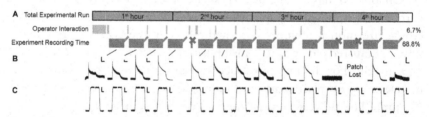

Fig. 9.5 Representative patcherBot$_{Pharma}$ run. (A) Timeline of experimental run. The breakdown of the time the operator interacted with the patcherBot$_{Pharma}$ and the duration of actual data collection is highlighted. The success of each experiment is noted with a check or cross. (B) Representative average GABA$_A$R responses (1 mM GABA, 1 s application) from all successful experiments. Scale bars indicate 20 pA and 0.5 s. (C) Open-tip position validation, showing the rising and falling exchange response to a jump to a solution containing 50% H$_2$O/50% wash solution. Scale bars indicate 200 pA and 20 ms. The rise and fall times (20%–80%, average ± SD (range)) for piezoelectric jumps were 3.06 ± 0.78 ms (1.30–4.11) and 3.56 ± 0.32 ms (2.27–6.55). *(Adapted with permission from R.E. Perszyk, M.C. Yip, O.L. McConnell, E.T. Wang, A. Jenkins, S.F. Traynelis, C.R. Forest. Automated intracellular pharmacological electrophysiology for ligand-gated ionotropic receptor and pharmacology screening. Mol. Pharmacol. 100 (1) (2021a) 73–82.)*

which equates to 72.0% of the total operation time. The data collected—GABA$_A$R rapid agonist jumps—upon analysis, were found to have similar activation, desensitization, and deactivation parameters as in the literature (Fig. 9.5B). During this representative run, the placement of the electrode at the interface of the solution manifold was reliable and had consistent solution exchange times (Fig. 9.5C).

1.5 patcherBot$_{Pharma}$ performance: A case study

The capacity of the patcherBot$_{Pharma}$ was tested with a case study (Fig. 9.6) where the concentration response of propofol (PRO), a widely used anesthetic, was measured on rapid solution application of agonist to GABA$_A$R in the stable cells line previously mentioned. Propofol is known to prolong the deactivation of GABA$_A$R, and it has been well characterized [19,20]. For this case study, the patcherBot$_{Pharma}$ was run with operator assistance (Fig. 9.6A and B), to optimize the time the patcherBot$_{Pharma}$ spent collecting biological data. The effect on GABA$_A$R deactivation was determined using an eight-point concentration response of propofol, split into two sets with a propofol-free control before and after drug application in each set (Fig. 9.6C). In four half-day recording sessions totaling 12.95 h of patcherBot$_{Pharma}$ operation, both concentration sets were recorded twice. In total, 42 cell recordings were attempted, obtaining 39 gigaseal patches, achieving 28 whole-cell conformations, with 24 successful outside-out patches, yielding 18 completed experimental datasets collected (including six incomplete), and resulting in 113 concentration data points (Fig. 9.6D and E). After triaging the recordings with unacceptable leak current, too small a response amplitude, or recording artifacts that disrupted accurate measurement, 71 data points remained, which were used to calculate the EC$_{50}$ of propofol actions on GABA$_A$R deactivation (EC$_{50}$ = 11.8 ± 4.6 μM, Fig. 9.6D and E).

During the entire case study, the operator interacted with the patcherBot$_{Pharma}$ for only 2.49 h of the entire 12.95 h of operation. Additionally, the patcherBot$_{Pharma}$ was in the process of collected experimental data for 9.07 h of the total time (Table 9.2). There were 1.39 h of other automated processing time (cleaning, other system processes, etc.). The total operator interaction time comprised of, primarily, cell selection, landing the electrode on the cell, and gigaseal formation. On average, during each iteration, patcherBot$_{Pharma}$ spent 1.99 min translating the electrode and cleaning the electrode. Additionally, the time the operator spent landing the electrode on the cell and establishing the whole-cell conformation totaled

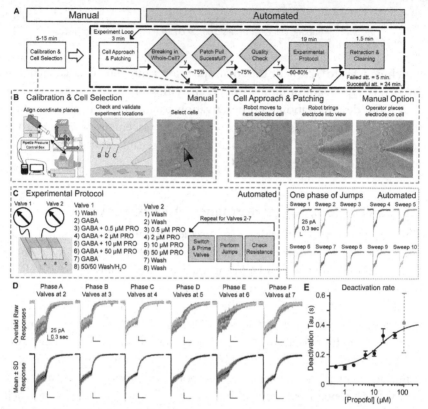

Fig. 9.6 Propofol concentration response case study. (A) A procedural flowchart illustrating the operation, timing, and success rate of the patcherBot$_{Pharma}$. The white boxes indicate manual steps and *gray boxes* indicate automated steps. After the initial calibration and cell selection step, the patcherBot$_{Pharma}$ iterates through the main cycle of the process and records from each cell in turn. Quality check points are placed periodically to determine if the current iteration should be terminated and proceed to the next one. (B) The details of the manual steps are shown: the calibration and cell selection step (*left*, typically 7–12 cells) whereby the alignment of the coordinate systems takes place, checking whether the saved locations of the solution manifold are indeed correct, and lastly the cell selection step. At the beginning of each loop, the cell approach and patching step (*right*) starts with the patcherBot$_{Pharma}$ moving the stage to the next selected cell and positioning the electrode above the cell (100 µm). The operator-assisted task includes lowering the electrode to the cell, and then either manually making the gigaseal and break-in or turning control over to the patcherBot$_{Pharma}$ for automated control of these steps. (C) The details of the experimental protocols of this case study include six sets of solutions used during each experimental set (two control and four propofol solutions per set, *left*). Each phase includes the collection of 10 replicates of the intended jump protocol (*right*) after the valves have been changed and primed. (D) The sweeps collected from a single experiment, showing all collected replicates (*top*) and the average response (±SD, shown by *shaded gray area*). The desensitization and deactivation are depicted on top of the averaged responses (*white line*). (E) The concentration response of propofol on the deactivation tau fitted with the Hill equation. The highest concentration of propofol (100 µM) was omitted due to the enhanced desensitization, which prevented accurate measurement of the deactivation rate. (*Adapted with permission from R.E. Perszyk, M.C. Yip, O.L. McConnell, E.T. Wang, A. Jenkins, S.F. Traynelis, C.R. Forest. Automated intracellular pharmacological electrophysiology for ligand-gated ionotropic receptor and pharmacology screening. Mol. Pharmacol. 100 (1) (2021a) 73–82.*)

Table 9.2 Breakdown of the patcherBot$_{Pharma}$ operation time in the propofol case study.

	Total operation	Electrode cleaning (robotic control)	Data collection (robotic control)	Patch establishment (operator control)	Nonrecording time[a]
Total time (percentage)	12.95 h	1.39 h (10.7%)	9.07 h (70.0%)	2.49 h (19.3%)	3.88 h (30%)
Time per cycle (successful cycle)		1.99 min	24.6 min	3.55 min	5.54 min
Time per data point (successful cycle)			4.1 mins[b] (~1 min solution change time) (1.67 min data collection)	2.1 min[c]	

[a]The nonrecording time is the time the robot is not performing the data collection protocol.
[b]Each data collection phase equals the solution change time plus the data collection time, however, the mean time per data point reflects the additional time needed to pull the patch and validate the jump at the end of the experiment averaged into the timing for each phase. Adapted with permission from Perszyk et al. [18].
[c]This rate represents the total time the operator spent interacting with the patcherBot$_{Pharma}$ during the entire experiment performance (cell selection, solution maintenance, electrode placement on the cell, and gigaseal formation).

~2–3 min, on average. If all steps were successful, the 24.6 min experimental data collection step would be initiated. If any quality control checks were not satisfied or if the patch broke down in the process of pulling the outside-out patch (1.73 min process), then the current patcherBot$_{Pharma}$ cycle would be terminated and the system would clean the electrode and commence the next attempt in less than 2 min. The 57% success rate of the case study (24/42 attempts were successful) did not hinder the overall performance of the patcherBot$_{Pharma}$ (Table 9.3). Theoretically, in 12.95 h, the patcherBot$_{Pharma}$ could have performed 25.8 experiments; this is only modestly higher than the 18 successfully experiments collected. This case study delivered data collection at the rate of 2.1 min (of the operator's effort) per data point. The theoretical minimum of this parameter is about twice as fast, collecting a data point every 0.97 min of the operator's time spend controlling the patcherBot$_{Pharma}$.

Table 9.3 Performance of the patcherBot$_{Pharma}$ in the propofol case study.

	Counts	Yield	Theoretical max[a]
Patch attempts	42		
Successfully established gigaseals	39	93%	
Whole-cell conformations obtained	28	72%	
Outside-out patches obtained (experiments started)	24	86%	
Successful experiments	18	75%	25.8
Total data points collected	113		154.8
Data points passed quality control	71		

[a]The theoretical maximum values were determined by taking the total operation time divided by the total time for one successful cycle. Adapted with permission from Perszyk et al. [18].

2 Discussion

The patcherBot$_{Pharma}$ has the ability to enhance the process of patch–clamp electrophysiology experimentation, allowing for more rapid data collection, requiring less training for operation, and making the data more consistent. The features introduced in the patcherBot$_{Pharma}$, namely precise and accurate electrode translations, solution handling, electrode cleaning, and rapid solution exchange, allow for ligand-gated ionotropic receptor pharmacological screening, which expands the range of experiments the patcherBot can perform. Nearly any in vitro pharmacological experiment involving ligand-gated or voltage-gated ion channels can be performed using the patcherBot$_{Pharma}$ [18]. Similar to a traditional patch-clamp rig, the patcherBot$_{Pharma}$ retains the ability to be retooled as needed, and has the option to be run autonomously or with operator assistance. For example, the patcherBot$_{Pharma}$ can apply compounds to the entire bath and record from adherent cells.

The patcherBot$_{Pharma}$ achieves giga-ohm resistance patches and can break in to the whole-cell patch conformation at a high rate (80%–100%). The newly developed methods to lift isolated patch-clamped cells and to excise outside-out patches have a high success rate (70%–90% yield). The patcher-Bot$_{Pharma}$ removes the need for an experimenter to manually guide the patch electrode to desired locations throughout the course of the full experiment and can allow for parallel tasks to be performed while the system conducts the programmed electrophysiology experiments. The patcherBot$_{Pharma}$

enables high efficiency data collection, allowing for the ability to rapidly collect a dataset, with eight-point concentration response curve of the effects of propofol on $GABA_AR$ deactivation, which would take weeks/months of recording, reducing it to approximately 13 h of recording.

Unlike other systems, the patcherBot$_{Pharma}$ has the normal capabilities of electrophysiology rigs, as it is comprised of typical components of a traditional electrophysiology rig that can be retooled as needed with various components. The example shown used a larger, three-barreled manifold, which enables solution exchange times in the low millisecond range ($\sim$1–2 ms); however, this could be reduced further ($<$1 ms) using pulled theta tubes or other different solution manifolds [21,22]. This would make it possible to study rapidly desensitizing receptors, which cannot be measured currently on commercially available multiwell high-throughput planar patch-clamp systems. The patcherBot$_{Pharma}$ running costs are low—primarily the cost of the cell culture preparation, salt solutions, and compounds being evaluated—and comparable to the cost of operating a traditional patch-clamp rig. Additionally, upgrading from a traditional electrophysiology rig can be feasible since the differences between a traditional and the patcherBot$_{Pharma}$ system are limited [18]. Furthermore, reduced effort and time on behalf of the user enables two other advantages. First, the highly skilled electrophysiologist may now spend more time focused on other tasks rather than manually controlling the patch-clamp rig through the tedious experimental steps of data collection. Second, the requirement for a highly skilled electrophysiologist is reduced, allowing for training and staffing of users who did not dedicate years of their research learning how to patch-clamp.

Facilitating pharmacological experimentation on ligand gated channels can not only increase productivity but also can reduce the chance of human bias when collecting data, as the experiment protocols are explicitly defined prior to experiment execution. Indeed, the reduction in human bias can play a critical role in validating results and conclusions across distinct research groups studying similar research areas. For instance, divergent results and misinterpretation of nonsignificant findings remain problematic in science, and the reproducibility of each other's works remains highly nonuniform [23].

The system reduces the burden of electrophysiology recording and allows single-cell electrophysiology to be applied to more labor-intensive questions (collecting multiple concentration data points or testing more constructs), allows for more complex experimental protocols, or increase the number of replicates and more controls to make more reproducible data.

Such increase in efficiency and throughput for single cells in physiological conditions (i.e., nonsuspended) would help rectify issues such as having low power in the experimental design, cited by many neuroscience studies [24]. Additionally, as the patcherBot$_{Pharma}$ experiment easily allows for full data logging, the data collected by the patcherBot$_{Pharma}$ may be more reproducible, as the experimental parameters are automatically documented and can be easily shared [25].

Looking forward, the patcherBot$_{Pharma}$ could be applied to more nuanced electrophysiology experimentation, e.g., single-channel recoding. Single-channel recording is even more laborious and tedious, as the electrodes require adding polymer near the tip to reduce capacitance and many patches must be pulled in the search for an excised patch that contains only a single channel. Thus, the patcherBot$_{Pharma}$ system has the capacity to provide greater benefit in reducing the need for the experimenter to perform this even more tedious, highly repetitive, and low yield type of patch–clamp electrophysiology. In drug development programs, such as large library screens and systematic structure–activity evaluation, in most cases, more general effects of compounds are typically tested with easier to perform assays. As a result, more subtle drug effects may be missed, such as conductance modification [26,27]. The patcherBot$_{Pharma}$ in these applications may lead to unexpected discoveries or help elucidate more complete mechanistic understanding of the tested compounds, which could lead to novel pharmacological strategies to treat diseases or to better engineer drugs with specific properties.

Furthermore, the patcherBot abilities could be augmented by using rapidly developing machine vision and image processing techniques. Improvements in the ability to correct for small manipulator errors that accrue over the many repetitive movements made throughout one cycle of the patcherBot$_{Pharma}$ process could have multiple benefits. In a recent paper [28], the authors detail the training of a computational neural net that can identify the tip of the recording electrode with improved time savings and greater precision than previous applied methods (such as those used with the patcherBot$_{Pharma}$). This machine learning method may improve electrode placement by circumventing the need for operator intervention in the identified key steps that lead to reduced yield in obtaining gigaseals and pulling outside-out patches. Furthermore, methods for identification of viable cells, potentially those that are transfected, through computational methods, such as the real-time machine learning method in [29], would enable agnostic cell selection and improve time savings in the cell selections phases of the patcherBot$_{Pharma}$ process.

Another area of advancement in the evolution of the patcherBot$_{Pharma}$ is to combine the ability for pharmacological screening with the ability to test more relevant biological or physiological activity, i.e., testing compounds on neuronal activity. Typically, only a few candidate compounds are carried forward from drug development to be tested in physiological tissues. The in vitro effects of compounds do not always translate into ex vivo or in vivo actions. Thus, accelerating the testing of pharmacological compounds in tissue could impart more precise evaluation of the physiological actions of novel drugs, in that perhaps structure–activity relationships could be conducted using biological actions instead of basic pharmacological properties. However, there would be numerous challenges to enabling a patcherBot system to conduct these more physiological experiments, including controlling various stimulation methods, tissue maintenance, and tissue loading and removal. The future of robotic experimentation can greatly enhance the effectiveness of many electrophysiologists through generating larger, more efficient, and unbiased datasets, and thus, developing a deeper understanding of pharmacology, physiology, and neuroscience.

References

[1] E. Neher, B. Sakmann, Single-channel currents recorded from membrane of denervated frog muscle fibres, Nature 260 (5554) (1976) 799–802.
[2] H.-J. Suk, E.S. Boyden, I. van Welie, Advances in the automation of whole-cell patch clamp technology, J. Neurosci. Methods 326 (2019), 108357.
[3] A. Auerbach, Y. Zhou, Gating reaction mechanisms for NMDA receptor channels, J. Neurosci. 25 (35) (2005) 7914–7923.
[4] S. Chakrapani, J.F. Cordero-Morales, V. Jogini, A.C. Pan, D.M. Cortes, B. Roux, E. Perozo, On the structural basis of modal gating behavior in K(+) channels, Nat. Struct. Mol. Biol. 18 (1) (2011) 67–74.
[5] D. Colquhoun, F.J. Sigworth, Fitting and statistical analysis of single-channel records, in: B. Sakmann, E. Neher (Eds.), Single-Channel Recording, Springer US, Boston, MA, 1995, pp. 483–587.
[6] H.-w. Ai, Fluorescent-protein-based probes: general principles and practices, Anal. Bioanal. Chem. 407 (1) (2015) 9–15.
[7] C. Deo, L.D. Lavis, Synthetic and genetically encoded fluorescent neural activity indicators, Curr. Opin. Neurobiol. 50 (2018) 101–108.
[8] C. Liu, T. Li, J. Chen, Role of high-throughput electrophysiology in drug discovery, Curr. Protoc. Pharmacol. 87 (1) (2019), e69.
[9] I. Mollinedo-Gajate, C. Song, T. Knopfel, Genetically encoded fluorescent calcium and voltage indicators, Handb. Exp. Pharmacol. 260 (2019) 209–229.
[10] A. Obergrussberger, T.A. Goetze, N. Brinkwirth, N. Becker, S. Friis, M. Rapedius, C. Haarmann, I. Rinke-Weiß, S. Stölzle-Feix, A. Brüggemann, M. George, N. Fertig, An update on the advancing high-throughput screening techniques for patch clamp-based ion channel screens: implications for drug discovery, Expert Opin. Drug Discov. 13 (3) (2018) 269–277.

[11] H.-b. Yu, M. Li, W.-p. Wang, X.-l. Wang, High throughput screening technologies for ion channels, Acta Pharmacol. Sin. 37 (1) (2016) 34–43.

[12] L.A. Annecchino, A.R. Morris, C.S. Copeland, O.E. Agabi, P. Chadderton, S.R. Schultz, Robotic automation of in vivo two-photon targeted whole-cell patch-clamp electrophysiology, Neuron 95 (5) (2017) 1048–1055.e3.

[13] L. Campagnola, M.B. Kratz, P.B. Manis, ACQ4: an open-source software platform for data acquisition and analysis in neurophysiology research, Front. Neuroinform. 8 (2014) 3.

[14] B.A. Suter, T. O'Connor, V. Iyer, L.T. Petreanu, B.M. Hooks, T. Kiritani, K. Svoboda, G.M. Shepherd, Ephus: multipurpose data acquisition software for neuroscience experiments, Front. Neural Circuits 4 (2010) 100.

[15] Q. Wu, A.A. Chubykin, Application of automated image-guided patch clamp for the study of neurons in brain slices, J. Vis. Exp. (125) (2017), 56010.

[16] I. Kolb, C.R. Landry, M.C. Yip, C.F. Lewallen, W.A. Stoy, J. Lee, A. Felouzis, B. Yang, E.S. Boyden, C.J. Rozell, C.R. Forest, PatcherBot: a single-cell electrophysiology robot for adherent cells and brain slices, J. Neural Eng. 16 (4) (2019), 046003.

[17] I. Kolb, W.A. Stoy, E.B. Rousseau, O.A. Moody, A. Jenkins, C.R. Forest, Cleaning patch-clamp pipettes for immediate reuse, Sci. Rep. 6 (2016) 35001.

[18] R.E. Perszyk, M.C. Yip, O.L. McConnell, E.T. Wang, A. Jenkins, S.F. Traynelis, C.R. Forest, Automated intracellular pharmacological electrophysiology for ligand-gated ionotropic receptor and pharmacology screening, Mol. Pharmacol. 100 (1) (2021) 73–82.

[19] S. Adodra, T.G. Hales, Potentiation, activation and blockade of GABAA receptors of clonal murine hypothalamic GT1-7 neurones by propofol, Br. J. Pharmacol. 115 (6) (1995) 953–960.

[20] B. Orser, L. Wang, P. Pennefather, J. MacDonald, Propofol modulates activation and desensitization of GABAA receptors in cultured murine hippocampal neurons, J. Neurosci. 14 (12) (1994) 7747–7760.

[21] N.G. Glasgow, J.W. Johnson, Whole-cell patch-clamp analysis of recombinant NMDA receptor pharmacology using brief glutamate applications, in: M. Martina, S. Taverna (Eds.), Patch-Clamp Methods and Protocols, Springer New York, New York, NY, 2014, pp. 23–41.

[22] D.M. MacLean, Constructing a rapid solution exchange system, in: G.K. Popescu (Ed.), Ionotropic Glutamate Receptor Technologies, Springer New York, New York, NY, 2016, pp. 165–183.

[23] O.A. Moody, S. Talwar, M.A. Jenkins, et al., Rigor, reproducibility, and in vitro cerebrospinal fluid assays: the devil in the details, Ann. Neurol. 81 (6) (2017) 904–907.

[24] K.S. Button, J.P.A. Ioannidis, C. Mokrysz, B.A. Nosek, J. Flint, E.S.J. Robinson, M.R. Munafò, Power failure: why small sample size undermines the reliability of neuroscience, Nat. Rev. Neurosci. 14 (5) (2013) 365–376.

[25] M.R. Munafò, B.A. Nosek, D.V.M. Bishop, K.S. Button, C.D. Chambers, N. Percie du Sert, U. Simonsohn, E.-J. Wagenmakers, J.J. Ware, J.P.A. Ioannidis, A manifesto for reproducible science, Nat. Hum. Behav. 1 (1) (2017) 0021.

[26] R.E. Perszyk, S.A. Swanger, C. Shelley, A. Khatri, G. Fernandez-Cuervo, M.P. Epplin, J. Zhang, P. Le, P. Bülow, E. Garnier-Amblard, P.K.R. Gangireddy, G.J. Bassell, H. Yuan, D.S. Menaldino, D.C. Liotta, L.S. Liebeskind, S.F. Traynelis, Biased modulators of NMDA receptors control channel opening and ion selectivity, Nat. Chem. Biol. 16 (2) (2020) 188–196.

[27] R.E. Perszyk, Z. Zheng, T.G. Banke, J. Zhang, L. Xie, M.J. McDaniel, B.M. Katzman, S.C. Pelly, H. Yuan, D.C. Liotta, S.F. Traynelis, The negative allosteric modulator EU1794-4 reduces Single-Channel conductance and Ca^{2+}

permeability of GluN1/GluN2A N-methyl-d-aspartate receptors, Mol. Pharmacol. 99 (5) (2021) 399.

[28] M.M. Gonzalez, C.F. Lewallen, M.C. Yip, C.R. Forest, Machine learning-based pipette positional correction for automatic patch clamp in vitro, eNeuro 8 (4) (2021), ENEURO.0051-21.2021.

[29] M.C. Yip, M.M. Gonzalez, C.R. Valenta, M.J.M. Rowan, C.R. Forest, Deep learning-based real-time detection of neurons in brain slices for in vitro physiology, Sci. Rep. 11 (2021) 6065.

CHAPTER 10

Automated cell aspiration for genetic and mechanical analysis

Mingzhu Sun[a,b], Huiying Gong[a,b], Yujie Zhang[a,b], Yaowei Liu[a,b], Yue Du[a,b], Qili Zhao[a,b], and Xin Zhao[a,b]
[a]Institute of Robotics and Automatic Information System, Tianjin Key Laboratory of Intelligent Robotics, Nankai University, Tianjin, China
[b]Institute of Intelligence Technology and Robotic Systems, Shenzhen Research Institute of Nankai University, Shenzhen, China

1 Introduction

In the past three decades, biological cell characterization has become a popular research topic in many subjects, such as micro/nanotechnology, biomechanics, biomedical engineering, and so on [1–3]. Researchers have designed various types of models and developed all sorts of sensors to measure the physical properties of single cells, such as mass, density, and traction force [4–7]. The measurement of cellular force enabled mechanical characterization of the cells [8, 9], while response testing of the cellular substances to different molecules can be used for gene diagnosis [10]. These cellular characteristics have provided us with explanations of cellular functionality [11, 12].

Genetic and mechanical analysis at the single-cell level needs the techniques of cell manipulation or cell surgery. A variety of techniques, including microfluidics [13], magnetic fields [14], optical tweezers [15], electrophoresis (EP) or dielectric electrophoresis (DEP) [16], acoustic manipulation [17], and cell aspiration by glass capillary micropipettes [18], have been introduced to manipulate single cells. Compared with the other noncontact methods, the cell aspiration system has the longest history. This kind of method is the most widely accepted and used technique due to its low requirements for equipment. Cell aspiration is generally realized by applying negative pressure to capillary micropipettes filled with a medium so that the whole cells or subcellular structures are aspirated into the micropipette. This technique is widely used in cell pick-up and positioning [19, 20], cell holding [21, 22], and cell transfer [23, 24], as well as in cell genetic and mechanical analysis [25–27].

Robotics for Cell Manipulation and Characterization
https://doi.org/10.1016/B978-0-323-95213-2.00002-8
223

Traditionally, cell aspiration depends heavily on skilled operators [28]. The operators operate multiple devices according to microscope images. They need to move the micropipettes to the appropriate positions and aspirate the target substances carefully by constantly controlling the pressure of the syringe connected to the micropipettes. They also need to perform experimental measurements based on microscopic images or transfer the substances in the pipettes for further analysis. Therefore, manual cell aspiration has high technical requirements, long operation time, and low efficiency, as a consequence of which, it is difficult to meet the requirements of high throughput. With the development of automation techniques in the past 20 years, automated cell aspiration systems have been proposed for genetic and mechanical analysis [29, 30]. Automation improves the accuracy and efficiency of biological operations and reduces the dependence of cell aspiration on labor.

In this chapter, we summarize the existing automation systems for genetic and mechanical analysis based on cell aspiration and discuss the major robotic techniques involved in developing these systems. The automation systems are categorized into two groups: automated cell aspiration for mechanical analysis and automated cell aspiration for genetic analysis. Finally, we will conclude this chapter and analyze the future trends of automated cell aspiration systems.

2 Automated cell aspiration for mechanical analysis

Ever-growing evidence indicates that the mechanical properties of biological cells are fundamental for cellular behavior and reflect the physiological and developmental state of the cells [31–34]. For example, Young's modulus of oocytes was used to study their maturation process, confirming that the stiffness of the zona pellucid would increase following fertilization [35]. A large number of studies have shown that cell mechanical properties may also be related to the progression of some diseases [36, 37]. For example, human red blood cells (RBCs) under different conditions, including in patients with thalassemia [38], burn-induced [39], and harboring *Plasmodium falciparum* [40], showed different shear modulus. The stiffness of metastatic cancer cells was much softer than the benign cells that line the body cavity [41, 42]. In addition, cell transformation and tumorigenicity were also related to cellular viscoelasticity [43, 44].

To characterize the mechanical properties of a single cell, the cells should be deformed under a known force. The force-deformation data then need to be measured accurately. Several techniques have been developed to measure the mechanical properties of the cells [45]. In atomic force microscopy (AFM) systems, the mechanical deformation is generated by the tip of a cantilever beam. The deflection of the cantilever is used to calculate the applied force [46–49]. In optical tweezer techniques, a laser beam captures high refractive index particles attached to cells to apply force to cells [50, 51]. As for magnetic tweezer systems, magnetic beads are attached to the cell surface, and the torsional torque generated by a magnetic field deforms the cells in magnetic twisted cytometry (MTC) [52, 53]. In the techniques of micropipette aspiration, the whole or part of the cells are aspirated in a glass capillary micropipette using the negative pressure generated by the injectors. The deformation of the cells is recorded under the applied pressure [54–57]. Compared with other techniques, micropipette aspiration provides a large range of aspiration force (from about 10 pn to hundreds of nn) with simple devices.

However, there are some problems with the traditional micropipette aspiration system. First, the suspended or adhered target cells are manually searched for and captured under the microscope, which is time-consuming. The negative pressure for aspirating cells must then be carefully controlled; however, the continuous evaporation of the liquid leads to the baseline drift of pressure, which makes it difficult to measure the pressure accurately. After acquiring the images of the cell deformation and the pressure data by the CCD camera and pressure sensor, respectively, a large number of cell deformation images need postprocessing. The manual measurement of cell deformation parameters is tedious and error-prone for human operators. Finally, the mechanical properties of the cells are quantified based on various models [58–62], and errors can grow very rapidly because of the low accuracy of offline manual measurement. Therefore, the key techniques for mechanical analysis based on automated cell aspiration include measured cell capture, pressure control in the micropipette, visual detection of cellular geometric parameters, and mechanical analysis based on cellular deformation.

2.1 Measured cell capture

The cells to be measured should be localized and the tip of the micropipette should be detected before cell aspiration. If the cells are scattered in a petri dish, it is necessary to select the cells in the field of view under the microscope by mouse clicking by the operator [63] or using image processing methods. It is also necessary to move the micromanipulators or the motorized stage of the microscope so that the cells are close to the tip of the micropipette. In batch cell measurement, it takes a lot of time to locate and capture the cells. Putting the cells in an inappropriate experimental environment for a long time will lead to changes in cell mechanical properties and inaccurate measurement results, due to the high sensitivity of the cells to the culture environment (oxygen content [64], temperature [65], osmotic pressure [66]). Therefore, improving the speed of target cell capture is important for biological applications.

It has been reported that microfluidic devices were combined with the micropipette aspiration to facilitate cell measurement. Guo et al. [67, 68] designed a microfluidic device for deforming single cells using precisely controlled pressure. The cells were introduced into the fluid and constrained in a funnel constriction. As shown in Fig. 10.1A, Li et al. [69] proposed a microfluidic device for quantifying cell mechanics at the single-cell level by combining the micropipette aspiration technique. In this device, the inlet was connected to a syringe pump filled with the cell suspension. The cell suspension was pumped continuously during the cell trapping. The cell deformation due to the micropipette aspiration was observed and used to measure the mechanical properties of single cells. After measurement, the aspirated cells can be released by withdrawing the cell suspension at the inlet. Using microfluidics effectively captures cells without manual operations.

Arranging the cells in order before the operation will eliminate the cell search time and realize the rapid measurement of batch cells. Dong et al. [70] developed a biochip, which consists of a regular microwell array. By introducing a path planning algorithm for micropipette movement, automated cell aspiration could be realized in the microwell array. As shown in Fig. 10.1B, Wang et al. [71] reported a vacuum-based cell holding device to fix batch zebrafish embryos in an array. The cells were batch immobilized on individual through-holes via a negative pressure, while extra cells were flushed off the device. Based on the sequence of the through-holes, cells can be operated continuously. Shakoor et al. [72] also designed a vacuum-based microfluidic chip to arrange cells in an array of polydimethylsiloxane (PDMS) channels. The negative pressure generated by the digital injector was applied to the cells. The cells then were held at the opening of the microfluidic channel. Up to 100 cells can be fixed in position in a one-dimensional array in a few minutes with a trap success rate of 80.6%.

The vacuum-based cell holding device needs to include specific devices and complex peripheral systems for each kind of cell with different sizes, which limits the application of these techniques [68, 73, 74]. As shown in

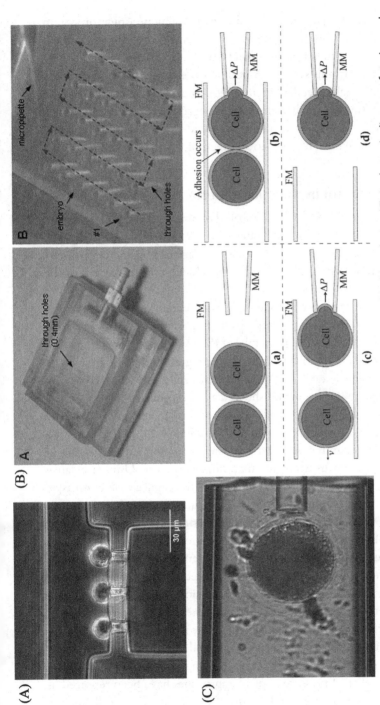

Fig. 10.1 Cell capture methods. (A) A microfluidic device. (B) A vacuum-based cell holding device. (C) A schematic diagram of storing and picking up cells one by one with a feeder micropipette. ((A) Reproduced from Y.-J. Li, Y.-N. Yang, H.-J. Zhang, C.-D. Xue, D.-P. Zeng, T. Cao, K.-R. Qin, A microfluidic micropipette aspiration device to study single-cell mechanics inspired by the principle of wheatstone bridge, Micromachines 10 (2) (2019) 131, copyright 2019 MDPI (Basel, Switzerland); (B) Reproduced from W. Wang, X. Liu, D. Gelinas, B. Ciruna, Y. Sun, A fully automated robotic system for microinjection of zebrafish embryos, PLoS One 2 (9) (2007) e862, copyright 2007; (C) Reproduced from Y. Liu, M. Cui, J. Huang, M. Sun, X. Zhao, Q. Zhao, Robotic micropipette aspiration for multiple cells, Micromachines 10 (5) (2019) 348, copyright 2019 MDPI (Basel, Switzerland).)

Fig. 10.1C, Liu et al. [75] developed a robotic micropipette aspiration system capable of storing multiple cells with a feeder micropipette [76] and picking up cells one by one to measure their mechanical properties with a measurement micropipette. In this system, the thinner measuring micropipette was inserted into the thicker feeder micropipette and held the target cell, the feeder micropipette then applied the proper negative pressure to draw back the adhesive cells. The robotic system was able to continuously measure more than 20 cells with a manipulation speed of 0.5 min/cell.

2.2 Pressure control in the micropipette

In the automated measurement system based on micropipette aspiration, the accuracy of the applied pressure will affect the accuracy of mechanical properties of the cells calculated using the models. Various pressure control systems and algorithms have been designed to achieve precise control of the pressure.

Microfluidic devices have been designed to accurately generate pressure. As shown in Fig. 10.2A, Li et al. [69] reported a microfluidic device to control the pressure based on the principle of the Wheatstone bridge circuit. With fixed resistances, the microfluidic device regulated the bridge pressure by controlling the flow rate. The pressure difference was linearly proportional to the volume flow rate, and showed a discrepancy of 10% at a maximum flow rate. The pressure model was proposed to accurately calculate the pressure value. As shown in Fig. 10.2B, Zhao et al. [35] proposed a method achieving quantitative control of the factual aspiration pressure in micropipette aspiration using a pneumatic microinjector. Using a balance pressure model, they quantified the influence of the capillary effect on aspiration pressure, which produced a significant pulling force and generated an extra aspiration pressure to cells. The final aspiration pressure was calibrated by both the output pressure and the capillary effect.

Control algorithms have been introduced to ensure the accuracy of pressure control. Shojaei-Baghini et al. [63] developed a closed-loop controlled pressure system for cell aspiration based on a proportional integral derivative (PID) controller. The aspiration pressure was generated by creating a height difference between reference and delta tanks. The positions of the two water tanks were controlled by a standard PID controller with pressure feedback from the differential pressure transducer, achieving a control error of ± 2 Pa. Moreover, a number of pressure control approaches have been reported in the literature for aspiration and positioning of a single cell in the

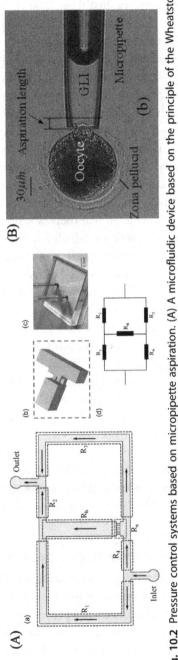

Fig. 10.2 Pressure control systems based on micropipette aspiration. (A) A microfluidic device based on the principle of the Wheatstone bridge circuit. (B) A pneumatic micropipette aspiration method using a balance pressure model. ((A) *Reproduced from Y.-J. Li, Y.-N. Yang, H.-J. Zhang, C.-D. Xue, D.-P. Zeng, T. Cao, K.-R. Qin, A microfluidic micropipette aspiration device to study single-cell mechanics inspired by the principle of wheatstone bridge, Micromachines 10 (2) (2019) 131, copyright 2019 MDPI (Basel, Switzerland); (B) Reproduced from Q. Zhao, M. Wu, M. Cui, Y. Qin, J. Yu, M. Sun, X. Zhao, X. Feng, A novel pneumatic micropipette aspiration method using a balance pressure model, Rev. Sci. Instrum. 84 (12) (2013) 123703, copyright 2013 AIP Publishing LLC.)*

micropipette. Anis et al. [23] presented a robotic manipulation system for living cell selection and transfer for the first time. The cells were aspirated into the micropipette based on vision-based feedback and closed-loop process control. The cell could be successfully aspirated, but it would disappear into the micropipette since the aspiration process was not well controlled. Lu et al. [77] proposed a micropipette pick-up method for partial cells and compared it with the whole-cell pick-up method. The operation time of partial cell pick-up (15 s) was shorter than that of the whole-cell (30 s), and the success rate reached 80%. However, it is easy for cells to be damaged using this method, and it is inevitable to lose cells in the process of micropipette movement. Zhang et al. [78] designed a closed-loop robust controller integrating cell dynamics to position the cell inside the micropipette. Human sperm cells were used as the controlled objects in the experiment. Since the head width of the sperm is only 3 μm, it was assumed that there was no relative motion between the sperm and the culture medium, which is not applicable to most kinds of cells. The average pick-up and positioning time was 10 s. Shan et al. [20] established a nonlinear dynamic model of cell motion inside the pipette by considering oil compressibility and connecting tube's deformation. Based on this model, an adaptive controller was designed to aspirate and control the position of the cell inside the micropipette accurately. This method could be applied to cells of different sizes, but the system exhibited an overshoot with the decrease of the cell size. Specifically, the smaller the cell size, the larger the overshoot. Sun et al. [19] considered the relative motion of the object and the fluid and established an overall dynamic model of microbead motion inside and outside the micropipette based on computational fluid dynamics (CFD). An adaptive sliding mode controller (ASMC) was then designed for microbead aspiration outside the micropipette and positioning inside the pipette based on the dynamic model. The positioning errors of the microbeads of different sizes converge to zero without overshoot, revealing the strong robustness of the cell aspiration and positioning system.

2.3 Visual detection of cellular geometric parameters

In automated measurement systems, image processing methods have been widely introduced to measure the geometric parameters of the cell deformation, which are used to calculate the mechanical properties of cells based on the related models. The geometric parameters mainly include the aspiration length of the cell inside the micropipette and the radius of the cell portion outside the micropipette during micropipette aspiration. To obtain the cellular geometric parameters, it would be natural to detect the tip of the micropipette first. The entire microscopic image is then divided into two regions of interest (ROIs): the region inside the pipette and the outer region near the pipette tip. The two regions are used to calculate the cellular aspiration length and radius, respectively.

Generally, the micropipette does not move during cell aspiration, so we only need to locate the micropipette tip before each experiment. Before micropipette detection, a low-pass Gaussian filter and median filter are used for denoising and smoothing of the microscopic images. The position of the micropipette tip is then obtained using contour detection or template matching methods. Liu et al. [28] used a gradient subtraction method to remove most of the edge information from the cell in the very first image frame and extract the micropipette edge information. After adaptive thresholding using the OTSU method [79], a morphological close operation was used to connect the outer edge and inner edge of each micropipette wall into a single contour. Finally, the horizontal image coordinate of the rightmost point in the contour was identified as the location of the micropipette tip. Shojaei-Baghini et al. [63] used an optimized rotation–invariant normalized cross-correlation (NCC) method [80] to track the micropipette tip visually. A template of the micropipette tip was selected when the micropipette was mounted on the micromanipulator at the beginning of the experiment. The template matching method was then performed to determine the position and angle information of the micropipette tip. Pu et al. [29] used OTSU thresholding for micropipette and cell segmentation, and then used the Shi-Tomasi corner detection algorithm [81] to locate the contact edge and two contact points between the micropipette and the cell.

According to the position of the micropipette tip, the outer region near the pipette tip was defined as an ROI to calculate the cellular radius. For spherical cells with regular shapes, the cell contour was obtained by the contour detection method. Canny algorithm [82] was widely used because of its simplicity and superior effect. The cellular radius was then fitted by the least square method or Hough circle detection after contour detection. For cells with rough surfaces, morphological operations were used to obtain the convex hull of the cell after cell ROI image thresholding. The obtained convex hulls were usually screened using an area threshold to avoid the influence of surrounding impurities. Finally, the cell radius was estimated by circle fitting methods.

When a cell was aspirated into the micropipette, an ROI was formed to include the inside of the micropipette. The aspiration length of the cell inside the micropipette could be calculated in this ROI. In [63], the ROI was first convolved with a standard gradient filter, for example, Sebel kernel, to eliminate the edges of the micropipette and enhance the contours of the cells. After binarization and morphology close operation, the beginning of the left-most edge in the ROI image was recognized as the protruded cell contour. The aspiration length of the cell was determined by the distance between the tracked protrusion and the micropipette opening. Kalman filter [83] was also introduced to track the real position of the aspiration surface of the cell for noise suppression.

2.4 Mechanical analysis based on cellular deformation

Finally, the mechanical properties of the cells are quantified based on cellular deformation according to the related models. In developing a mechanical model, the cellular structure should be taken into account as well as cell mechanical response [84]. Obviously, different choices of models and experimental parameters would lead to different mechanical properties. Various models have been proposed for solid-like cells and liquid-like cells, respectively.

For solid-like cells, such as endothelial cells [85, 86], and chondrocytes [87], the aspiration pressure only makes a part of the cells enter the micropipette for some distance, so cells show the characteristics of a solid. The aspiration length of the cell varies with the aspiration pressure. Various half-space models, layered models, and shell models are employed to measure Young's modulus (elastic characterization) and viscoelastic properties based on length-pressure data. Sato et al. [86] developed a homogenous half-space model to analyze Young's modulus and viscoelastic properties of the endothelial cells. The model treated the cell as a half-space and did not account for the finite boundaries of the cell. The mechanical properties were calculated according to the inner diameter of the micropipette, the increment of the aspiration pressure, and the aspiration lengths. This model was linear, based on the assumption of small deformation. Khalilian et al. [58] analyzed Young's modulus of the zona pellucida by using the theoretical models of the oocyte as an elastic incompressible half-space (half-space model), an elastic compressible bilayer (layered model), or an elastic compressible shell (shell model). Experimental results showed that incorporation of the layered geometry of the ovum and the compressibility of the zona pellucida in the layered and shell models could more accurately characterize the elasticity. Liu et al. [28] measured the viscoelastic properties of interstitial cells with the aid of a homogenous half-space linear viscoelastic model. The key elastic constants in the model were experimentally obtained.

For liquid-like cells, such as neutrophils [59,62] and erythrocytes [88], the cells are aspirated into the micropipette by the negative pressure completely, showing liquid-like properties. Both theory and experiment showed that the leading edge of the liquid-like cells moved at a nearly constant velocity during most of the aspiration process and with a rapid acceleration at the end. The viscosity of the cytoplasm can be characterized according to the liquid drop or shell-liquid models. Tsai et al. [60, 84] simplified the neutrophil as a system of three components: the cortical membrane, a lobular segmented nucleus, and the cytoplasm, and developed a numerical simulation to calculate cell viscosity based on the Newtonian liquid drop model. The cytoplasmic viscosity was determined as a function of the ratio of the initial cell size to the pipette radius, the cortical tension, aspiration pressure, and the cell aspiration time. Cortical shell-liquid core models, which described the flow of a Newtonian liquid from either a hemisphere or a spherical segment into a cylinder, were used in [59,62]. The radius of the cell portion outside the pipette and the aspiration length of the cell inside the micropipette were calculated to quantify the viscosity in these models.

3 Automated cell aspiration for genetic analysis

Cell aspiration is also widely used for single-cell surgery at the subcellular level. The operators penetrate the micropipette into the cell and extract the subcellular components and/or organelles from the cell [89], which are then analyzed by high-throughput sequencing technology [90]. Such techniques are beneficial in studying diseases and their causes [24, 91, 92], mRNA analysis [93, 94], polar body biopsy [95,96], embryo biopsy [97, 98], and cloning [99, 100].

There are four main steps in cell aspiration-based genetic analysis: first, align the micropipette with the subcellular components inside the cell; then penetrate the cell with the micropipette tip; next, extract the target cellular material from the cell; finally, analyze the cellular material. Compared with cell-level aspiration, we should accurately detect the subcellular structures to be analyzed in subcellular level operations, facilitating the positioning of the micropipette inside the cells. In addition, the volume of the material extracted from the cells also needs to be accurately controlled to reduce the damage to the cells. Therefore, the key techniques for genetic analysis based on automated cell aspiration include visual detection of subcellular structures and aspiration volume control from cells.

3.1 Visual detection of subcellular structures

In cell aspiration-based genetic analysis, it is often necessary to locate the target subcellular component first. The positioning result is used as visual feedback of cell orientation, as well as to judge whether the aspiration is complete. Various visual detection algorithms have been designed and implemented for different subcellular structures.

For subcellular components visible under the bright field of a microscope, such as the blastomeres in embryo biopsy, some classical image processing algorithms are used to identify the subcellular components based on the shape or texture features. Paranawithana et al. [101] utilized Canny edge detection and circle Hough transformation to detect the blastomere closest to the tip of the micropipette, and used the Kanade Lucas Tomasi (KLT) algorithm [102] to track the motion of the targeted features when the blastomere was being aspirated into the pipette. Wong et al. [103] proposed a blastomere-tracking algorithm, which segmented and tracked the position of the extracted blastomeres in the image using image subtraction during blastomere retrieval. Moreover, Jang et al. [104] proposed a blastomere instance segmentation algorithm based on Amodal instance segmentation, which aimed to recover the complete silhouette of an object even when the object is not fully visible. The testing results implied that this method outperformed previous state-of-the-art methods, showing the great potential of deep learning in subcellular structure detection.

For subcellular components that are partially visible due to being blocked, such as the polar body, visual detection becomes more complex. The polar body, which contains a copy of the genetic information of an oocyte or embryo, needs to be located before enucleation and polar body biopsy. Polar body detection approaches can be divided into two main categories: traditional image processing approaches and learning-based detection approaches. The polar body is identified by image processing based on thresholding segmentation, image texture, or contour curvature [105–108]. These traditional image processing approaches are convenient to implement and integrate into the micromanipulation system for online applications, but the detection results are susceptible to the shape, size, and focus of the polar body. As for learning-based approaches, as shown in Fig. 10.3A, Chen et al. [109] proposed an algorithm that combined a histogram of oriented gradient

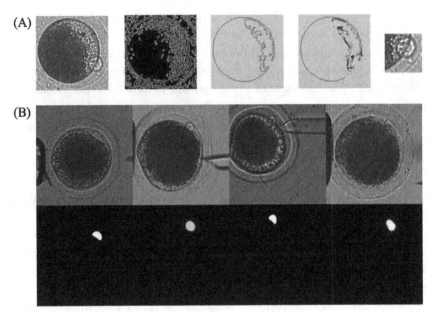

Fig. 10.3 Visual detection methods of subcellular structures. (A) A polar body detection algorithm that combined HOG feature and curvature of cytoplasmic contour. (B) A polar body segmentation algorithm using an improved U-net. *((A) Reproduced from D. Chen, M. Sun, X. Zhao, Oocytes polar body detection for automatic enucleation, Micromachines 7 (2) (2016) 27, copyright 2016 MDPI (Basel, Switzerland); (B) Reproduced from Y. Wang, Y. Liu, M. Sun, X. Zhao, Deep-learning-based polar-body detection for automatic cell manipulation, Micromachines 10 (2) (2019) 120, copyright 2019 MDPI (Basel, Switzerland).)*

(HOG) feature and curvature of cytoplasmic contour for polar body detection in automatic enucleation. The detection problem was interpreted as image classification for the first time. The SVM classifier was utilized to determine the position of the polar body. The method achieved a success rate of 96% for various types of polar bodies, but it did not consider the deformation and defocusing of the polar body. Dai et al. [110] applied a classical deep neural network to robust polar body detection. The network demonstrated an accuracy of 97.6% in 0.2 s on mouse oocytes. As shown in Fig. 10.3B, Wang et al. [111] proposed an improved U-net for polar body segmentation, in which inception-V3 architecture was integrated to enhance the network's attention to small targets. The method performed well in porcine oocyte rotation with an accuracy of 98.7% in 0.10 s on the GPU server. Gong et al. [112] further improved this algorithm by combining image segmentation and visual tracking, so that the polar body could be detected accurately and robustly during cell rotation. Deep learning-based approaches demonstrate effective and robust detection results.

For invisible subcellular components under the bright field microscope, such as mitochondria and nuclei, visual detection is performed after fluorescent staining. The fluorescence images are first converted to gray images or HSV space and denoised by a Gaussian filter or median filter. The target subcellular structures are then detected using thresholding and contour detection methods [30].

3.2 Aspiration volume control from cells

Generally, using a micropipette with a larger inner diameter or increasing the aspiration volume of the target materials from the cells may increase the success rate of micropipette aspiration, but such operations would potentially cause irreversible damage to the cells [103]. Therefore, it is necessary to accurately control the aspiration volume of the materials from cells, considering the cell activity.

When the subcellular components are visible under a bright field microscope or fluorescence microscope, the position of the target component could be used as visual feedback to determine whether the aspiration is completed. However, aspiration control becomes more difficult due to the lack of feedback when the position of the subcellular components cannot be clearly determined. For example, the invisibility of the oocyte nucleus in the bright field currently forces operators to blindly aspirate the nucleus out in oocyte enucleation, usually causing large cytoplasm losses and poor developmental competencies of cloned embryos. Zhao et al. [113] determined the relative positions between the injection pipette and nucleus through a finite element modeling of nucleus aspiration and derived the trajectory of the injection pipette inside the oocyte. The volume of removed cytoplasm was then measured online according to the edge of the aspirated cytoplasm in the injection pipette. The experimental results showed that the system was capable of reducing cytoplasm loss by 60% at the same level of enucleation success rate.

4 Conclusion and outlook

Cell aspiration has been utilized in biomedical engineering for a long time. This kind of method is the most widely accepted and used technique for genetic and mechanical analysis due to its low requirements for equipment. In this chapter, we summarize the existing automated cell aspiration systems for genetic and mechanical analysis that have been developed so far. The automation systems are categorized into two groups: automated cell aspiration for mechanical analysis and automated cell aspiration for genetic analysis. For mechanical analysis, four key techniques are reviewed in detail, including measured cell capture, pressure control in the micropipette, visual detection of cellular geometric parameters, and mechanical analysis based on cellular deformation. For genetic analysis, two key techniques are described, including visual detection of subcellular structures and aspiration volume control from cells.

Although many techniques have been proposed, challenges still exist in genetic and mechanical analysis based on automated cell aspiration. For example, application of micropipette aspiration for adherent cells. Most of the techniques of cell aspiration are aimed at suspended cells. The mechanical properties of the adherent cells are also important, such as the adhesion force of the adherent cells. More effort should be spent on automated measurement methods for adherent cells. Another challenge is micropipette aspiration of single cells for multibiophysical property characterization. At present, some systems have realized the characterization of both mechanical and electrical properties of the same single cell [29, 73]. The micropipette aspiration technique can be combined with microfluidic devices and electronic devices to measure a variety of biophysical parameters and improve efficiency. Until now, most techniques of cell aspiration are still in the theory research stage. In the future, continuous efforts should be focused on making full use of the advantages of automated measurement, so that the big data obtained by cell aspiration can be used in the research of biological and biomedical mechanisms.

References

[1] Z. Wang, R. Samanipour, K.I. Koo, K. Kim, Organ-on-a-chip platforms for drug delivery and cell characterization: a review, Sens. Mater. 27 (6) (2015) 487–506.

[2] J. Carmen, S.R. Burger, M. McCaman, J.A. Rowley, Developing assays to address identity, potency, purity and safety: cell characterization in cell therapy process development, Regen. Med. 7 (1) (2012) 85–100.

[3] Y. Shen, T. Fukuda, State of the art: micro-nanorobotic manipulation in single cell analysis, Robot. Biomimetics 1 (1) (2014) 1–13.

[4] D.-H. Kim, P.K. Wong, J. Park, A. Levchenko, Y. Sun, Microengineered platforms for cell mechanobiology, Annu. Rev. Biomed. Eng. 11 (2009) 203–233.

[5] J. Rajagopalan, M.T.A. Saif, MEMS sensors and microsystems for cell mechanobiology, J. Micromech. Microeng. 21 (5) (2011) 054002.

[6] N. Cermak, S. Olcum, F.F. Delgado, S.C. Wasserman, K.R. Payer, M.A. Murakami, S.-M. Knudsen, R.J. Kimmerling, M.M. Stevens, Y. Kikuchi, High-throughput

measurement of single-cell growth rates using serial microfluidic mass sensor arrays, Nat. Biotechnol. 34 (10) (2016) 1052–1059.

[7] W.H. Grover, A.K. Bryan, M. Diez-Silva, S. Suresh, J.M. Higgins, S.R. Manalis, Measuring single-cell density, Proc. Natl. Acad. Sci. USA 108 (27) (2011) 10992–10996.

[8] J.J. Norman, V. Mukundan, D. Bernstein, B.L. Pruitt, Microsystems for biomechanical measurements, Pediatr. Res. 63 (5) (2008) 576–583.

[9] X.R. Zheng, X. Zhang, Microsystems for cellular force measurement: a review, J. Micromech. Microeng. 21 (5) (2011) 054003.

[10] R. Todd, D.H. Margolin, Challenges of single-cell diagnostics: analysis of gene expression, Trends Mol. Med. 8 (6) (2002) 254–257.

[11] E. Armingol, A. Officer, O. Harismendy, N.E. Lewis, Deciphering cell-cell interactions and communication from gene expression, Nat. Rev. Genet. 22 (2) (2021) 71–88.

[12] I.C. Macaulay, C.P. Ponting, T. Voet, Single-cell multiomics: multiple measurements from single cells, Trends Genet. 33 (2) (2017) 155–168.

[13] T. Luo, L. Fan, R. Zhu, D. Sun, Microfluidic single-cell manipulation and analysis: methods and applications, Micromachines 10 (2) (2019) 104.

[14] J. Dobson, Remote control of cellular behaviour with magnetic nanoparticles, Nat. Nanotechnol. 3 (3) (2008) 139–143.

[15] A. Banerjee, S. Chowdhury, S.K. Gupta, Optical tweezers: autonomous robots for the manipulation of biological cells, IEEE Robot. Autom. Mag. 21 (3) (2014) 81–88.

[16] T.Z. Jubery, S.K. Srivastava, P. Dutta, Dielectrophoretic separation of bioparticles in microdevices: a review, Electrophoresis 35 (5) (2014) 691–713.

[17] D. Ahmed, A. Ozcelik, N. Bojanala, N. Nama, A. Upadhyay, Y. Chen, W. Hanna-Rose, T.J. Huang, Rotational manipulation of single cells and organisms using acoustic waves, Nat. Commun. 7 (1) (2016) 11085.

[18] H. Wang, F. Zhou, Y. Guo, L.A. Ju, Micropipette-based biomechanical nanotools on living cells, Eur. Biophys. J. 51 (2) (2022) 119–133.

[19] M. Sun, Y. Yao, X. Zhao, L. Li, H. Gong, J. Qiu, Y. Liu, X. Zhao, Precise aspiration and positioning control based on dynamic model inside and outside the micropipette, IEEE Trans. Autom. Sci. Eng. 20 (1) (2023) 385–393.

[20] G. Shan, Z. Zhang, C. Dai, X. Wang, L.-T. Chu, Y. Sun, Model-based robotic cell aspiration: tackling nonlinear dynamics and varying cell sizes, IEEE Robot. Autom. Lett. 5 (1) (2019) 173–178.

[21] Q. Zhao, M. Sun, M. Cui, J. Yu, Y. Qin, X. Zhao, Robotic cell rotation based on the minimum rotation force, IEEE Trans. Autom. Sci. Eng. 12 (4) (2014) 1504–1515.

[22] C. Zhao, Y. Liu, M. Sun, X. Zhao, Robotic cell rotation based on optimal poking direction, Micromachines 9 (4) (2018) 141.

[23] Y.H. Anis, M.R. Holl, D.R. Meldrum, Automated selection and placement of single cells using vision-based feedback control, IEEE Trans. Autom. Sci. Eng. 7 (3) (2010) 598–606.

[24] J.J. Tokar, C.N. Stahlfeld, J.M. Sperger, D.J. Niles, D.J. Beebe, J.M. Lang, J.W. Warrick, Pairing microwell arrays with an affordable, semiautomated single-cell aspirator for the interrogation of circulating tumor cell heterogeneity, SLAS Technol. 25 (2) (2020) 162–176.

[25] L.M. Lee, A.P. Liu, The application of micropipette aspiration in molecular mechanics of single cells, J. Nanotechnol. Eng. Med. 5 (4) (2014) 040902.

[26] R.M. Hochmuth, Micropipette aspiration of living cells, J. Biomech. 33 (1) (2000) 15–22.

[27] P. Actis, Sampling from single cells, Small Methods 2 (3) (2018) 1700300.

[28] X. Liu, Y. Wang, Y. Sun, Cell contour tracking and data synchronization for real-time, high-accuracy micropipette aspiration, IEEE Trans. Autom. Sci. Eng. 6 (3) (2009) 536–543.

[29] H. Pu, N. Liu, J. Yu, Y. Yang, Y. Sun, Y. Peng, S. Xie, J. Luo, L. Dong, H. Chen, Micropipette aspiration of single cells for both mechanical and electrical characterization, IEEE Trans. Biomed. Eng. 66 (11) (2019) 3185–3191.

[30] A. Shakoor, M. Xie, T. Luo, J. Hou, Y. Shen, J.K. Mills, D. Sun, Achieving automated organelle biopsy on small single cells using a cell surgery robotic system, IEEE Trans. Biomed. Eng. 66 (8) (2018) 2210–2222.

[31] C.F. Guimarães, L. Gasperini, A.P. Marques, R.L. Reis, The stiffness of living tissues and its implications for tissue engineering, Nat. Rev. Mater. 5 (5) (2020) 351–370.

[32] M. Mak, F. Spill, R.D. Kamm, M.H. Zaman, Single-cell migration in complex microenvironments: mechanics and signaling dynamics, J. Biomech. Eng. 138 (2) (2016) 021004.

[33] S. Alt, P. Ganguly, G. Salbreux, Vertex models: from cell mechanics to tissue morphogenesis, Philos. Trans. R. Soc. B Biol. Sci. 372 (1720) (2017) 20150520.

[34] L.Z. Yanez, J. Han, B.B. Behr, R.A.R. Pera, D.B. Camarillo, Human oocyte developmental potential is predicted by mechanical properties within hours after fertilization, Nat. Commun. 7 (1) (2016) 10809.

[35] Q. Zhao, M. Wu, M. Cui, Y. Qin, J. Yu, M. Sun, X. Zhao, X. Feng, A novel pneumatic micropipette aspiration method using a balance pressure model, Rev. Sci. Instrum. 84 (12) (2013) 123703.

[36] S. Suresh, J. Spatz, J.P. Mills, A. Micoulet, M. Dao, C.T. Lim, M. Beil, T. Seufferlein, Connections between single-cell biomechanics and human disease states: gastrointestinal cancer and malaria, Acta Biomater. 1 (1) (2005) 15–30.

[37] I.V. Pivkin, Z. Peng, G.E. Karniadakis, P.A. Buffet, M. Dao, S. Suresh, Biomechanics of red blood cells in human spleen and consequences for physiology and disease, Proc. Natl. Acad. Sci. USA 113 (28) (2016) 7804–7809.

[38] A.M. Dondorp, K.T. Chotivanich, S. Fucharoen, K. Silamut, J. Vreeken, P.A. Kager, N.J. White, Red cell deformability, splenic function and anaemia in thalassaemia, Br. J. Haematol. 105 (2) (1999) 505–508.

[39] S.B. Zaets, T.L. Berezina, D.-Z. Xu, Q. Lu, J. Ricci, D. Cohen, P. Ananthakrishnan, E.A. Deitch, G.W. Machiedo, Burn-induced red blood cell deformability and shape changes are modulated by sex hormones, Am. J. Surg. 186 (5) (2003) 540–546.

[40] J.P. Mills, M. Diez-Silva, D.J. Quinn, M. Dao, M.J. Lang, K.S.W. Tan, C.T. Lim, G. Milon, P.H. David, O. Mercereau-Puijalon, Effect of plasmodial RESA protein on deformability of human red blood cells harboring *Plasmodium falciparum*, Proc. Natl. Acad. Sci. USA 104 (22) (2007) 9213–9217.

[41] S.E. Cross, Y.-S. Jin, J. Rao, J.K. Gimzewski, Nanomechanical analysis of cells from cancer patients, Nat. Nanotechnol. 2 (12) (2007) 780–783.

[42] L.M. Rebelo, J.S. de Sousa, J. Mendes Filho, M. Radmacher, Comparison of the viscoelastic properties of cells from different kidney cancer phenotypes measured with atomic force microscopy, Nanotechnology 24 (5) (2013) 055102.

[43] K. Onwudiwe, J. Hu, J. Obayemi, V. Uzonwanne, C. Ani, C. Nwazojie, C. Onyekanne, T. Ezenwafor, O. Odusanya, W. Soboyejo, Actin cytoskeletal structure and the statistical variations of the mechanical properties of non-tumorigenic breast and triple-negative breast cancer cells, J. Mech. Behav. Biomed. Mater. 119 (2021) 104505.

[44] E.M. Darling, S. Zauscher, J.A. Block, F. Guilak, A thin-layer model for viscoelastic, stress-relaxation testing of cells using atomic force microscopy: do cell properties reflect metastatic potential? Biophys. J. 92 (5) (2007) 1784–1791.

[45] P.-H. Wu, D.R.-B. Aroush, A. Asnacios, W.-C. Chen, M.E. Dokukin, B.L. Doss, P. Durand-Smet, A. Ekpenyong, J. Guck, N.V. Guz, A comparison of methods to assess cell mechanical properties, Nat. Methods 15 (7) (2018) 491–498.

[46] E. Spedden, C. Staii, Neuron biomechanics probed by atomic force microscopy, Int. J. Mol. Sci. 14 (8) (2013) 16124–16140.

[47] S. Vichare, S. Sen, M.M. Inamdar, Cellular mechanoadaptation to substrate mechanical properties: contributions of substrate stiffness and thickness to cell stiffness measurements using AFM, Soft Matter. 10 (8) (2014) 1174–1181.

[48] Y. Abidine, A. Giannetti, J. Revilloud, V.M. Laurent, C. Verdier, Viscoelastic properties in cancer: from cells to spheroids, Cells 10 (7) (2021) 1704.

[49] S. Kasas, P. Stupar, G. Dietler, AFM contribution to unveil pro-and eukaryotic cell mechanical properties, in: Seminars in Cell & Developmental Biology, vol. 73, Elsevier, 2018, pp. 177–187.

[50] J.P. Mills, L. Qie, M. Dao, C.T. Lim, S. Suresh, Nonlinear elastic and viscoelastic deformation of the human red blood cell with optical tweezers, Mol. Cell. Biochem. 1 (3) (2004) 169.

[51] Y. Li, C. Wen, H. Xie, A. Ye, Y. Yin, Mechanical property analysis of stored red blood cell using optical tweezers, Colloids Surf. B Biointerfaces 70 (2) (2009) 169–173.

[52] V.M. Laurent, S. Hénon, E. Planus, R. Fodil, M. Balland, D. Isabey, F.O. Gallet, Assessment of mechanical properties of adherent living cells by bead micromanipulation: comparison of magnetic twisting cytometry vs optical tweezers, J. Biomech. Eng. 124 (4) (2002) 408–421.

[53] Y. Zhang, F. Wei, Y.-C. Poh, Q. Jia, J. Chen, J. Chen, J. Luo, W. Yao, W. Zhou, W. Huang, Interfacing 3D magnetic twisting cytometry with confocal fluorescence microscopy to image force responses in living cells, Nat. Protoc. 12 (7) (2017) 1437–1450.

[54] B. González-Bermúdez, G.V. Guinea, G.R. Plaza, Advances in micropipette aspiration: applications in cell biomechanics, models, and extended studies, Biophys. J. 116 (4) (2019) 587–594.

[55] V.K. Chivukula, B.L. Krog, J.T. Nauseef, M.D. Henry, S.C. Vigmostad, Alterations in cancer cell mechanical properties after fluid shear stress exposure: a micropipette aspiration study, Cell Health Cytoskelet. 7 (2015) 25.

[56] K. Wang, X.H. Sun, Y. Zhang, T. Zhang, Y. Zheng, Y.C. Wei, P. Zhao, D.Y. Chen, H.A. Wu, W.H. Wang, Characterization of cytoplasmic viscosity of hundreds of single tumour cells based on micropipette aspiration, R. Soc. Open Sci. 6 (3) (2019) 181707.

[57] M.-J. Oh, F. Kuhr, F. Byfield, I. Levitan, Micropipette aspiration of substrate-attached cells to estimate cell stiffness, J. Vis. Exp. (67) (2012) e3886.

[58] M. Khalilian, M. Navidbakhsh, M.R. Valojerdi, M. Chizari, P.E. Yazdi, Estimating Young's modulus of zona pellucida by micropipette aspiration in combination with theoretical models of ovum, J. R. Soc. Interface 7 (45) (2010) 687–694.

[59] E. Evans, A. Yeung, Apparent viscosity and cortical tension of blood granulocytes determined by micropipet aspiration, Biophys. J. 56 (1) (1989) 151–160.

[60] M.A. Tsai, R.E. Waugh, P.C. Keng, Cell cycle-dependence of HL-60 cell deformability, Biophys. J. 70 (4) (1996) 2023–2029.

[61] F. Guilak, J.R. Tedrow, R. Burgkart, Viscoelastic properties of the cell nucleus, Biochem. Biophys. Res. Commun. 269 (3) (2000) 781–786.

[62] D. Needham, R.M. Hochmuth, Rapid flow of passive neutrophils into a 4 μm pipet and measurement of cytoplasmic viscosity, J. Biomech. Eng. 112 (3) (1990) 269–276.

[63] E. Shojaei-Baghini, Y. Zheng, Y. Sun, Automated micropipette aspiration of single cells, Ann. Biomed. Eng. 41 (6) (2013) 1208–1216.

[64] U. Waldenström, A.-B. Engström, D. Hellberg, S. Nilsson, Low-oxygen compared with high-oxygen atmosphere in blastocyst culture, a prospective randomized study, Fertil. Steril. 91 (6) (2009) 2461–2465.

[65] A. Lourens, H. Van den Brand, R. Meijerhof, B. Kemp, Effect of eggshell temperature during incubation on embryo development, hatchability, and posthatch development, Poult. Sci. 84 (6) (2005) 914–920.

[66] E. Popova, M. Bader, A. Krivokharchenko, Effect of culture conditions on viability of mouse and rat embryos developed in vitro, Genes 2 (2) (2011) 332–344.

[67] Q. Guo, S.J. Reiling, P. Rohrbach, H. Ma, Microfluidic biomechanical assay for red blood cells parasitized by *Plasmodium falciparum*, Lab Chip 12 (6) (2012) 1143–1150.

[68] Q. Guo, S. Park, H. Ma, Microfluidic micropipette aspiration for measuring the deformability of single cells, Lab Chip 12 (15) (2012) 2687–2695.

[69] Y.-J. Li, Y.-N. Yang, H.-J. Zhang, C.-D. Xue, D.-P. Zeng, T. Cao, K.-R. Qin, A microfluidic micropipette aspiration device to study single-cell mechanics inspired by the principle of wheatstone bridge, Micromachines 10 (2) (2019) 131.

[70] H. Dong, W. Wang, Z. Wang, L. Zhou, L. Liu, Automated micropipette aspiration of cell using resistance-based voltage feedback control, in: 7th IEEE International Conference on Nano/Molecular Medicine and Engineering, IEEE, 2013, pp. 142–146.

[71] W. Wang, X. Liu, D. Gelinas, B. Ciruna, Y. Sun, A fully automated robotic system for microinjection of zebrafish embryos, PLoS One 2 (9) (2007) e862.

[72] A. Shakoor, T. Luo, S. Chen, M. Xie, J.K. Mills, D. Sun, A high-precision robot-aided single-cell biopsy system, in: 2017 IEEE International Conference on Robotics and Automation (ICRA), IEEE, 2017, pp. 5397–5402.

[73] J. Chen, Y. Zheng, Q. Tan, Y.L. Zhang, J. Li, W.R. Geddie, M.A.S. Jewett, Y. Sun, A microfluidic device for simultaneous electrical and mechanical measurements on single cells, Biomicrofluidics 5 (1) (2011) 014113.

[74] L.M. Lee, J.W. Lee, D. Chase, D. Gebrezgiabhier, A.P. Liu, Development of an advanced microfluidic micropipette aspiration device for single cell mechanics studies, Biomicrofluidics 10 (5) (2016) 054105.

[75] Y. Liu, M. Cui, J. Huang, M. Sun, X. Zhao, Q. Zhao, Robotic micropipette aspiration for multiple cells, Micromachines 10 (5) (2019) 348.

[76] L.S. Mattos, E. Grant, R. Thresher, K. Kluckman, Blastocyst microinjection automation, IEEE Trans. Inf. Technol. Biomed. 13 (5) (2009) 822.

[77] Z. Lu, X. Zhang, C. Leung, N. Esfandiari, R.F. Casper, Y. Sun, Robotic ICSI (intracytoplasmic sperm injection), IEEE Trans. Biomed. Eng. 58 (7) (2011) 2102–2108.

[78] X.P. Zhang, C. Leung, Z. Lu, N. Esfandiari, R.F. Casper, Y. Sun, Controlled aspiration and positioning of biological cells in a micropipette, IEEE Trans. Biomed. Eng. 59 (4) (2012) 1032–1040.

[79] N. Otsu, A threshold selection method from gray-level histograms, IEEE Trans. Syst. Man Cybern. 9 (1) (1979) 62–66.

[80] J.-C. Yoo, T.H. Han, Fast normalized cross-correlation, Circuits Syst. Signal Process. 28 (6) (2009) 819–843.

[81] J. Shi, Good features to track, in: 1994 Proceedings of IEEE Conference on Computer Vision and Pattern Recognition, IEEE, 1994, pp. 593–600.

[82] J. Canny, A computational approach to edge detection, IEEE Trans. Pattern Anal. Mach. Intell. (6) (1986) 679–698.

[83] P. Zarchan, Progress in Astronautics and Aeronautics: Fundamentals of Kalman Filtering: A Practical Approach, vol. 208, AIAA, 2005.

[84] M.A. Tsai, R.S. Frank, R.E. Waugh, Passive mechanical behavior of human neutrophils: power-law fluid, Biophys. J. 65 (5) (1993) 2078–2088.

[85] D.P. Theret, M.J. Levesque, M. Sato, R.M. Nerem, L.T. Wheeler, The application of a homogeneous half-space model in the analysis of endothelial cell micropipette measurements, J. Biomech. Eng. 110 (3) (1988) 190–199.

[86] M. Sato, D.P. Theret, L.T. Wheeler, N. Ohshima, R.M. Nerem, Application of the micropipette technique to the measurement of cultured porcine aortic endothelial cell viscoelastic properties, J. Biomech. Eng. 112 (3) (1990) 263–268.

[87] W.R. Jones, H.P. Ting-Beall, G.M. Lee, S.S. Kelley, R.M. Hochmuth, F. Guilak, Alterations in the Young's modulus and volumetric properties of chondrocytes isolated from normal and osteoarthritic human cartilage, J. Biomech. 32 (2) (1999) 119–127.

[88] D.E. Discher, D.H. Boal, S.K. Boey, Simulations of the erythrocyte cytoskeleton at large deformation. II. Micropipette aspiration, Biophys. J. 75 (3) (1998) 1584–1597.

[89] J. Liu, J. Wen, Z. Zhang, H. Liu, Y. Sun, Voyage inside the cell: microsystems and nanoengineering for intracellular measurement and manipulation, Microsyst. Nanoeng. 1 (1) (2015) 1–15.

[90] J.A. Reuter, D.V. Spacek, M.P. Snyder, High-throughput sequencing technologies, Mol. Cell 58 (4) (2015) 586–597.

[91] B.K. Chen, D. Anchel, Z. Gong, R. Cotton, R. Li, Y. Sun, D.P. Bazett-Jones, Nanodissection and sequencing of DNA at single sub-nuclear structures, Small 10 (16) (2014) 3267–3274.

[92] P. Actis, M.M. Maalouf, H.J. Kim, A. Lohith, B. Vilozny, R.A. Seger, N. Pourmand, Compartmental genomics in living cells revealed by single-cell nanobiopsy, ACS Nano 8 (1) (2014) 546–553.

[93] T. Osada, H. Uehara, H. Kim, A. Ikai, mRNA analysis of single living cells, J. Nanobiotechnol. 1 (1) (2003) 1–8.

[94] K. Lee, Y. Cui, L.P. Lee, J. Irudayaraj, Quantitative imaging of single mRNA splice variants in living cells, Nat. Nanotechnol. 9 (6) (2014) 474–480.

[95] A. Kuliev, S. Rechitsky, Polar body-based preimplantation genetic diagnosis for Mendelian disorders, Mol. Hum. Reprod. Basic Sci. Reprod. Med. 17 (5) (2011) 275–285.

[96] M. Montag, Polar body biopsy, in: In Vitro Fertilization, Springer, Cham, 2019, pp. 583–590.

[97] J.J. Tarín, A.H. Handyside, Embryo biopsy strategies for preimplantation diagnosis, Fertil. Steril. 59 (5) (1993) 943–952.

[98] D. Cimadomo, A. Capalbo, F.M. Ubaldi, C. Scarica, A. Palagiano, R. Canipari, L. Rienzi, The impact of biopsy on human embryo developmental potential during preimplantation genetic diagnosis, BioMed Res. Int. 2016 (2016) 7193075.

[99] T. Wakayama, A.C.F. Perry, M. Zuccotti, K.R. Johnson, R. Yanagimachi, Full-term development of mice from enucleated oocytes injected with cumulus cell nuclei, Nature 394 (6691) (1998) 369–374.

[100] D. Iuso, M. Czernik, F. Zacchini, G. Ptak, P. Loi, A simplified approach for oocyte enucleation in mammalian cloning, Cell. Reprogram. (Formerly "Cloning and Stem Cells") 15 (6) (2013) 490–494.

[101] I. Paranawithana, W.-X. Yang, U.-X. Tan, Tracking extraction of blastomere for embryo biopsy, in: 2015 IEEE International Conference on Robotics and Biomimetics (ROBIO), IEEE, 2015, pp. 380–384.

[102] C. Tomasi, T. Kanade, Shape and motion from image streams under orthography: a factorization method, Int. J. Comput. Vis. 9 (2) (1992) 137–154.

[103] C.Y. Wong, J.K. Mills, Cell extraction automation in single cell surgery using the displacement method, Biomed. Microdevices 21 (3) (2019) 1–11.

[104] W.-D. Jang, D. Wei, X. Zhang, B. Leahy, H. Yang, J. Tompkin, D. Ben-Yosef, D. Needleman, H. Pfister, Learning vector quantized shape code for Amodal blastomere instance segmentation, arXiv preprint arXiv:2012.00985 (2020).

[105] C. Leung, Z. Lu, X.P. Zhang, Y. Sun, Three-dimensional rotation of mouse embryos, IEEE Trans. Biomed. Eng. 59 (4) (2012) 1049–1056.

[106] L. Feng, B. Turan, U. Ningga, F. Arai, Three dimensional rotation of bovine oocyte by using magnetically driven on-chip robot, in: 2014 IEEE/RSJ International Conference on Intelligent Robots and Systems, IEEE, 2014, pp. 4668–4673.

[107] Y.L. Wang, X. Zhao, Q.L. Zhao, G.Z. Lu, Illumination intensity evaluation of microscopic image based on texture information and application on locating polar body in oocytes, in: Proceedings of the China Automation Conference, Beijing, China, 2011, pp. 7–10.

[108] Z. Wang, Y. Hu, J. Wei, W.T. Latt, Visual servoed robotic mouse oocyte rotation, IEEE Trans. Biomed. Eng. 67 (8) (2019) 2389–2396.

[109] D. Chen, M. Sun, X. Zhao, Oocytes polar body detection for automatic enucleation, Micromachines 7 (2) (2016) 27.

[110] C. Dai, Z. Zhang, Y. Lu, G. Shan, X. Wang, Q. Zhao, C. Ru, Y. Sun, Robotic manipulation of deformable cells for orientation control, IEEE Trans. Robot. 36 (1) (2019) 271–283.

[111] Y. Wang, Y. Liu, M. Sun, X. Zhao, Deep-learning-based polar-body detection for automatic cell manipulation, Micromachines 10 (2) (2019) 120.

[112] H. Gong, L. Li, J. Qiu, Y. Yao, Y. Liu, M. Cui, Q. Zhao, X. Zhao, M. Sun, Automatic cell rotation based on real-time detection and tracking, IEEE Robot. Autom. Lett. 6 (4) (2021) 7909–7916.

[113] Q. Zhao, J. Qiu, Z. Feng, Y. Du, Y. Liu, Z. Zhao, M. Sun, M. Cui, X. Zhao, Robotic label-free precise oocyte enucleation for improving developmental competence of cloned embryos, IEEE Trans. Biomed. Eng. 68 (8) (2020) 2348–2359.

CHAPTER 11

Cell characterization by nanonewton force sensing

Xiaowei Jin[a], Jordan Rosenbohm[a], Grayson Minnick[a],
Amir M. Esfahani[b], Bahareh Tajvidi Safa[a], and Ruiguo Yang[a,c]
[a]Department of Mechanical and Materials Engineering, University of Nebraska-Lincoln, Lincoln, NE, United States
[b]Department of Bioengineering, School of Medicine, Johns Hopkins University, Baltimore, MD, United States
[c]Nebraska Center for Integrated Biomolecular Communication, University of Nebraska-Lincoln, Lincoln, NE, United States

1 Introduction

Characterization of the mechanical behavior of living cells and how the behavior is modified by cell microenvironment have long been of interest to researchers in biological sciences and bioengineering. On the one hand, cell mechanical properties quantified using different techniques are considered as markers for underlying disease conditions, such as malaria [1] and cancer [2]. A wide spectrum of techniques has been developed or fitted to precisely measure the mechanical characteristics of living cells, including elastic modulus, loss modulus, and stress relaxation time [3]. On the other hand, mechanobiology represents a new group of studies that connects mechanical stimulation and cell biology to decipher the influence of external cell environments on cell behavior [4]. In this regard, studies often look at how cells sense and transduce mechanical forces, both extracellular and intracellular, and ultimately regulate their behavior. Both perspectives are gaining new meaning in the current age of tissue fabrication by three-dimensional (3D) printing, and in association with multiomics in single cell resolution.

In this chapter, we will examine the working principles and applications of the techniques for mechanical characterization of living cells by nano-newton force sensing. This will be discussed in two categories: techniques that measure cell-generated forces externally and methods that sense forces within the cell. The first group of technologies includes conventional ones such as micropipette aspiration, optical and magnetic tweezers, atomic force microscopy, and traction microscopy. New techniques made possible by micro-electromechanical systems (MEMS) microfabrication and two

Robotics for Cell Manipulation and Characterization
https://doi.org/10.1016/B978-0-323-95213-2.00014-4
245

photon polymerization are also included. These will be discussed in Section 2. For the second category, discussion will be focused on Förster resonance energy transfer (FRET) sensors and DNA tensional sensors that are engineered to living cells. These techniques utilize well-calibrated optical detection to quantify the subnanonewton forces in molecules within the cell to provide an assessment of the intracellular tensional status. These will be discussed in Section 3.

2 Techniques for sensing cell-generated forces

2.1 Traction force microscopy

The actomyosin complex in cells generates traction forces that are transmitted to the extracellular matrix (ECM) via focal adhesions (FAs) allowing cells to sense and respond to their microenvironment. According to mechanobiology studies, cell-ECM interactions regulate cellular behaviors such as proliferation [5], migration [6–8], morphogenesis [9,10], cancer progression [11,12], cell–cell interactions [13], and determining stem cell fate [5,14]. Traction force microscopy (TFM) techniques have been developed to quantify traction forces and reveal the regulatory system that controls cellular behavior.

Two approaches are used to compute in-plane traction forces: hydrogels with embedded fluorescent beads and post array detectors (mPDAs). The first approach employs isotropic linear elastic hydrogels, such as polyacrylamide (PAA) [15], polyethylene glycol (PEG) [16], and polydimethylsiloxane (PDMS) [17], with tunable stiffness (Fig. 11.1A). The posts used in the second method are also fabricated from linear elastic materials, such as PDMS [18],

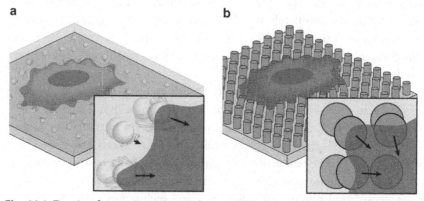

Fig. 11.1 Traction force microscopy techniques. (A) Gel embedded with fluorophore beads to track the displacement and the traction force. (B) Micropost array with a cell adhered to the tips which apply traction force to the posts.

and their stiffness can be adjusted according to the experiment (Fig. 11.1B). Cells cause bead (or post) displacement by exerting traction forces on the hydrogel (or posts). Particle tracking/image velocimetry (PTV/PIV) [19], correction-based PTV [19], and feature-based registration techniques [20] can be used to detect displacements. Force maps are reconstructed from bead displacements using the Boussinesq solution (either with Bayesian [21] or Fourier transform [20,22]). The 2D TFM methods are straightforward and informative; however, they cannot quantify out-of-plane forces, nor do they account for the time- and strain-dependent properties of ECM in vivo.

3D TFM methods are proposed to compute both tangential and normal forces using z-stack images from confocal microscopy. In these techniques, cells are either seeded on a 2D hydrogel while measuring 3D forces (2.5D TFM) [23], or embedded in a 3D matrix [24], mimicking the natural environment of the cells. The difficulty in calculating traction forces due to the nonlinear mechanical properties of the gel and the requirement for intensive image processing are the drawbacks of the 3D approaches [25]. Living tissues display time-dependent behavior (viscoelasticity), nonlinear elasticity, plasticity, and anisotropy based on the forces they are subjected to and their biological condition. Gels with adjustable viscoelastic characteristics are utilized to investigate the influence of time- and strain-dependent features of ECM on mechanotransduction [26]. 3D traction force microscopy has also employed in viscoelastic gels [27].

TFM experiments have been effective in measuring cell-produced forces; nevertheless, to improve the accuracy of the results, additional experimental and computational methodologies are required. The existing methods are not high throughput, and the data processing techniques do not account for the viscoelastic characteristics of cells and the substrate. In recent years, many attempts have been made to improve the accuracy of the results in 3D techniques using super-resolution stimulated emission depletion (STED) microscopy [28] and 3D super-resolution fluorescent structured illumination microscopy (SIM) [29], which have provided fine details in the reconstructed force fields. Improving the resolution and throughput of TFM methods allows them to remain the standard tool for studying cell-generated forces in the mechanobiology field.

2.2 Atomic force microscopy

Atomic force microscopy (AFM) has been widely used in measuring cell mechanical properties [30]. In this method, a laser is focused on the tip of a cantilever and reflected to a sensor (Fig. 11.2A). When the cantilever

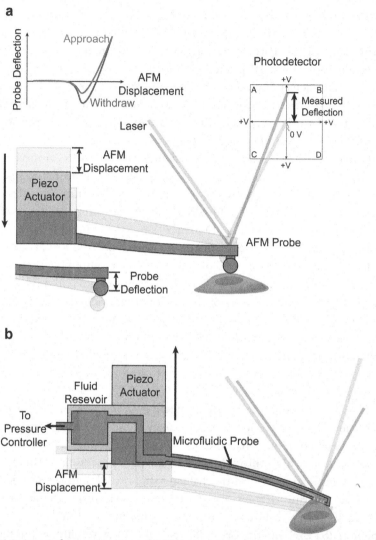

Fig. 11.2 Atomic force microscopy. (A) AFM concept and schematic show a laser beam emitted to a cantilever beam and reflected to a photodetector. The cantilever deflection results in laser beam reflection change to the photodetector. (B) FluidFM working principle is similar to that of AFM but includes a fluidic microchannel to manipulate cells and beads.

contacts a cell, the force between the cantilever tip and the cell membrane results in bending of the cantilever, altering the point of the laser reflection on the sensor. Based on the stiffness of the cantilever, the curve of the cantilever deflection for approach and withdrawal is plotted and analyzed to calculate the cell stiffness and Young's modulus (Fig. 11.2B) [31]. Furthermore, different cantilever geometries with different stiffness are designed for different applications to precisely measure biological samples' properties [32].

While this method was initially adopted in mechanobiology research to capture the mechanical properties of cells, other applications emerged. For example, researchers functionalized the cantilever tip with different proteins, such as ligands, to measure cell-ECM or cell–cell adhesion strength [33–35]. In addition, introducing different modes of measurement can make it a diagnostic platform for cancer detection [36]. In a modified version of AFM, known as fluid force microscopy (fluidFM), researchers developed a fluidic microchannel inside the cantilever to aspirate fluids to adhere to cells or beads (Fig. 11.2B) [37,38]. This method has a few advantages over conventional AFM. For example, in AFM, after using a functionalized beam in multiple experiments, one should stop the experiments and repeat the coating process. Contrastingly, in fluidFM, thousands of beads can be functionalized and used to run several experiments. This platform can also be used to pick and place a cell in the desired location and make monoclonals. Although this method has many advantages, there are difficulties in detaching cells or beads from the cantilever opening, even with high pressure. Furthermore, due to the delicate fabrication method and the material, it is expensive.

2.3 Optical and magnetic tweezers

Optical and magnetic tweezers are a group of methods in which specific microbeads are inserted into the cells or attached to the cell membrane. In optical tweezers, dielectric beads attach to the cells. A laser is focused on the bead, which is attracted to the focal point of the laser under force exerted by changes in momentum of refracted photons. The magnitude of the force depends on how the laser is focused on the bead (Fig. 11.3A) [39,40]. In this method, the refractive index of the cell should be larger than the media. An alternative to this method is the optical stretcher, in which the laser is divergent, and no beads are needed [41,42]. In this method, two laser lights are directed to different sides of the cells. As the photons interact with each other, a force is applied to the cells (Fig. 11.3B). The advantages of this method over optical tweezers include the application of greater force and no physical changes to the cells. In magnetic tweezers, micro magnets are attached to the cell membrane. An external magnetic field will cause the magnet to move and apply a force to the cell (Fig. 11.3C) [43]. With this method, cyclic loads and torque can be applied [44]. However, because of the relationship between the magnet size and the amount of force, this method is limited to very small forces.

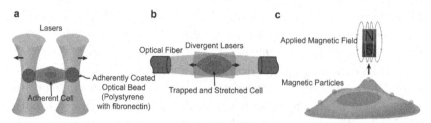

Fig. 11.3 Optical and magnetic techniques. (A) Optical tweezer concept to stretch a single cell with a small amount of force. (B) Optical stretcher working principle to stretch a cell without adding optical beads. (C) Magnetic tweezer is used to apply force and torque to the cells.

2.4 Microfabricated structures from MEMS

Many approaches have been developed and used to quantify cellular forces at various levels [30], but microscale structures, in addition to their higher precision and resolution, provide a better physiological environment. With the recent advances in microfabrication and the use of biocompatible materials, researchers have invented microdevices to study the mechanics of cells at small scales. Through increasing force and displacement resolutions, microstructure-based technologies have improved the precision, efficiency, and consistency of cell manipulation and characterization activities. Microstructure devices have been regarded as ideal instruments for cell manipulation and characterization because of their unique qualities, such as size matching to single cells and the ability to generate/measure microscale movements and forces [45,46]. These devices are intended to study the cell–cell and cell-ECM junctions in different regimes including tension, compression, and traction [47].

The first generation of these microdevices was adapted from MEMS accelerometers. Most systems for measuring these forces use two beams, one of which serves as the actuator and the other as the sensor. They can provide various resolutions and precisions based on actuation and sensing approaches. The comb drive structure of MEMS devices enables the application of tension through electrostatic forces and the study of the effect of stretch on cells' behavior (Fig. 11.4A) [48]. Other researchers have adapted this idea to make spring-like plates. In one device, cells are cultured on two close plates and stretched using a needle and a micromanipulator. By tracking the displacement of these plates, force (stress) and strain can be derived [49]. Fabrication of these devices requires cleanroom processes, such as deposition and etching, which limits creativity due to complex fabrication

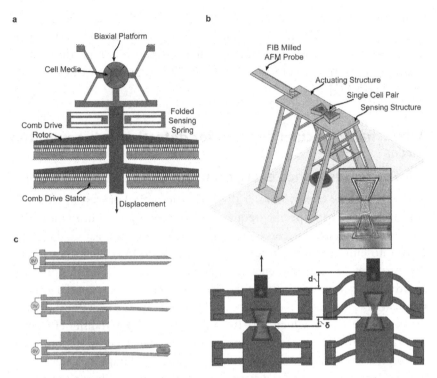

Fig. 11.4 Microstructure techniques. (A) MEMS comb drive biaxial tensile tester used to stretch a single cell. (B) Single cell junction micro tensile tester that allows the investigation of mechanical properties of a junction between a pair of cells subjected to stretch. (C) Microgripper principle to squish and compress a single cell to interrogate the cell mechanical characteristics during compression.

processes. The invention of nano/micro 3D printers helped with the fabrication of the next generation of these microdevices. In a recent study, a two-photon polymerization printer was used to fabricate a biocompatible platform that can measure the forces at a single junction between a single cell pair, while being submerged in liquid media (Fig. 11.4B) [50].

While stretch and tension are important for cell activities, understanding the compression mechanics and deformability of cells is also important for specific types of cells such as red blood cells since they need to move through channels of different sizes [51]. Microgrippers and microtweezers have been widely developed during the last several decades. Researchers have used electromagnetic, electrothermal, optomechanical, and electrostatic methods to actuate the structure and apply force to cells [52–54]. These devices can grasp single cells, apply static and cyclic compression to the cells, and capture their mechanical properties (Fig. 11.4C). These methods can measure rupture force, stiffness, Young's modulus, viscosity, and deformability. Integrating these devices with microfluidic chips makes it a high-throughput screening platform to compare the mechanical properties of healthy and diseased cells [55].

3 Internal molecular sensors for detecting cell-generated forces

3.1 Förster resonance energy transfer (FRET) tension sensor

Förster resonance energy transfer (FRET) is an optical process in which an emissive energy with specific frequency, produced by one fluorophore that is in its excited state, transfers to another fluorophore, stimulating the fluorophore to emit its characteristic energy (Fig. 11.5) [56]. The fluorophore that provides the emissive energy is called the donor, while the one that absorbs the transferred energy is the acceptor. This energy transfer process is distance dependent, and only happens between two adjacent fluorophores, the distance between which is 1–10 nm [57]. The energy transfer efficiency is inversely proportional to the sixth power of distance between donor and acceptor, allowing FRET to be used as a nanoscale technique to detect intramolecular or intermolecular distances. By connecting the donor and acceptor with an elastic linker peptide, such as F40 (TSMod), HP35, sstFRET, or stFRET [57], the spatial distance reading from FRET can be used to evaluate force changes. Thereby, different types of FRET force biosensors are widely used to detect cellular force changes on the order of several piconewtons (pNs).

Fluorescent proteins (FPs)-based FRET force biosensors are one of the most popular biosensors, and achieve cellular force sensing via reconstructing

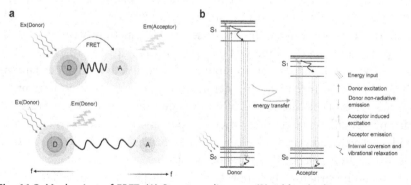

Fig. 11.5 Mechanism of FRET. (A) Structure diagram. (B) Jablonski diagram.

the gene expression of functional proteins and directly fusing the FRET pairs to the force sensing protein domains. Depending on the characteristics of FPs, the FRET pairs can be classified as CFPs/YFPs, GFPs/RFPs, FFPs/IFPs, LSS FPs, dark FPs, ptFPs, multicolor, and Homo FRET pairs [56]. CFPs/YFPs are the most popular FRET pairs, due to their high quantum yield and low pH sensitivity. By inserting these FRET pairs into live cells and detecting the fluorescent intensity changes between donor and acceptor, the real-time force response can be mapped.

There are several methods to measure forces in cells via FRET changes, including spectral imaging FRET (siFRET), acceptor photobleaching FRET (apFRET), fluorescence lifetime imaging FRET (FLIM-FRET), and sensitized emission FRET (seFRET) [56]. Among these methods, siFRET, apFRET, and FLIM-FRET are used to detect the FRET efficiency, while seFRET only relates the fluorescent signal changes; therefore, seFRET is not suitable for directly measuring the quantitative force in cells and is only used for investigating relative force changes. However, the three quantitative methods have their own disadvantages. siFRET is only suitable for intramolecular biosensors due to needing a known ratio between donor and acceptor fluorophores, while apFRET is not compatible with dynamic research on proteins in live cells because of the damage to the acceptor from photobleaching. FLIM-FRET needs a long image acquisition time and is sensitive to the microenvironment of cells. Although seFRET has limited capacity in evaluating force magnitude, it is the easiest and fastest way to obtain relative tension changes in cells. In short, FRET biosensors with stable, real-time, and wash-free fluorescent signals are a very useful technique in exploring the force regulation in cells. It is also a technology that requires many skills to implement and the selection of FPs and measuring methods need to be carefully considered, based on the experimental purposes.

Currently, many types of FRET tension sensors are inserted into cells to investigate the intracellular force response. Most FRET tension sensors are engineered into force-sensing and transduction proteins located on adhesions, such as cell–cell adhesions and cell-ECM adhesions, where force changes happen. Depending on the proteins where FRET pairs are integrated, FRET tension sensors are classified as junctional adhesion FRET sensors and focal adhesion FRET sensors. Junctional adhesion FRET sensors explore mechanotransduction on cell–cell contacts via adhesion proteins, such as cadherin, catenin, desmoplakin, and desmoglein. These tension sensors are mainly used for exploring how cells sense forces from their neighbors, how cells respond to mechano-disruption, and how force equilibrium is maintained in the junction. An EcadTSMod tension sensor, built by inserting the TSMod,whose structure is mTFP-elastic linker-mEYFP between the fourth residues of the juxtamembrane domain (V742) and the fifth residues of the juxtamembrane domain (K743) of canine E-cadherin cDNA, was engineered into MDCK cells and used to show that E-cadherin directly bore the pN tensile forces from actomyosin. The researchers also found that this tension was constitutively applied to E-cadherin at the junction, whether E-cadherin clustered on the cell–cell junction or not [58].

A similar tension sensor was engineered into drosophila to study the tension difference between the front and back of migrating border cells. Compared to the back of the migrating border cell clusters, the E-cadherin FRET ratio results displayed greater tension at the front without increasing the number of E-cadherin molecules, indicating differences in the strength of adhesive bonds and/or actomyosin contractility between the back and front during migration (Fig. 11.6A–C) [59]. Besides the tension sensing in epithelial cells, TSMod sensors have also been engineered into endothelial cells to explore tension sensing under fluid shear stress (FSS). By inserting TSMod between the p120 catenin and the β–catenin binding domains of VE-cadherin and the second ITIM NAD exon 15 domain of PECAM-1 in HUVECs, VE-cadherin and PECAM-1 FRET tension sensors were developed. VE-cadherin FRET sensors displayed a rapid decrease in tension across VE-cadherin under FSS, while the PECAM-1 tension sensor indicated an increase in tension across the junctional PECAM-1. Interestingly, FSS has no effect on nonjunctional PECAM-1 [60]. Besides monolayer tension studies, the VE-cadherin tension sensor was also engineered into zebrafish and used to find that VE-cadherin tension decreased during maturation of dorsal aorta (DA) junctions in zebrafish. Furthermore, VE-cadherin tension in mature endothelial cells has no alteration after a complete loss of blood flow, which implied a minor influence of fluid flow on the tension in mature DA junctions compared to the tension regulated by actomyosin [61] (Fig. 11.6D–F).

In addition to cadherin, FRET sensors are also widely used in catenin, desmoplakin, and desmoglein to monitor the mechanical force responses in cells, especially mechanotransduction between junctional proteins and the cytoskeleton. An αE-catenin tension sensor was made by inserting a TSMod between the N-terminal β-catenin binding domain of αE–catenin and its F-actin binding domain in Caco-2 cells to monitor the force sensing of αE-catenin during actomyosin contractility. The researchers found that mDia1 played a crucial role on the contractility of zonula adhesion, which

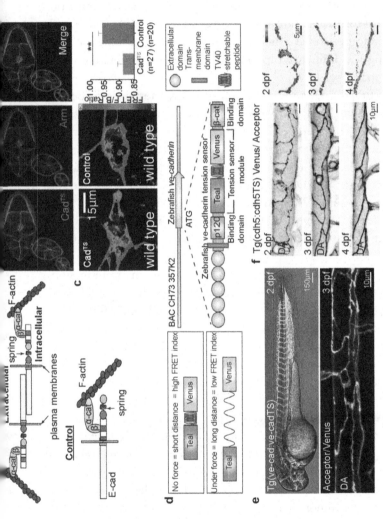

Fig. 11.6 Cadherin tension sensor. (A) Schematic of E-cadherin tension sensor and control. (B) E-cadTS expressed in drosophila with arm colocalization. (C) Lower FRET F/B ratio of E-cadTS displayed greater tension per molecule at the front of the E-cadherin cluster. (D) Schematic of VE-cadherin tension sensor. (E) Venus expressed throughout the blood vasculature in zebrafish at 2 dpf (top) and at cell–cell junctions of the dorsal aorta (DA) at 3 dpf. (F) Increased FRET index reported a significant decrease in the tension across the junction during the mature of the DA junction. (*Source: D. Cai, S.C. Chen, M. Prasad, L. He, X. Wang, V. Choesmel-Cadamuro, J.K. Sawyer, G. Danuser, D.J. Montell, Mechanical feedback through E-cadherin promotes direction sensing during collective cell migration, Cell 157(5) (2014) 1146–59; A.K. Lagendijk, G.A. Gomez, S. Baek, D. Hesselson, W.E. Hughes, S. Paterson, D.E. Conway, H.-G. Belting, M. Affolter, K.A. Smith, M.A. Schwartz, A.S. Yap, B.M. Hogan, Live imaging molecular changes in junctional tension upon VE-cadherin in zebrafish, Nat. Commun. 8(1) (2017); Copyright permission file: Copyright Permission-1, Copyright Permission-2.*)

determined the morphological and physiological performance of epithelia in tissue barriers [62]. Similarly, an mTFP1-F40-meYFP module was inserted after the rod domain and prior to Pro1946 and Thr1354 to fabricate DPI-TS and DPII-TS individually in MDCK and MEK cells, respectively, to directly reveal the loadbearing characteristics of desmosomes. The results revealed that there was no tension across DPI and DPII during collective migration and desmosome formation. However, both DPI and DPII temporarily became mechanically loaded when cells were exposed to external stress and this loading was influenced by the magnitude and orientation of the external stress [63] (Fig. 11.7).

Instead of studying DPs to understand the force response of desmosomes, a desmoglein-2 (DSG-2) FRET tension sensor was constructed by inserting a TSMod module between the intracellular anchor (IA) domain and the intracellular catenin-binding site (ICS) in MDCK cells. DSG-2 experienced more tension loading in 3D MDCK acini than the 2D monolayers [64]. This Dsg-2 tension sensor was also engineered into intestinal epithelial cells (IECs) and used to show that the loss of Dsc2 increased the force loading on Dsg2, indicating the important role of Dsc2 in mediating the mechanical tension in desmosomes in IECs during junction formation [65].

Vinculin and talin are two main force-sensing proteins in focal adhesions. Vinculin, a protein whose recruitment within focal adhesions is highly force dependent, is widely used to study force transmission and focal adhesion formation during cell processes such as migration, differentiation, and proliferation. A vinculin tension sensor (VTS) was developed by connecting monomeric teal fluorescent protein (mTFP1) and Venus (A206K), with an elastic domain between the vinculin head and tail, to directly report the distance between two fluorophores and measure the tension across the vinculin. VTS was then engineered into Vinculin$^{-/-}$ cells subjected to tension during the assembling and enlarging of adhesions. The sensor disclosed that the process of recruiting vinculin to focal adhesions was separated from the force transmission across the vinculin [66]. This VTS was further engineered into breast tumor cells and indicated that the decreased focal adhesion tension in tumor cells reduced their potential migration [67] (Fig. 11.8A–B). Moreover, VTS was also used in MSCs to explore the function of TRPV4-mediated Ca^{2+} signaling on force loading of focal adhesions. Researchers found that the activation of TRPV4 increased tension across vinculin, indicating that tension within focal adhesions in MSCs can be regulated by calcium via TRPV4 [68].

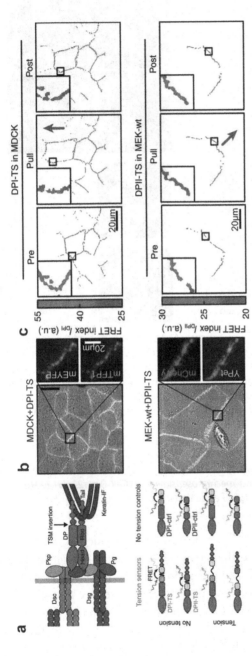

Fig. 11.7 Desmoplakin (DP) tension sensor. (A) Schematic of DP junctional localization (top), and DP no tension sensors and DP tension sensors and DP no tension controls (bottom). (B) DPI-TS and DPII-TS expressed in MDCK cells and MEK cells, respectively. (C) The decreasing FRET index of both DPI-TS and DPII-TS during the pulling process revealed the mechanical loading of both DP isoforms in response to external stress. (*Source: A.J. Price, A.-L. Cost, H. Ungewiß, J. Waschke, A.R. Dunn, C. Grashoff, Mechanical loading of desmosomes depends on the magnitude and orientation of external stress, Nat. Commun. 9(1) (2018); Copyright permission file: Copyright Permission-3.*)

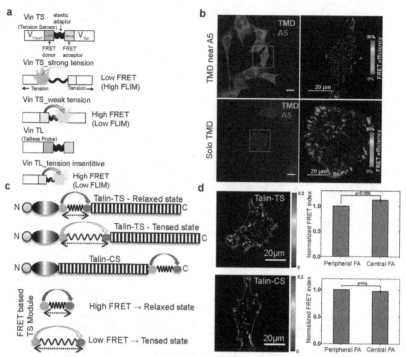

Fig. 11.8 Focal adhesion tension sensor. (A) Schematic of vinculin tension sensor (VinTS) and vinculin tailless probe (VinTL). (B) Increased VinTS FRET efficiency in TMD cocultured with A5 visualized the potential inhibition of A5 on tumor cell migration. (C) Schematic of talin tension sensor (talin-TS) and no tension control (talin-CS). (D) Lower talin FRET index expressed in peripheral FA, indicating high mechanical loading on talin localized at the cell edges. *(Source: F. Li, A. Chen, A. Reeser, Y. Wang, Y. Fan, S. Liu, X. Zhao, R. Prakash, D. Kota, B.-Y. Li, H. Yokota, J. Liu, Vinculin force sensor detects tumor-osteocyte interactions, Sci. Rep. 9(1) (2019); A. Kumar, M. Ouyang, K. Van den Dries, E.J. McGhee, K. Tanaka, M.D. Anderson, A. Groisman, B.T. Goult, K.I. Anderson, M.A. Schwartz, Talin tension sensor reveals novel features of focal adhesion force transmission and mechanosensitivity, J. Cell Biol. 213(3) (2016) 371–83; Copyright permission file: Copyright Permission-4, Copyright Permission-5.)*

Talin, whose activity is closely associated with vinculin, is an adhesion plaque protein that connects the integrin binding head domain and an actin binding rod domain. Hence, talin tension sensors are usually constructed by inserting the tension sensor module between those two domains. Talin1 tension sensor built in NIH 3T3 cells showed the different force response of talin on substrates with different stiffness. Meanwhile, the sensor also displayed gradient tension distribution on talin, where higher tension is located in the center and lower tension at the periphery, in assembling adhesions;

however, the gradient tension distribution could not be observed in mature adhesions [69–71] (Fig. 11.8C–D). Recently, a similar tension sensor module was engineered into talin2 in papillary collecting duct cells without expression of both talin1 and talin2 to study the relationship between those isoforms. Researchers reported that cells only expressing talin2 failed to lock talin on the membrane, form mature focal adhesions, and spread [72]. In addition to inserting talin tension sensors in cells, a drosophila with talin tension sensors was developed by inserting the YPet-elastic linker-mCherry between talin head and rod domains (F40-TS, TS, stTS) at the analogous position in drosophila. This model revealed that not all talin molecules responded to mechanotransduction at any time and mature muscles experienced less tension within talin [73].

3.2 DNA-based tension sensors

To probe the forces on the piconewton scale that are exerted on and exerted by cells, researchers have employed DNA-based tension sensors (Fig. 11.9). These sensors, in general, consist of four functional moieties. The central element is the force-sensing module made from either single-stranded DNA (ssDNA) or double-stranded DNA (dsDNA), in which the hydrogen bonds between nucleotides are ruptured under tension. The tension required for rupture, called the tension tolerance, is related to the total number of base pairs and the content of guanine (G) and cytosine (C). Next, a fluorescent reporter, which is either a fluorophore-quencher pair or a FRET pair, is used to observe if the base pairs within the DNA force-sensing

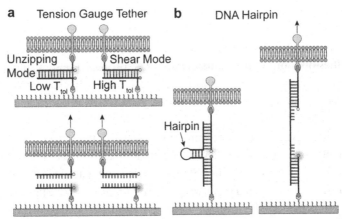

Fig. 11.9 Working mechanism of DNA-based tension sensors. (A) Tension gauge tether. (B) DNA hairpin.

module have been broken. The final two moieties include an anchoring unit, such as biotin-streptavidin, to immobilize the sensor to a surface, and a ligand specific to a protein, such as integrins, on the cell to incorporate the probe to the desired mechanotransduction pathway to be studied [74].

3.2.1 Tension gauge tethers

Tension gauge tethers (TGTs) were the first type of DNA-based tension sensor used in the study of cell forces. The sensor was first designed to limit the force that cells could transmit to their ECM, and was later modified to include a fluorophore for force readouts [75,76]. These sensors use dsDNA, and, in addition to the number of nucleotides and GC content, the tension tolerance can be controlled by the relative position of the connection points between the dsDNA strand to the anchoring unit and ligand moiety. Lower rupture forces are obtained when placed on the same end of the dsDNA strand (unzipping mode), and higher when on the opposite ends of the dsDNA strand (shear mode) (Fig. 11.9A). By controlling these parameters, researchers have developed TGTs with rupture forces on the order of a few piconewtons to 100 pN. An important consideration when using this type of tension sensor is that the process of dsDNA rupture is irreversible, and, therefore, cannot be used for multiple measurements.

In one study, TGTs were used to investigate the influence of epidermal growth factor receptor (EGFR) on the formation of integrin-based focal adhesions (Fig. 11.10A). The TGTs presented a ligand to bind to the integrin heterodimer $\alpha V \beta 3$ with a tension tolerance of either 12 pN or 56 pN. Cos-7 cells were cultured on surfaces functionalized with these TGTs for an hour with and without stimulation of EGFR via exposure to epidermal growth factor (EGF). It was found that stimulation with EGF resulted in enhanced cell spreading, formation of more mature and radially oriented focal adhesions, and higher signal from TGT probes. These results demonstrate that growth factors such as EGFR can act as "mechano-organizers," regulating the formation of tension within focal adhesions during focal adhesion maturation [77].

An advancement of this technology was the development of a method for amplifying signal from TGTs using a hybridization chain reaction (HCR) (Fig. 11.10B). In this method, called mechano-HCR, the TGT contains the initiator for HCR, and when a force large enough to rupture the TGT is achieved, the initiator is exposed, initiating HCR and amplifying the fluorescence signal. Mechano-HCR was used to investigate contractile forces of platelets when exposed to different chemical compounds. From the data, the IC50 (half-maximal inhibitory concentration) of the drugs were

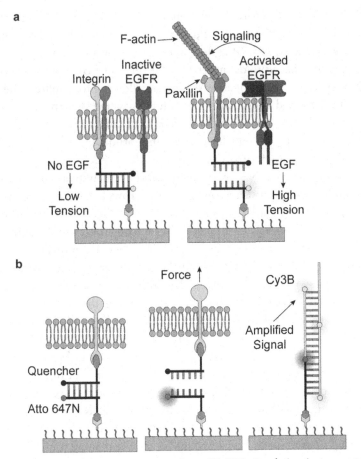

Fig. 11.10 Studies with tension gauge tethers. (A) EGF stimulation increases tension across integrins in Cos-7 cells. (B) Hybridization chain reaction increases signal from TGTs, allowing for calculation of IC_{50} of different chemical compounds with a plate reader.

calculated, which were in close agreement with values from literature. Importantly, the results of the assay can be quantified in a plate reader due to the amplification of the signal, which significantly decreases the time to quantify the results compared to analyzing microscopy images and allows for high-throughput screening [78].

3.2.2 DNA hairpin force probes

The first type of reversible probe for sensing forces from adhesive junctions used polymers, such as poly-ethylene glycol [79–81], engineered polypeptides [82–84], and proteins [85] for the force sensing module. However,

DNA-based reversible force probes offer distinct advantages, including having a high signal to noise ratio and a greater force sensitivity. DNA hairpin force probes employ ssDNA as the force-sensing module, which is in a stem-loop structure that unfolds under tension (Fig. 11.9B). Unlike rupture of dsDNA in TGTs, this process is reversible, allowing for multiple measurements. This method offers a distinct advantage over sensors that employ proteins, polypeptides, or synthetic polymers as the force-sensing module, due to the binary nature of stem-loop unfolding, producing a clear signal when tension levels are higher than the tension tolerance.

A hairpin-based DNA tensioner platform was implemented within a microfluidic device consisting of arrays of microwells to screen the mechanical property heterogeneity of single living cells (Fig. 11.11A–C) [86]. The DNA tension sensor was assembled by hybridizing three DNA fragments, F, Q, and H, to form a hairpin structure, which, when under mechanical forces that exceed its binding force, will transform to an open conformation, leading to increased distance between the fluorophore (Cye5) and quencher (Fig. 11.11D). Mechanical characterization using the tensioner platform showed that paclitaxel-resistant Hela cells (HeLa-T) exhibited higher fluorescence intensity than paclitaxel-sensitive HeLa cells, suggesting reduced force at the hairpin sensor (Fig. 11.11E). The study further used the tensioner platform to screen for mechanical heterogeneity of tumor cells from lung cancer patients. It was found that tissue tumor cells exhibited significantly lower forces than pleural effusion tumor cells (Fig. 11.11F). The device can screen single cell arrays in a throughput larger than 10,000 cells, suggesting a potential for high-throughput drug screening using force-based DNA tensioner platforms.

In another study, a DNA hairpin force probe was developed that is locked when unfolded when in the presence of locking DNA strand, storing mechanical information. Following which, an unlocking strand can be used to restore force sensitivity. This probe is useful for studying bonds that are typically short-lived, such as the interaction of T cell receptors (TCRs) with antigens. This probe was used to study the interaction between OT-1 naïve $CD8^+$ cells and the cognate antigen peptide-major histocompatibility complex (pMCH) N4 (Fig. 11.12A). TCR-pMHC tension signals were weak in real time without the use of the locking strand; but with the locking strand, accumulation of signal is observed, allowing for visualization of these short-lived interactions [87].

In addition to studying cell-ECM adhesions, DNA hairpin probes have recently been adapted for use to detect forces within cell–cell adhesions. In one such report, a probe called DNA-based membrane tension ratiometric

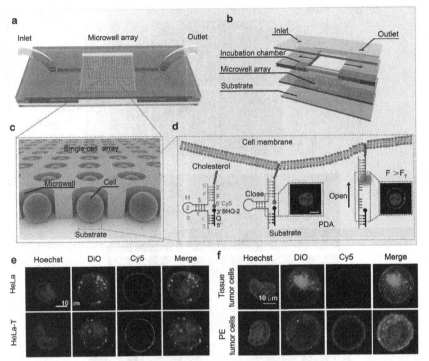

Fig. 11.11 Studies with DNA hairpins. (A) Schematic of the microfluidic device that uses DNA tensioner for mechanical characterization of single cells. (B) Inlet and outlet setup of the device. (C) Microwell array for single cell confinement. (D) Schematic of the DNA tensioner assembly on the cell membrane. (E) Imaging of HeLa-T and HeLa cells with the DNA tensioner in the *red* (*gray* in print version) channel. (F) Imaging of tissue and pleural effusion (PE) tumor cells with the DNA tensioner platform. *(Source: X. Hang, S. He, Z. Dong, Y. Li, Z. Huang, Y. Zhang, H. Sun, L. Lin, H. Li, Y. Wang, High-throughput DNA tensioner platform for interrogating mechanical heterogeneity of single living cells, Small 18(12) (2022) 2106196; Copyright permission file: Copyright Permission-6.)*

probe, or DNAMeter, was designed for studying tension forces within E–cadherin (Fig. 11.12B). This probe features two self-assembled DNA hairpins with tension thresholds of 4.4 pN and 8.1 pN, each with distinct fluorophore–quencher pairs, allowing for tracking of different ranges of forces, along with a domain to bind to E–cadherin and a cholesterol anchor to integrate into the cell membrane. With this probe, it was shown that both large (8.1 pN) and medium (4.4 pN) forces reduce when calcium in the environment is chelated, which is required for E–cadherin trans–dimers to form. In addition, the researchers investigated E–cadherin forces during cell migration, and found that the number of E–cadherins experiencing tension increases with distance from the leading edge up to about 15 cell lengths, after which it remains at a constant level [88].

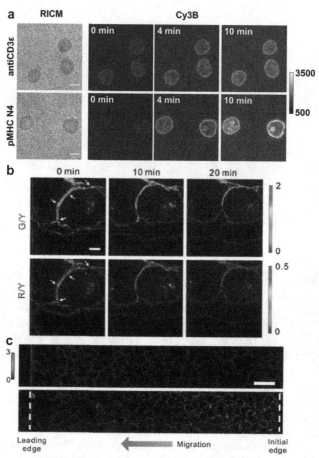

Fig. 11.12 Studies with DNA hairpins. (A) Use of a locking mechanism to record tension history allows for visualizing short lived forces exerted by immune cells. (B) DNAMeter was used to show decreased forces in E-cadherin when calcium is removed from the environment. (C) Distribution of intercellular forces in migrating cells, with low forces at the leading edge and higher forces in the middle. *(Source: R. Ma, A.V. Kellner, Y. Hu, B.R. Deal, A.T. Blanchard, K. Salaita, DNA tension probes to map the transient piconewton receptor forces by immune cells, J. Vis. Exp. (169) (2021); B. Zhao, N. Li, T. Xie, Y. Bagheri, C. Liang, P. Keshri, Y. Sun, M. You, Quantifying tensile forces at cell–cell junctions with a DNA-based fluorescent probe, Chem. Sci. 11(32) (2020) 8558–8566; Copyright permission file: Copyright Permission-7, Copyright Permission-8.)*

4 Concluding remarks

It is widely accepted that external and internal forces play critical roles in regulating cell behavior. This will place an ever increasing demand on techniques to characterize the mechanics of living cells with ever increasing resolution requirements. The advancement in 3D fabrication of complex designs with biocompatible materials allows more precise measurements in many sensitive cell types. Furthermore, 4D printing introduces bioactive materials into the designs to probe cell behaviors in 3D environments actively. Despite these promises, current techniques need to be significantly improved in throughput to meet the demands of industrial scale screening and to provide batch processing of single cell measurement. It is worth noting that, although microfluidic devices have been employed to provide high-throughput external measurement of the mechanics of living cells, because the characterization is often performed on cells in an unnatural suspended state, rather than adherent cells, often by monitoring cell deformability as cells flow through well-defined microfluidic channels, these devices are not included in this chapter. They play a more important role in multiomics characterizations than in mechanical characterizations.

FRET is an effective and precise method to monitor and measure the force response in cells, no matter whether studying the cell activities or understanding the biophysical mechanism behind diseases. It reveals intramolecular or intermolecular force loading and provides force distribution at different locations within the cell with superior resolution. However, since this technique evaluates the force response by detecting the fluorescent intensity changes between donor and acceptor fluorophores, it puts a high demand on microscopy configurations and postimaging data processing. With the advances in DNA nanotechnology, DNA-based tension systems are promising in delivering piconewton level detection. However, in order for them to be widely adopted, limitations such as nuclease degradation, complications in DNA sensor modification, and short working duration need to be addressed.

References

[1] S. Suresh, J. Spatz, J.P. Mills, A. Micoulet, M. Dao, C.T. Lim, M. Beil, T. Seufferlein, Connections between single-cell biomechanics and human disease states: gastrointestinal cancer and malaria, Acta Biomater. 1 (1) (2005) 15–30.

[2] S.E. Cross, Y.-S. Jin, J. Rao, J.K. Gimzewski, Nanomechanical analysis of cells from cancer patients, Nat. Nanotechnol. 2 (12) (2007) 780–783.

[3] C. Lim, E. Zhou, S. Quek, Mechanical models for living cells—a review, J. Biomech. 39 (2) (2006) 195–216.

[4] T. Iskratsch, H. Wolfenson, M.P. Sheetz, Appreciating force and shape-the rise of mechanotransduction in cell biology, Nat. Rev. Mol. Cell Biol. 15 (12) (2014) 825–833.

[5] A.E. Miller, P. Hu, T.H. Barker, Feeling things out: bidirectional signaling of the cell–ECM interface, implications in the mechanobiology of cell spreading, migration, proliferation, and differentiation, Adv. Healthc. Mater. 9 (8) (2020) 1901445.

[6] C.-M. Lo, H.-B. Wang, M. Dembo, Y.-L. Wang, Cell movement is guided by the rigidity of the substrate, Biophys. J. 79 (1) (2000) 144–152.

[7] T.R. Morin Jr., S.A. Ghassem-Zadeh, J. Lee, Traction force microscopy in rapidly moving cells reveals separate roles for ROCK and MLCK in the mechanics of retraction, Exp. Cell Res. 326 (2) (2014) 280–294.

[8] R. Alert, X. Trepat, Physical models of collective cell migration, Annu. Rev. Condens. Matter Phys. 11 (2020) 77–101.

[9] F. Bosveld, I. Bonnet, B. Guirao, S. Tlili, Z. Wang, A. Petitalot, R. Marchand, P.-L. Bardet, P. Marcq, F. Graner, Mechanical control of morphogenesis by fat/dachsous/four-jointed planar cell polarity pathway, Science 336 (6082) (2012) 724–727.

[10] K.M. Yamada, J.W. Collins, D.A. Cruz Walma, A.D. Doyle, S.G. Morales, J. Lu, K. Matsumoto, S.S. Nazari, R. Sekiguchi, Y. Shinsato, Extracellular matrix dynamics in cell migration, invasion and tissue morphogenesis, Int. J. Exp. Pathol. 100 (3) (2019) 144–152.

[11] Y.A. Miroshnikova, J.K. Mouw, J.M. Barnes, M.W. Pickup, J.N. Lakins, Y. Kim, K. Lobo, A.I. Persson, G.F. Reis, T.R. McKnight, Tissue mechanics promote IDH1-dependent HIF1α–tenascin C feedback to regulate glioblastoma aggression, Nat. Cell Biol. 18 (12) (2016) 1336–1345.

[12] D.T. Butcher, T. Alliston, V.M. Weaver, A tense situation: forcing tumour progression, Nat. Rev. Cancer 9 (2) (2009) 108–122.

[13] S.R. Polio, S.E. Stasiak, R.R. Jamieson, J.L. Balestrini, R. Krishnan, H. Parameswaran, Extracellular matrix stiffness regulates human airway smooth muscle contraction by altering the cell-cell coupling, Sci. Rep. 9 (1) (2019) 1–12.

[14] A.J. Engler, S. Sen, H.L. Sweeney, D.E. Discher, Matrix elasticity directs stem cell lineage specification, Cell 126 (4) (2006) 677–689.

[15] M. Dembo, Y.-L. Wang, Stresses at the cell-to-substrate interface during locomotion of fibroblasts, Biophys. J. 76 (4) (1999) 2307–2316.

[16] N.Q. Balaban, U.S. Schwarz, D. Riveline, P. Goichberg, G. Tzur, I. Sabanay, D. Mahalu, S. Safran, A. Bershadsky, L. Addadi, Force and focal adhesion assembly: a close relationship studied using elastic micropatterned substrates, Nat. Cell Biol. 3 (5) (2001) 466–472.

[17] K.H. Vining, D.J. Mooney, Mechanical forces direct stem cell behaviour in development and regeneration, Nat. Rev. Mol. Cell Biol. 18 (12) (2017) 728–742.

[18] J. Fu, Y.-K. Wang, M.T. Yang, R.A. Desai, X. Yu, Z. Liu, C.S. Chen, Mechanical regulation of cell function with geometrically modulated elastomeric substrates, Nat. Methods 7 (9) (2010) 733–736.

[19] J.P. Butler, I.M. Tolic-Nørrelykke, B. Fabry, J.J. Fredberg, Traction fields, moments, and strain energy that cells exert on their surroundings, Am. J. Phys. Cell Phys. 282 (3) (2002) C595–C605.

[20] Z. Yang, J.-S. Lin, J. Chen, J.H. Wang, Determining substrate displacement and cell traction fields—a new approach, J. Theor. Biol. 242 (3) (2006) 607–616.

[21] U.S. Schwarz, N.Q. Balaban, D. Riveline, A. Bershadsky, B. Geiger, S. Safran, Calculation of forces at focal adhesions from elastic substrate data: the effect of localized force and the need for regularization, Biophys. J. 83 (3) (2002) 1380–1394.

[22] M. Lekka, K. Gnanachandran, A. Kubiak, T. Zieliński, J. Zemła, Traction force microscopy—measuring the forces exerted by cells, Micron 150 (2021), 103138.

[23] W.R. Legant, C.K. Choi, J.S. Miller, L. Shao, L. Gao, E. Betzig, C.S. Chen, Multi-dimensional traction force microscopy reveals out-of-plane rotational moments about focal adhesions, Proc. Natl. Acad. Sci. 110 (3) (2013) 881–886.

[24] J. Steinwachs, C. Metzner, K. Skodzek, N. Lang, I. Thievessen, C. Mark, S. Münster, K.E. Aifantis, B. Fabry, Three-dimensional force microscopy of cells in biopolymer networks, Nat. Methods 13 (2) (2016) 171–176.

[25] S.S. Hur, J.H. Jeong, M.J. Ban, J.H. Park, J.K. Yoon, Y. Hwang, Traction force microscopy for understanding cellular mechanotransduction, BMB Rep. 53 (2) (2020) 74.

[26] E.E. Charrier, K. Pogoda, R. Li, C.Y. Park, J.J. Fredberg, P.A. Janmey, A novel method to make viscoelastic polyacrylamide gels for cell culture and traction force microscopy, APL Bioeng. 4 (3) (2020), 036104.

[27] J. Toyjanova, E. Hannen, E. Bar-Kochba, E.M. Darling, D.L. Henann, C. Franck, 3D viscoelastic traction force microscopy, Soft Matter 10 (40) (2014) 8095–8106.

[28] H. Colin-York, D. Shrestha, J.H. Felce, D. Waithe, E. Moeendarbary, S.J. Davis, C. Eggeling, M. Fritzsche, Super-resolved traction force microscopy (STFM), Nano Lett. 16 (4) (2016) 2633–2638.

[29] H. Colin-York, Y. Javanmardi, L. Barbieri, D. Li, K. Korobchevskaya, Y. Guo, C. Hall, A. Taylor, S. Khuon, G.K. Sheridan, Spatiotemporally super-resolved volumetric traction force microscopy, Nano Lett. 19 (7) (2019) 4427–4434.

[30] R. Yang, J.A. Broussard, K.J. Green, H.D. Espinosa, Techniques to stimulate and interrogate cell-cell adhesion mechanics, Extreme Mech. Lett. 20 (2018) 125–139.

[31] D. Johnson, N. Hilal, W.R. Bowen, Basic principles of atomic force microscopy, in: Atomic Force Microscopy in Process Engineering: An Introduction to Afm for Improved Processes and Products, 2009, pp. 1–30.

[32] T. Bouzid, A.M. Esfahani, B.T. Safa, E. Kim, V. Saraswathi, J.K. Kim, R. Yang, J.Y. Lim, Rho/ROCK mechanosensor in adipocyte stiffness and traction force generation, Biochem. Biophys. Res. Commun. 606 (2022) 42–48.

[33] L. Wildling, B. Unterauer, R. Zhu, A. Rupprecht, T. Haselgrubler, C. Rankl, A. Ebner, D. Vater, P. Pollheimer, E.E. Pohl, P. Hinterdorfer, H.J. Gruber, Linking of sensor molecules with amino groups to amino-functionalized AFM tips, Bioconjug. Chem. 22 (6) (2011) 1239–1248.

[34] V.T. Moy, E.L. Florin, H.E. Gaub, Adhesive forces between ligand and receptor measured by Afm, Biophys. J. 66 (2) (1994) A340.

[35] D.K. Deda, B.B.S. Pereira, C.C. Bueno, A.N. da Silva, G.A. Ribeiro, A.M. Amarante, E.F. Franca, F.L. Leite, The use of functionalized AFM tips as molecular sensors in the detection of pesticides, Mater Res-Ibero-Am. J. 16 (3) (2013) 683–687.

[36] S. Prasad, A. Rankine, T. Prasad, P. Song, M.E. Dokukin, N. Makarova, V. Backman, I. Sokolov, Atomic force microscopy detects the difference in Cancer cells of different neoplastic aggressiveness via machine learning, Adv. Nanobiomed. Res. 1 (8) (2021).

[37] M.G.A. Meister, P. Behr, P. Studer, J. Vörös, P. Niedermann, J. Bitterli, J. Polesel-Maris, M. Liley, H. Heinzelmann, T. Zambelli, FluidFM: Combining Atomic Force Microscopy and Nanofluidics in a Universal Liquid Delivery System for Single Cell Applications and Beyond, 2009.

[38] O. Guillaume-Gentil, E. Potthoff, D. Ossola, C.M. Franz, T. Zambelli, J.A. Vorholt, Force-controlled manipulation of single cells: from AFM to FluidFM, Trends Biotechnol. 32 (7) (2014) 381–388.

[39] C.J. Bustamante, Y.R. Chemla, S. Liu, M.D. Wang, Optical tweezers in single-molecule biophysics, Nat. Rev. Methods Primers 1 (2021).

[40] J. Derganc, S.Z. Jokhadar, Probing cell mechanics at small deformations with lateral indentation using optical tweezers, Eur. Biophys. J. Biophy. 50 (Suppl 1) (2021) 133.

[41] J. Guck, R. Ananthakrishnan, H. Mahmood, T.J. Moon, C.C. Cunningham, J. Käs, The optical stretcher: a novel laser tool to micromanipulate cells, Biophys. J. 81 (2) (2001) 767–784.

[42] C.T. Mierke, The role of the optical stretcher is crucial in the investigation of cell mechanics regulating cell adhesion and motility, Front. Cell Dev. Biol. 7 (2019).

[43] R. Sarkar, V.V. Rybenkov, A guide to magnetic tweezers and their applications, Front. Phys-Lausanne 4 (2016).

[44] J. Lipfert, J.W.J. Kerssemakers, T. Jager, N.H. Dekker, Magnetic torque tweezers: measuring torsional stiffness in DNA and RecA-DNA filaments, Nat. Methods 7 (12) (2010) 977–U54.

[45] P. Pan, W.H. Wang, C.H. Ru, Y. Sun, X.Y. Liu, MEMS-based platforms for mechanical manipulation and characterization of cells, J. Micromech. Microeng. 27 (12) (2017).

[46] X.Y. Liu, Y. Sun, W.H. Wang, B.M. Lansdorp, Vision-based cellular force measurement using an elastic microfabricated device, J. Micromech. Microeng. 17 (7) (2007) 1281–1288.

[47] A.M. Esfahani, G. Minnick, J. Rosenbohm, H. Zhai, X. Jin, B. Tajvidi Safa, J. Brooks, R. Yang, Microfabricated platforms to investigate cell mechanical properties, Med. Device Novel Technol. 13 (2022), 100107.

[48] N. Scuor, P. Gallina, H. Panchawagh, R. Mahajan, O. Sbaizero, V. Sergo, Design of a novel MEMS platform for the biaxial stimulation of living cells, Biomed. Microdevices 8 (3) (2006) 239–246.

[49] V. Mukundan, B.L. Pruitt, MEMS electrostatic actuation in conducting biological media, J. Microelectromech. Syst. 18 (2) (2009) 405–413.

[50] A.M. Esfahani, J. Rosenbohm, B.T. Safa, N.V. Lavrik, G. Minnick, Q. Zhou, F. Kong, X. Jin, E. Kim, Y. Liu, Y. Lu, J.Y. Lim, J.K. Wahl 3rd, M. Dao, C. Huang, R. Yang, Characterization of the strain-rate-dependent mechanical response of single cell-cell junctions, Proc. Natl. Acad. Sci. U. S. A. 118 (7) (2021).

[51] M. Cauchi, I. Grech, B. Mallia, P. Mollicone, N. Sammut, Analytical, numerical and experimental study of a horizontal electrothermal MEMS microgripper for the deformability characterisation of human red blood cells, Micromachines-Basel 9 (3) (2018).

[52] J.L. Walker, L.H.C. Patterson, E. Rodriguez-Mesa, K. Shields, J.S. Foster, M.T. Valentine, A.M. Doyle, K.L. Foster, Controlled single-cell compression with a high-throughput MEMS actuator, J. Microelectromech. Syst. 29 (5) (2020) 790–796.

[53] R. Parreira, E. Ozelci, M.S. Sakar, Investigating tissue mechanics in vitro using untethered soft robotic microdevices, Front. Robot. AI 8 (2021).

[54] B. Barazani, S. Warnat, T. Hubbard, A.J. MacIntosh, Mechanical characterization of individual brewing yeast cells using microelectromechanical systems (MEMS): cell rupture force and stiffness, J. Am. Soc. Brew. Chem. 75 (3) (2017) 236–243.

[55] D. Pekin, G. Perret, Q. Rezard, J.C. Gerbedoen, S. Meignan, D. Collard, C. Lagadec, M.C. Tarhan, Subcellular imaging during single cell mechanical characterization, Proc. IEEE Micr. Elect. (2020) 62–65.

[56] B.T. Bajar, E.S. Wang, S. Zhang, M.Z. Lin, J. Chu, A guide to fluorescent protein FRET Pairs, Sensors (Basel) 16 (9) (2016).

[57] L.S. Fischer, S. Rangarajan, T. Sadhanasatish, C. Grashoff, Molecular force measurement with tension sensors, Annu. Rev. Biophys. 50 (2021) 595–616.

[58] N. Borghi, M. Sorokina, O.G. Shcherbakova, W.I. Weis, B.L. Pruitt, W.J. Nelson, A.R. Dunn, E-cadherin is under constitutive actomyosin-generated tension that is increased at cell-cell contacts upon externally applied stretch, Proc. Natl. Acad. Sci. U. S. A. 109 (31) (2012) 12568–12573.

[59] D. Cai, S.C. Chen, M. Prasad, L. He, X. Wang, V. Choesmel-Cadamuro, J.K. Sawyer, G. Danuser, D.J. Montell, Mechanical feedback through E-cadherin promotes direction sensing during collective cell migration, Cell 157 (5) (2014) 1146–1159.

[60] D.E. Conway, M.T. Breckenridge, E. Hinde, E. Gratton, C.S. Chen, M.A. Schwartz, Fluid shear stress on endothelial cells modulates mechanical tension across VE-cadherin and PECAM-1, Curr. Biol. 23 (11) (2013) 1024–1030.

[61] A.K. Lagendijk, G.A. Gomez, S. Baek, D. Hesselson, W.E. Hughes, S. Paterson, D.E. Conway, H.-G. Belting, M. Affolter, K.A. Smith, M.A. Schwartz, A.S. Yap, B.M. Hogan, Live imaging molecular changes in junctional tension upon VE-cadherin in zebrafish, Nat. Commun. 8 (1) (2017).

[62] B.R. Acharya, S.K. Wu, Z.Z. Lieu, R.G. Parton, S.W. Grill, A.D. Bershadsky, G.A. Gomez, A.S. Yap, Mammalian diaphanous 1 mediates a pathway for E-cadherin to

stabilize epithelial barriers through junctional contractility, Cell Rep. 18 (12) (2017) 2854–2867.

[63] A.J. Price, A.-L. Cost, H. Ungewiß, J. Waschke, A.R. Dunn, C. Grashoff, Mechanical loading of desmosomes depends on the magnitude and orientation of external stress, Nat. Commun. 9 (1) (2018).

[64] S.R. Baddam, P.T. Arsenovic, V. Narayanan, N.R. Duggan, C.R. Mayer, S.T. Newman, D.A. Abutaleb, A. Mohan, A.P. Kowalczyk, D.E. Conway, The desmosomal cadherin desmoglein-2 experiences mechanical tension as demonstrated by a FRET-based tension biosensor expressed in living cells, Cell 7 (7) (2018).

[65] A. Raya-Sandino, A.C. Luissint, D.H.M. Kusters, V. Narayanan, S. Flemming, V. Garcia-Hernandez, L.M. Godsel, K.J. Green, S.J. Hagen, D.E. Conway, C.A. Parkos, A. Nusrat, Regulation of intestinal epithelial intercellular adhesion and barrier function by desmosomal cadherin desmocollin-2, Mol. Biol. Cell 32 (8) (2021) 753–768.

[66] C. Grashoff, B.D. Hoffman, M.D. Brenner, R. Zhou, M. Parsons, M.T. Yang, M.A. McLean, S.G. Sligar, C.S. Chen, T. Ha, M.A. Schwartz, Measuring mechanical tension across vinculin reveals regulation of focal adhesion dynamics, Nature 466 (7303) (2010) 263–266.

[67] F. Li, A. Chen, A. Reeser, Y. Wang, Y. Fan, S. Liu, X. Zhao, R. Prakash, D. Kota, B.-Y. Li, H. Yokota, J. Liu, Vinculin force sensor detects tumor-osteocyte interactions, Sci. Rep. 9 (1) (2019).

[68] C.L. Gilchrist, H.A. Leddy, L. Kaye, N.D. Case, K.E. Rothenberg, D. Little, W. Liedtke, B.D. Hoffman, F. Guilak, TRPV4-mediated calcium signaling in mesenchymal stem cells regulates aligned collagen matrix formation and vinculin tension, Proc. Natl. Acad. Sci. 116 (6) (2019) 1992–1997.

[69] A. Kumar, M. Ouyang, K. Van den Dries, E.J. McGhee, K. Tanaka, M.D. Anderson, A. Groisman, B.T. Goult, K.I. Anderson, M.A. Schwartz, Talin tension sensor reveals novel features of focal adhesion force transmission and mechanosensitivity, J. Cell Biol. 213 (3) (2016) 371–383.

[70] A. Kumar, J.K. Placone, A.J. Engler, Understanding the extracellular forces that determine cell fate and maintenance, Development 144 (23) (2017) 4261–4270.

[71] A. Kumar, K.L. Anderson, M.F. Swift, D. Hanein, N. Volkmann, M.A. Schwartz, Local tension on Talin in focal adhesions correlates with F-actin alignment at the nanometer scale, Biophys. J. 115 (8) (2018) 1569–1579.

[72] E.S. Rangarajan, M.C. Primi, L.A. Colgan, K. Chinthalapudi, R. Yasuda, T. Izard, A distinct talin2 structure directs isoform specificity in cell adhesion, J. Biol. Chem. 295 (37) (2020) 12885–12899.

[73] S.B. Lemke, T. Weidemann, A.-L. Cost, C. Grashoff, F. Schnorrer, A small proportion of Talin molecules transmit forces at developing muscle attachments in vivo, PLoS Biol. 17 (3) (2019), e3000057.

[74] Q. Tian, P. Keshri, M. You, Recent developments in DNA-based mechanical nanodevices, Chem. Commun. (2022).

[75] X. Wang, T. Ha, Defining single molecular forces required to activate integrin and notch signaling, Science 340 (6135) (2013) 991–994.

[76] Y. Wang, X. Wang, Integrins outside focal adhesions transmit tensions during stable cell adhesion, Sci. Rep. 6 (1) (2016) 1–9.

[77] T.C. Rao, V.P.-Y. Ma, A. Blanchard, T.M. Urner, S. Grandhi, K. Salaita, A.L. Mattheyses, EGFR activation attenuates the mechanical threshold for integrin tension and focal adhesion formation, J. Cell Sci. 133 (13) (2020) jcs238840.

[78] Y. Duan, R. Glazier, A. Bazrafshan, Y. Hu, S.A. Rashid, B.G. Petrich, Y. Ke, K. Salaita, Mechanically triggered hybridization chain reaction, Angew. Chem. Int. Ed. 60 (36) (2021) 19974–19981.

[79] D.R. Stabley, C. Jurchenko, S.S. Marshall, K.S. Salaita, Visualizing mechanical tension across membrane receptors with a fluorescent sensor, Nat. Methods 9 (1) (2012) 64–67.

[80] Y. Liu, K. Yehl, Y. Narui, K. Salaita, Tension sensing nanoparticles for mechano-imaging at the living/nonliving interface, J. Am. Chem. Soc. 135 (14) (2013) 5320–5323.

[81] Y. Chang, Z. Liu, Y. Zhang, K. Galior, J. Yang, K. Salaita, A general approach for generating fluorescent probes to visualize piconewton forces at the cell surface, J. Am. Chem. Soc. 138 (9) (2016) 2901–2904.

[82] M. Morimatsu, A.H. Mekhdjian, A.S. Adhikari, A.R. Dunn, Molecular tension sensors report forces generated by single integrin molecules in living cells, Nano Lett. 13 (9) (2013) 3985–3989.

[83] M. Morimatsu, A.H. Mekhdjian, A.C. Chang, S.J. Tan, A.R. Dunn, Visualizing the interior architecture of focal adhesions with high-resolution traction maps, Nano Lett. 15 (4) (2015) 2220–2228.

[84] S.J. Tan, A.C. Chang, S.M. Anderson, C.M. Miller, L.S. Prahl, D.J. Odde, A.R. Dunn, Regulation and dynamics of force transmission at individual cell-matrix adhesion bonds, Sci. Adv. 6 (20) (2020) eaax0317.

[85] K. Galior, Y. Liu, K. Yehl, S. Vivek, K. Salaita, Titin-based nanoparticle tension sensors map high-magnitude integrin forces within focal adhesions, Nano Lett. 16 (1) (2016) 341–348.

[86] X. Hang, S. He, Z. Dong, Y. Li, Z. Huang, Y. Zhang, H. Sun, L. Lin, H. Li, Y. Wang, High-throughput DNA tensioner platform for interrogating mechanical heterogeneity of single living cells, Small 18 (12) (2022) 2106196.

[87] R. Ma, A.V. Kellner, Y. Hu, B.R. Deal, A.T. Blanchard, K. Salaita, DNA tension probes to map the transient piconewton receptor forces by immune cells, J. Vis. Exp 169 (2021).

[88] B. Zhao, N. Li, T. Xie, Y. Bagheri, C. Liang, P. Keshri, Y. Sun, M. You, Quantifying tensile forces at cell–cell junctions with a DNA-based fluorescent probe, Chem. Sci. 11 (32) (2020) 8558–8566.

CHAPTER 12

Cellular mechanical measurement by magnetic micro/nanorobots

Jintian Wang, Xian Wang, and Yu Sun
Department of Mechanical and Industrial Engineering, University of Toronto, Toronto, ON, Canada

1 Introduction

1.1 Cell mechanics

Cells generate, sustain, and respond to mechanical forces within their environment via an intrinsic system of physical and biochemical modules to maintain appropriate biological functions. For example, cells act on the extracellular matrix (ECM), by generating cell traction forces (CTFs) to maintain cell shape, migrating within tissues, reorganizing ECM, and communicating with neighboring cells [1]. Such CTFs play an important role in a wide range of biological processes, such as embryogenesis, angiogenesis, inflammation, wound healing, and metastasis [2]. At the same time, cells withstand diverse mechanical cues such as plasma membrane tension, shear stress, hydrodynamic pressure, and forces from homotypic and heterotypic cell interactions, and adapt accordingly via different mechanotransduction mechanisms (Fig. 12.1) [3–5]. Abnormal mechanical behaviors of cells have been shown to be highly correlated with the progression of diseases like cancers [6], malaria [7], sickle cell anemia [8], and atherosclerosis [9]. It has been confirmed that the stiffness of metastatic cancer cells is more than 70% lower than that of benign cells [10]. Researchers are striving to accurately measure mechanical properties of cells and build a realistic and predictive model to provide new perspectives of disease-related cell mechanics.

1.2 Techniques measuring cell mechanics: State of art

Cells present rich molecular dynamics that ultimately combine to drive macroscopic cellular processes. Specifically, cell mechanics of interest include molecular interactions, traction forces and adhesion strength, intrinsic materials properties such as stiffness and surface tension, and the capability to respond to applied stresses and strains. Accordingly, different techniques

Robotics for Cell Manipulation and Characterization
https://doi.org/10.1016/B978-0-323-95213-2.00012-0

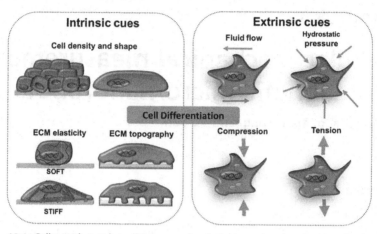

Fig. 12.1 Cells withstand multiple mechanical cues in the human body and corresponding cellular signaling pathways and behaviors are regulated by the mechanical interplay of cell-cell and cell-matrix interactions. Abnormal mechanical behaviors due to disrupted mechanosensing and mechanotransduction lead to disease progression [3]. *(Reproduced with permission: 2021, Frontiers.)*

for cell mechanics measurement have been developed based on approaches in physics and engineering [11].

Forster resonance energy transfer (FRET) utilizes the fluorescent molecules to investigate molecular interactions under tension [12]. In FRET measurement, the fluorescent molecule being excited by the incident light is referred as the donor and the molecule emitting the photon is referred as the acceptor. The excited electron will then activate a cascade of photon emission of both donor and receptor. Furthermore, the probability of FRET between fluorophores is a decreasing function of the distance between them (Fig. 12.2A) [13]. Thus, the distance between the donor and acceptor molecules can be determined by tracking the fluorescent emission of both molecules. As changes in FRET efficiency only indicate the distance change of fluorophore molecules, an elastic linker between two molecules needs to be calibrated with the tension–distance relationship. The FRET probes, when genetically inserted into vinculin, a focal adhesion protein, have revealed that the vinculin in protruding regions of the cell is stretched more than that in the retracting region [17]. Since the FRET signal is most sensitive to forces from 1 to 6 pN, various probes relying on different linkers have been inserted into vinculin [18], E-cadherin [19], PECAM-1 [20], and β-spectrin [21] to investigate tension at focal adhesions and cell-cell junctions.

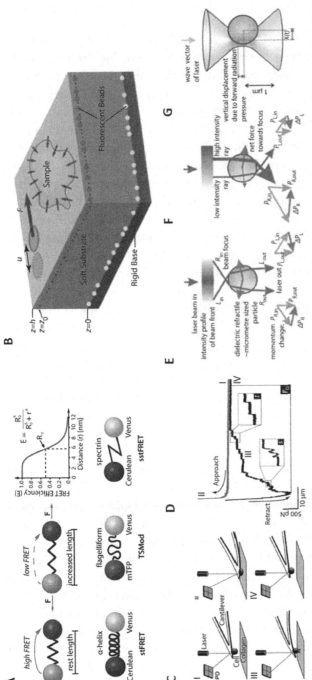

Fig. 12.2 Different techniques measuring cells mechanics. (A) The working principles of FRET-based biosensors: the efficiency of the FRET signal is dependent on the distance between two elastically linked fluorophores. Different linker molecules have been designed to connect FRET molecule pairs [13]. (B) A typical TFM on a soft elastic substrate commonly made of PAA or PDMS [14]. (C) The general principle of AFM: a laser reflected from the back of a cantilever and onto a photodiode so that deflection of cantilever can be recorded. (D) A representative cell adhesion experiment where force from a cell was measured as the cantilever was retracted from the surface [15]. (E) The optical tweezers have the force balance due to the refraction of light through a bead centered in the trap. (F) The bead off-center is feeling the restoring force pushing it back to focus. (G) The general shape of an optical trap is shown [16]. *(Panel A: Reproduced with permission. 2015, Springer Nature. Panel B: Reproduced with permission. 2014, Royal Society of Chemistry. Panel D: Reproduced with permission. 2010, Nature America. Panel G: Reproduced with permission. 2013, The Royal Society.)*

Traction force microscopy (TFM) techniques have been employed to measure the traction force of a cell being exerted on an external substrate, which needs to be deformable and elastic [22]. After cell adhesion, the traction force can be reflected by the small but measurable deformation that remains in the linear regime of the elastic behavior of the substrate material [23]. Elastic films made of polymers such as polyacrylamide (PAA) [24,25] and polydimethylsiloxane silicone (PDMS) [14] are coated with extracellular matrix (ECM) protein for cell adhesion (Fig. 12.2B) [26]. The ratio of polymer and crosslinker determines the film mechanical properties, which can be measured simply by macroscopic approaches and can be tuned to a broad range of physiological stiffness [27]. Tracking fluorescent microspheres embedded in films, 2D strain field and stress map [24,28] can be constructed with a spatial resolution of $0.5\,\mu m$ and on the order of $10-10^5\,Pa$, respectively [28,29]. Furthermore, the normal components of traction forces can be measured by tracking displacements of beads in 3D [30].

The atomic force microscope (AFM), originally developed as an imaging technique for measuring surface topography, has been used to measure adhesion and force generation of cells owning to its high sensitivity and controllable small distance mobility [31–33]. The cantilever deflection during cell contact can be directly converted into a measure of force using Hooke's law, $F = kx$, in the simplest form (Fig. 12.2C and D) [15,34]. The cell attached on the AFM tip is first compressed to substrate surface to form adhesion and was monitored during retraction to measure single bonds breaking and total adhesive force [15,34,35]. On the other hand, if holding cells under constant tension, height, or stiffness with feedback loop control, AFM can be used to probe strain, stress, and stiffness of cells. For example, Webster et al. have found that cells changed traction rate to adapt to the apparent stiffness change of the AFM tip [36].

Optical tweezers, since first being reported in 1986, have become one of the standard tools for measuring biomechanical interactions [37]. Particles trapped in the focus of a tightly focused laser beam endure forces proportional to the displacement from the focal point of the trap (Fig. 12.2E–G) [16]. The most common object used in the trap generated at the focal point is the plastic bead, which is conjugated with biochemical molecules on the surface. Dai et al. trapped IgG coated beads to bind with cell membranes and retracted the beads using optical tweezers to identify the tension of the cytoskeleton connected membrane surpassing the pure membrane tension [38].

While the optical tweezers use a high-powered laser, which can potentially induce thermal and photochemical damage to the cells, magnetic tweezers employing micro/nanorobots represent a less invasive technique, directly applying forces to cells to extract information about their mechanical properties [39]. Actuated by different setups of external magnetic field with temporal control, magnetic beads can apply up to 100 nN at ranges of 10 μm to measure the cellular rheological properties depending on both magnitude and duration of the deformation [40]. With targeted functionalization, magnetic beads can bind to specific cellular structures of interest to investigate the response and apply forces to proteins such as integrins, cadherins, nesprins, and actin stress fibers [41–44]. Furthermore, magnetic swarms of micro/nanorobots can be actuated via controlled fields to investigate the responses of hundreds to thousands of cells simultaneously, or be combined with fluorescent microscopy to measure mechanical responses of cells with high signal-to-noise ratio. For example, Wang et al. used magnetic swarms functionalized with fluorescent dyes to reveal the pH gradient from the leading edge to the trailing edge in migrating cells and perinuclear calcium increase under mechanical stimulation mediated by the mechanosensitive PIEZO1 ion channels (Fig. 12.3) [45].

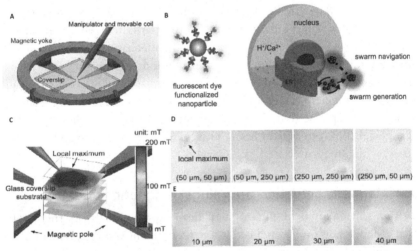

Fig. 12.3 Magnetic microrobotic swarms used for intracellular measurement with enhanced signal-to-noise ratio [45]. (A) A five-pole magnetic tweezer was developed with four stationary poles and one mobile pole controlled by a micromanipulator and is installed on a microscope for fluorescent imaging. (B) The intracellular measurement is conducted by fluorescent dye-coated magnetic nanoparticles that were formed into swarms and directed by magnetic tweezers. (C) Magnetic field simulation of the five-pole magnetic system showing a 3D magnetic field generation and a local maximum field strength in multiple 2D planes. (D and E) Swarm position and size control were achieved by the five-pole magnetic system [45]. *(Reproduced with permission. 2022, American Chemical Society.)*

2 Principles of magnetic actuation

2.1 Magnetic materials and nanoparticles

To be actuated by the external magnetic field, the magnetic micro/nanor-obots need to be made of either paramagnetic or ferromagnetic materials to exhibit induced magnetic moment. When the size is reduced to a critical threshold, the ferromagnetic magnetic material shows superparamagnetic behavior and presents single domain-like induced magnetic moment [46]. The loss of ferromagnetism is ascribed to the loss of domain wall motion when the formation of domain walls becomes energetically unfavorable below a critically small size [47]. The composition of magnetic materials includes cobalt, iron, and nickel, whereas iron oxide has been mostly used to fabricate magnetic micro/nanorobots due to superparamagnetic behaviors and high induced magnet moment by the external field.

Isotropic magnetic beads or nanoparticles are the most common forms of micro/nanorobots used for magnetic actuation to perform cellular and intra-cellular measurement, although anisotropic magnetic nanorods have also been used to measure the viscosity and viscoelastic properties of biological samples [48]. Magnetic beads with sizes from a few nanometers to tens of micrometers can bind to targets including proteins, intracellular organelles, viruses, and whole cells. When measuring the mechanical properties of cel-lular and intracellular structures, beads from hundreds of nanometers to sev-eral microns have generally been used for optimized cell internalization and generating sufficient forces under magnetic actuation. In addition, magnetic beads are encapsulated in biocompatible matrix or polymers like polyethyl-ene glycol (PEG) to avoid causing intracellular inflammation and oxidative stress, which may affect cell viability [49].

2.2 Magnetic actuation on a single magnetic nanoparticle

Within a magnetic field, the magnetic bead experiences the magnetic force that is dependent on the induced magnetic moment and the gradient of the magnetic field as follows:

$$F = m\nabla B \tag{12.1}$$

where F is the magnetic force, m is the magnetic moment, and ∇B is the gradient of the flux density of the magnetic field.

The induced magnetic moment of a particle is determined by the exter-nal field, the magnetic permeability of the material, and the total volume of the magnetic material within the field as follows:

$$m = \frac{3V}{\mu_0}\left(\frac{\mu - \mu_0}{\mu + 2\mu_0}\right)B \tag{12.2}$$

where B is the magnetic flux density, μ is the magnetic permeability of the bead, and μ_0 is the permeability of air.

It should be noted that the magnetic moment m scales down with volume, and thus, the magnetic force scales down with the dimension of the magnetic bead by a factor of three. To generate sufficient forces to stimulate and measure intracellular structures, superparamagnetic iron oxide nanoparticles are commonly used for high induced magnetic moment under external magnetic field. During cellular measurement, the beads in contact with a cell are modeled based on Hertz model to relate the magnetic force and cell deformation as follows [36]:

$$F_{elastic} = \frac{4}{3}\frac{E}{1 - v_s^2}R^{\frac{1}{2}}d^{\frac{3}{2}} \tag{12.3}$$

where E is the Young's modulus of the samples, v_s is the Poisson's ratio of the sample, R is the radius of the microbead, and d is the displacement of the bead.

In addition to elastic deformation, viscoelastic behaviors should be considered, with a viscoelastic model using springs and dashpots. In such a model, the stress-strain relationship is as follows:

$$\sigma = E\varepsilon + \eta\frac{d\varepsilon}{t} \tag{12.4}$$

where σ is stress, ε is strain, $\frac{d\varepsilon}{t}$ is strain rate, E is the elastic modulus, and η is the viscosity.

The corresponding forces from both elastic and viscous parts are then related with bead displacement and speed as follows:

$$F_{elastic} + F_{viscous} = \frac{4}{3}ER^{\frac{1}{2}}d^{\frac{3}{2}} + \eta\frac{v}{l}\pi Rd = F_{mag} \tag{12.5}$$

When placed in a static magnetic field, magnetic beads feel the magnetic torque due to tendency to align with self-magnetic moment with the direction of external field. The magnetic torque exerting on the bead is as follows:

$$T = m \times B = |m||B|\sin(\theta) \tag{12.6}$$

where θ is the angle between m and B and the torque reaches the maximum when m and B are perpendicular, while it reaches the minimum when m and B are aligned.

The magnetic torque effect is more pronounced in magnetic nanoparticles with shape anisotropy such as nanorods, whose magnetic moments are prone to be along their long axis. When applying magnetic torque to cells, the model relating the torque and local elastic properties of the cell was developed as follows:

$$g = \frac{T}{d} \qquad (12.7)$$

where g is a complex elastic modulus, and d is the particle displacement.

Furthermore, the elastic modulus g' (the real part of g), the loss modulus g'' (the imaginary part of g), and the loss tangent n (ratio of g'/g'') can be calculated for the magnetic nanoparticles so that a geometric factor α of the cell can be derived as follows:

$$g = \alpha g' \qquad (12.8)$$

The geometric factor α is an approximation and can be calculated using finite element analysis of cell deformation to model a representative bead-cell geometry, such as a bead-cytoskeleton dynamics [50].

3 Magnetic systems for magnetic actuation

Early magnetic tweezers utilized permanent magnets with sharp tips to apply forces to magnetic micro/nanorobots and achieved force control by adjusting the displacement of magnets [51]. However, the permanent magnet system lacks accurate field and force control due to dependency on magnet position, material type, lifetime, and shape [52]. Consequently, a magnetic system composed of an electromagnetic coil on a ferromagnetic core with a sharp tip was developed so that field generation could be delicately tuned by the applied current (Fig. 12.4A) [55–57]. The field strength produced also depended on the core material of the needle, size of the solenoid (i.e., the material and the number of turns), and the distance between the tip and the magnetic particles. Typically, forces with magnitudes of up to 100 nN at ranges of 10 μm are feasible [58].

The application of single-pole magnetic tweezers was limited by its one-directional attractive forces and, since 2005, people have been developing multipole magnetic tweezers for multidirectional manipulation [59]. Although capable of navigating in multiple directions, the magnetic micro/nanorobots under multipole magnetic tweezers were controlled in an open-loop manner that, with the small inertia and fast dynamics of

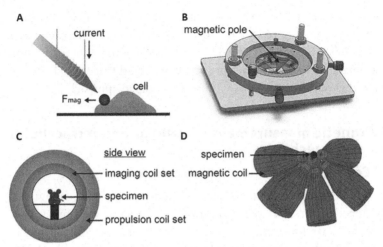

Fig. 12.4 (A) A typical single-pole magnetic tweezer pulling the microbead attached on a cell [52]. (B) A typical multipole magnetic tweezer navigating the magnetic micro/nanorobot [53]. (C) MRI-based system [54] and (D) MiniMag system [55]. *(Reproduced with permission. 2020, American Chemical Society.)*

magnetic micro/nanorobots, posed difficulties for position control. In addition, the force generation by applying current to each coil was built on empirical experiences instead of guidance from a force model. Therefore, efforts were made by Zhang et al. to develop a model for multipole magnetic tweezers based on magnetic monopole superposition and approximation. With the developed force model, the nonlinearity of the magnetic force on the particle with respect to the applied currents was characterized and stabilization of the particle was achieved by implementing a feedback control loop [60]. Wang et al. further related magnetic forces to electric currents applied to each coil and the particle position; and achieved force control of the magnetic nanorobots inside a single cell (Fig. 12.4B) [53].

In addition to magnetic tweezers, other electromagnetic systems such as an MRI-based system [54], OctoMag [61], MiniMag [55], and DeltaMag [62] have been developed to generate field gradient through coils for controlled magnetic micro/nanorobots actuation and movement (Fig. 12.4C and D). In the MRI-based system, a new magnetic resonance navigation (MRN) platform was developed by upgrading both hardware and software to achieve simultaneous deep-body image feedback and closed-loop control of the micro/nanobots [54]. Both the OctoMag and MiniMag systems were designed to endow micro/nanorobots with multiple locomotion, including rotating, pulling, and stick-slip motion; whereas, the OctoMag focused more on optimizing the manipulability of the magnetic field, and the Mini-Mag was designed to restrict the locations of the electromagnetic coils to a single hemisphere [55,61]. Finally, the DeltaMag has utilized three mobile electromagnetic coils to create larger effective workspace and the system also achieved closed-loop control with a vision-based approach [62].

4 Magnetic measurement of cellular and intracellular structure mechanics

4.1 Cell mechanics and rheological properties

Cells exhibit intrinsic viscoelastic properties and, thus, show adaptive behaviors to external mechanical stimulation based on the timescale of measurement. Therefore, the elastic and viscous behaviors can be revealed by measuring both magnitude and duration of deformation and relaxation in mechanical testing. To measure the mechanical properties of a single cell, magnetic micro/nanorobots are functionalized to attached to the cell surface [69]. Under external field actuation, the particles can apply a force or torque on the cell surface, which is concomitantly transmitted to the cellular structures through the cytoskeleton. For example, Aermes et al. used 4.5 μm superparamagnetic beads coated with fibronectin to couple to cells or collagen matrices and pulled them with a single-pole magnetic tweezer to apply physiological force up to 5 nN. The specific local linear and nonlinear viscoelastic behaviors of samples were recorded and described by a power law (Fig. 12.5A) [63].

In addition to the viscoelastic response, Bonakdar et al. identified the vital role of cell plasticity in the incomplete shape recovery of cells by using a 4 μm magnetic bead applying alternating step-force pulse of 10 nN and quantifying the irreversible deformation. They extended the viscoelastic power-law response model with a plastic element to predict the cell behavior more accurately under cyclic loading and further revealed the energy dissipation mechanism that protects the cell against mechanical damage (Fig. 12.5B) [64]. Employing magnetic measurement, long-term cell characterization during cell migration, mitosis, cancer progression, and stem cell maturation can also be established. Zhao et al. combined a single pole magnetic tweezer with a micropillar soft substrate to measure the contractile forces generated by cells and stiffness of the surrounding extracellular matrix simultaneously during tissue morphogenesis. The measurement lasted for up to 15 days and it was found that the mechanical properties of the microtissues increased quickly initially but remained stable subsequently (Fig. 12.5C–E) [65].

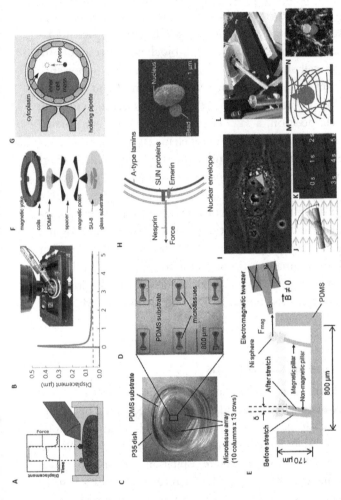

Fig. 12.5 Magnetic measurement of cell mechanics. (A) A single-pole magnetic tweezer pulling a superparamagnetic bead attached to collagen matrix [63]. (B) A single-pole magnetic tweezer measuring cell plastic deformation [64]. (C–E) Magnetic tweezer combined with a micropillar substrate for long-term cell traction force and microtissue stiffness measurement [65]. (F–G) A 3D magnetic tweezer measuring cytoplasm properties in a mouse embryo [66]. (H) Measurement of nuclear mechanical properties [44]. (I–K) Magnetic nanorods were rotated to measure nuclear interior rheological properties [67]. (L–N) NdFeB-based magnetic tweezers have applied forces up to 2000 pN to measure MT strength [68]. *(Panel A: Reproduced with permission. 2020, AIP Publishing. Panels F–G: Springer Nature. Panel B: Reproduced with permission. 2016, Springer Nature. Panels C–E: Reproduced with permission. 2014, Springer Nature. Panels I–K: Reproduced with permission. 2017, IEEE. Panel H: Reproduced with permission. 2014, Springer Nature. Panels L–N: Reproduced with permission. 2013, Elsevier.)*

Osborne et al. utilized magnetic tweezers to characterize the stiffness changes of epithelial cells during TGF-β-induced EMT (epithelial-to-mesenchymal transition) over time and revealed a functional connection between attenuated stiffness and stiffening response and increased invasion capacity after TGF-β-induced EMT [70].

Cytoplasm of living cells encompasses large macromolecular complexes, endomembranes, and an entangled cytoskeletal network [71]. The rheological properties of cytoplasm, such as viscosity and elasticity, impact fundamental cellular processes from the kinetics of biochemical reactions to vesicular transport and cell shape control [72–74]. Thus, the importance of rheological properties to cell physiology was recognized and has been studied since 1950 [75]. By actuating magnetic microrobots in cytoplasm and monitoring particle position, a viscoelastic model can be constructed using Navier-Stokes equation or a Kelvin-Voigt element [48,76]. For example, Wang et al. have actuated a 5 μm magnetic microrobot with a 3D magnetic tweezer to apply forces up to 120 pN with a resolution of 4 pN inside a mouse embryo. The cytoplasm viscosity was measured to be eight times that of water (Fig. 12.5F and G) [66]. In a recent study, Xie et al. utilized magnetic tweezers to displace and rotate mitotic spindles in living embryos and uncovered the active role of cytoplasm of imparting retractive forces that move organelles back to original positions. As the forces were found independent of cytoskeletal force generators, it was implied that bulk cytoplasm material properties are factors that regulate division positioning and cellular organization [77].

4.2 Mechanical properties of nucleus

The cell nucleus is the largest organelle that regulates the activity of the cell and is where hereditary materials area stored. Abnormal mechanical behaviors of the nucleus are known to be related to diseases like Hutchinson-Gilford progeria syndrome [78], Pelger-Huët anomaly [79], and cancers [80]. While tethered systems like AFM and micropipette aspiration can only measure isolated nuclei out of intact cells, an untethered magnetic tweezer system (single-pole and multipole) was able to measure mechanical properties of nuclei inside intact cells. Such measurement avoids altering the nucleus properties during the isolation process, where the nucleus integration with cytoskeleton and the communication with cytoplasmic environment via ions are interrupted. Specifically, Guilluy et al. attached a magnetic nanoparticle on a nuclear envelope protein, nesprin, to apply an oscillating force of 35 pN to the cell nucleus and found the existence of mechanotransduction on the nuclear envelope (Fig. 12.5H) [44]. Wang et al. performed mechanical measurement on different locations of the nuclear envelope and revealed the polarity of the nuclear mechanics in migrating cells [53]. Furthermore, Celedon et al. probed the nuclear interior with a magnetic nanorod and found significant decrease of nuclear viscosity and elasticity by lamin A/C knockout (Fig. 12.5I–K) [67].

4.3 Mechanical properties of cytoskeleton, motor proteins, and DNA strands

Magnetic microrobots and micromanipulations have also been used to measure mechanical properties of cytoskeleton, motor proteins, and DNA strands. To understand how molecular forces applied on the cytoskeleton regulate cell mitosis, Garzon-Coralin et al. used a single-pole magnetic tweezer to measure forces for a cell to maintain the position and orientation of the spindle during division. In detail, a 1 μm magnetic bead was injected into a *C. elegans* embryo cell and exerted forces up to 200 pN on mitotic spindles. The mechanism for spindle placement was identified with involvement of the astral microtubules acting as a force-generating machinery, whose stiffness was strong enough to quench thermal fluctuation, but at the same time, low enough to allow force generators to fine-tune the position of spindle to secure asymmetric cell division [76].

Yang et al. developed neodymium iron boron (NdFeB)-based magnetic tweezers to measure microtubules (MT) network strength with a force application up to 2000 pN. The MT network was identified as enthalpic viscoelastic solids dominated by filament bending instead of thermal undulations (Fig. 12.5L–N) [68]. If the magnetic force or torque is applied to balance the motion of motor proteins, such as kinesin, dynein, and rotary motor protein, the driving force or torque of an isolated motor protein can be measured. Collective behaviors of multiple kinesin motors were analyzed by Fallesen et al., and a minimal load-sharing model was revealed, as total forces can be scaled by the average number of motors [81].

Finally, mechanical measurements of DNA strands have been done for decades using magnetic micro/nanorobots. Through mechanical stretching (force) or twisting (torque) of the DNA strands, the unwinding forces, DNA binding force to adhesion molecules, molecular forces in DNA replication, and the forces on torsional constraints have been quantified [82]. In addition, Szczelkun et al. have utilized magnetic tweezers to apply torque to R-loops, a site-specific hybridization of RNA component of the DNA-targeting

CRISPR-Cas systems and an invading DNA, to investigate R-loop formation mechanism. They investigated both Cascade- and Cas9-based CRISPR-Cas systems and founded that Cascade had higher torque stability than Cas9 by using a conformational locking step [83].

5 Summary and outlook

The development of magnetic micro/nanorobots has enabled a wide range of cell mechanics measurements. Different cellular and intracellular structures have been characterized in different aspects via magnetic micro/nanorobots actuated by various magnetic systems. With controlled force or torque exertion, cell mechanics measurement, including cell mechanical and rheological properties, and mechanical properties of nucleus, cytoskeleton, motor proteins, and DNA strands can be quantified. It can be said that magnetic micro/nanorobots have become an important tool for cell mechanics measurement and will continue to be employed to discover novel cell mechanics. However, certain limitations of magnetic micro/nanorobots should be taken into consideration.

It has long been known that the cellular internalization of the microbeads is dependent on particle size, shape, surface charge, coating, and hydrophobicity. Once having undergone enzyme-mediated degradation in lysosomes after internalization, the magnetic micro/nanorobots may induce cytotoxicity by releasing reactive iron species and oxygen species [84]. Such cytotoxicity may disturb the cellular and intracellular measurement, especially those in a long-term manner. Currently, it is still unclear how the degradation of the magnetic micro/nanorobots may affect cell metabolism and what destiny these particles may have in the long run. However, in most cases surface functionalization with biocompatible materials or polymers is mandatory to reduce the cytotoxicity of these micro/nanorobots.

The research into cell mechanics is expected in the future to develop more quantitative description and more realistic cell models that explain the complex interactions that actuate cells. With such models, greater insight into mechanisms can be gained, and testable predictions can be made. Therefore, magnetic micro/nanorobot systems are expected to expand to more currently less-explored cellular and intracellular structures and achieve more accurate measurements. The next generation of micro/nanorobots should focus on actuation setups with higher position accuracy and higher force output, novel material compositions, more shapes and sizes, and adaptive functionalization for new targets.

References

[1] J.H.C. Wang, Cell traction forces (CTFs) and CTF microscopy applications in musculoskeletal research, Oper. Tech. Orthop. 20 (2010) 106–109.

[2] C.P. Heisenberg, Y. Bellaïche, Forces in tissue morphogenesis and patterning, Cell 153 (2013) 948–962.

[3] J. Petzold, E. Gentleman, Intrinsic mechanical cues and their impact on stem cells and embryogenesis, Front. Cell Dev. Biol. 9 (2021).

[4] D.H. Kim, K.W. Pak, J. Park, A. Levchenko, Y. Sun, Microengineered platforms for cell mechanobiology, Annu. Rev. Biomed. Eng. 11 (2009) 203–233.

[5] N.V. Bukoreshtliev, K. Haase, A.E. Pelling, Mechanical cues in cellular signalling and communication, Cell Tissue Res. 352 (2013) 77–94.

[6] S. Suresh, Biomechanics and biophysics of cancer cells, Acta Biomater. 3 (2007) 413–438.

[7] Y.K. Park, et al., Refractive index maps and membrane dynamics of human red blood cells parasitized by Plasmodium falciparum, Proc. Natl. Acad. Sci. U. S. A. 105 (2008) 13730–13735.

[8] G.Y.H. Lee, C.T. Lim, Biomechanics approaches to studying human diseases, Trends Biotechnol. 25 (2007) 111–118.

[9] L. Gong, et al., Micro-tweezers and force microscopy techniques for single-cell mechanobiological analysis, in: Handbook of Single-Cell Technologies, Springer, 2022, pp. 1011–1032.

[10] S.E. Cross, Y.S. Jin, J. Rao, J.K. Gimzewski, Nanomechanical analysis of cells from cancer patients, Nat. Nanotechnol. 2 (2007) 780–783.

[11] A.E. Pelling, M.A. Horton, An historical perspective on cell mechanics, Pflugers Arch. 456 (2008) 3–12.

[12] F.M. Fazal, S.M. Block, Optical tweezers study life under tension, Nat. Photon. 5 (2011) 318–321.

[13] A.L. Cost, P. Ringer, A. Chrostek-Grashoff, C. Grashoff, How to measure molecular forces in cells: a guide to evaluating genetically-encoded FRET-based tension sensors, Cell. Mol. Bioeng. 8 (2015) 96–105.

[14] R.W. Style, et al., Traction force microscopy in physics and biology, Soft Matter 10 (2014) 4047–4055.

[15] J. Friedrichs, J. Helenius, D.J. Muller, Quantifying cellular adhesion to extracellular matrix components by single-cell force spectroscopy, Nat. Protoc. 5 (2010) 1353–1361.

[16] M.C. Leake, The physics of life: one molecule at a time, Philos. Trans. R. Soc. B Biol. Sci. 368 (2013), https://doi.org/10.1098/rstb.2012.0248.

[17] C. Grashoff, et al., Measuring mechanical tension across vinculin reveals regulation of focal adhesion dynamics, Nature 466 (2010) 263–266.

[18] J.M. Leerberg, et al., Tension-sensitive actin assembly supports contractility at the epithelial zonula adherens, Curr. Biol. 24 (2014) 1689–1699.

[19] D. Cai, et al., Mechanical feedback through E-cadherin promotes direction sensing during collective cell migration, Cell 157 (2014) 1146–1159.

[20] D.E. Conway, et al., Fluid shear stress on endothelial cells modulates mechanical tension across VE-cadherin and PECAM-1, Curr. Biol. 23 (2013) 1024–1030.

[21] M. Krieg, A.R. Dunn, M.B. Goodman, Mechanical control of the sense of touch by β-spectrin, Nat. Cell Biol. 16 (2014) 224–233, https://doi.org/10.1038/ncb2915.

[22] T. Lecuit, P.F. Lenne, E. Munro, Force generation, transmission, and integration during cell and tissue morphogenesis, Annu. Rev. Cell Dev. Biol. 27 (2011) 157–184.

[23] B.D. Hoffman, C. Grashoff, M.A. Schwartz, Dynamic molecular processes mediate cellular mechanotransduction, Nature 475 (2011) 316–323.

[24] S.V. Plotnikov, B. Sabass, U.S. Schwarz, C.M. Waterman, High-resolution traction force microscopy, Methods Cell Biol. 123 (2014) 367–394.

[25] Y. Aratyn-Schaus, M.L. Gardel, Transient frictional slip between integrin and the ECM in focal adhesions under myosin II tension, Curr. Biol. 20 (2010) 1145–1153.

[26] Y. Aratyn-Schaus, P.W. Oakes, J. Stricker, S.P. Winter, M.L. Gardel, Preparation of complaint matrices for quantifying cellular contraction, J. Vis. Exp. 3–9 (2010), https://doi.org/10.3791/2173.

[27] M.T. Frey, A. Engler, D.E. Discher, J. Lee, Y.L. Wang, Microscopic methods for measuring the elasticity of gel substrates for cell culture: microspheres, microindenters, and atomic force microscopy, Methods Cell Biol. 83 (2007) 47–65.

[28] B. Sabass, M.L. Gardel, C.M. Waterman, U.S. Schwarz, High resolution traction force microscopy based on experimental and computational advances, Biophys. J. 94 (2008) 207–220.

[29] J. Stricker, B. Sabass, U.S. Schwarz, M.L. Gardel, Optimization of traction force microscopy for micron-sized focal adhesions, J. Phys. Condens. Matter 22 (2010), https://doi.org/10.1088/0953-8984/22/19/194104.

[30] J. Toyjanova, et al., 3D viscoelastic traction force microscopy, Soft Matter 10 (2014) 8095–8106.

[31] Y.F. Dufrêne, A.E. Pelling, Force nanoscopy of cell mechanics and cell adhesion, Nanoscale 5 (2013) 4094–4104.

[32] A.V. Taubenberger, D.W. Hutmacher, D.J. Muller, Single-cell force spectroscopy, an emerging tool to quantify cell adhesion to biomaterials, Tissue Eng. Part B Rev. 20 (2014) 40–55.

[33] J.J. Heinisch, et al., Atomic force microscopy—looking at mechanosensors on the cell surface, J. Cell Sci. 125 (2012) 4189–4195.

[34] J. Friedrichs, et al., A practical guide to quantify cell adhesion using single-cell force spectroscopy, Methods 60 (2013) 169–178.

[35] F. Li, S.D. Redick, H.P. Erickson, V.T. Moy, Force measurements of the $\alpha5\beta1$ integrin-fibronectin interaction, Biophys. J. 84 (2003) 1252–1262.

[36] K.D. Webster, A. Crow, D.A. Fletcher, An AFM-based stiffness clamp for dynamic control of rigidity, PLoS One 6 (2011), https://doi.org/10.1371/journal.pone.0017807.

[37] A. Ashkin, et al., Observation of a single-beam gradient force optical trap for dielectric particles, Opt. Lett. 11 (5) (1986) 288–290.

[38] J. Dai, M.P. Sheetz, Membrane tether formation from blebbing cells, Biophys. J. 77 (1999) 3363–3370.

[39] W.H. Goldmann, Mechanosensation: a basic cellular process, Prog. Mol. Biol. Transl. Sci. 126 (2014) 75–102.

[40] P. Kollmannsberger, C.T. Mierke, B. Fabry, Nonlinear viscoelasticity of adherent cells is controlled by cytoskeletal tension, Soft Matter 7 (2011) 3127–3132.

[41] G.F. Weber, M.A. Bjerke, D.W. DeSimone, A mechanoresponsive cadherin-keratin complex directs polarized protrusive behavior and collective cell migration, Dev. Cell 22 (2012) 104–115.

[42] S. Sugita, T. Adachi, Y. Ueki, M. Sato, A novel method for measuring tension generated in stress fibers by applying external forces, Biophys. J. 101 (2011) 53–60.

[43] N. Batra, et al., Mechanical stress-activated integrin $\alpha5\beta1$ induces opening of connexin 43 hemichannels, Proc. Natl. Acad. Sci. U. S. A. 109 (2012) 3359–3364, https://doi.org/10.1073/pnas.1115967109.

[44] C. Guilluy, et al., Isolated nuclei adapt to force and reveal a mechanotransduction pathway in the nucleus, Nat. Cell Biol. 16 (2014) 376–381.

[45] X. Wang, et al., Microrobotic swarms for intracellular measurement with enhanced signal-to-noise ratio, ACS Nano 16 (7) (2022) 10824–10839.

[46] D.J. Dunlop, Superparamagnetic and single-domain threshold sizes in magnetite, J. Geophys. Res. 78 (1973) 1780–1793.

[47] P. Fischer, A. Ghosh, Magnetically actuated propulsion at low Reynolds numbers: towards nanoscale control, Nanoscale 3 (2011) 557–563, https://doi.org/10.1039/c0nr00566e.

[48] J.F. Berret, Local viscoelasticity of living cells measured by rotational magnetic spectroscopy, Nat. Commun. 7 (2016) 1–9.

[49] D.B. Cochran, et al., Suppressing iron oxide nanoparticle toxicity by vascular targeted antioxidant polymer nanoparticles, Biomaterials 34 (2013) 9615–9622.

[50] B. Fabry, et al., Scaling the microrheology of living cells, Phys. Rev. Lett. 87 (2001) 1–4.

[51] R. Sarkar, V.V. Rybenkov, A guide to magnetic tweezers and their applications, Front. Physiol. 4 (2016) 48.

[52] X. Wang, et al., Magnetic measurement and stimulation of cellular and intracellular structures, ACS Nano 14 (2020) 3805–3821.

[53] X. Wang, et al., Intracellular manipulation and measurement with multipole magnetic tweezers, Sci. Robot. 4 (28) (2019), eaav6180.

[54] M. Vonthron, V. Lalande, G. Bringout, C. Tremblay, S. Martel, A MRI-based integrated platform for the navigation of microdevices and microrobots, in: IEEE International Conference on Intelligent Robots and Systems, 2011, pp. 1285–1290, https://doi.org/10.1109/IROS.2011.6048274.

[55] S. Schuerle, S. Erni, M. Flink, B.E. Kratochvil, B.J. Nelson, Three-dimensional magnetic manipulation of micro- and nanostructures for applications in life sciences, IEEE Trans. Magn. 49 (2013) 321–330.

[56] J.K. Fisher, et al., Three-dimensional force microscope: a nanometric optical tracking and magnetic manipulation system for the biomedical sciences, Rev. Sci. Instrum. 76 (5) (2005), 053711.

[57] J.K. Fisher, et al., Thin-foil magnetic force system for high-numerical-aperture microscopy, Rev. Sci. Instrum. 77 (2) (2006), 023702.

[58] P. Kollmannsberger, B. Fabry, High-force magnetic tweezers with force feedback for biological applications, Rev. Sci. Instrum. 78 (11) (2007), 114301.

[59] A.H.B. de Vries, B.E. Krenn, R. van Driel, J.S. Kanger, Micro magnetic tweezers for nanomanipulation inside live cells, Biophys. J. 88 (2005) 2137–2144.

[60] Z. Zhang, K. Huang, C.H. Menq, Design, implementation, and force modeling of quadrupole magnetic tweezers, IEEE/ASME Trans. Mechatron. 15 (2010) 704–713.

[61] M.P. Kummer, et al., Octomag: an electromagnetic system for 5-DOF wireless micromanipulation, IEEE Trans. Robot. 26 (2010) 1006–1017.

[62] L. Yang, X. Du, E. Yu, D. Jin, L. Zhang, DeltaMag: an electromagnetic manipulation system with parallel mobile coils, in: Proceedings of the IEEE International Conference on Robotics and Automation 2019-May, 2019, pp. 9814–9820.

[63] C. Aermes, A. Hayn, T. Fischer, C.T. Mierke, Environmentally controlled magnetic nano-tweezer for living cells and extracellular matrices, Sci. Rep. 10 (2020) 1–16.

[64] N. Bonakdar, et al., Mechanical plasticity of cells, Nat. Mater. 15 (2016) 1090–1094.

[65] R. Zhao, T. Boudou, W.G. Wang, C.S. Chen, D.H. Reich, Magnetic approaches to study collective three-dimensional cell mechanics in long-term cultures (invited), J. Appl. Phys. 115 (2014), https://doi.org/10.1063/1.4870918.

[66] A.T.M. Tweezer, et al., Short papers navigation and measurement, IEEE Trans. Robot. 34 (2018) 240–247.

[67] A. Celedon, C.M. Hale, D. Wirtz, Magnetic manipulation of nanorods in the nucleus of living cells, Biophys. J. 101 (2011) 1880–1886.

[68] Y. Yang, M.T. Valentine, Determining the structure–mechanics relationships of dense microtubule networks with confocal microscopy and magnetic tweezers-based microrheology, Methods Cell Biol. 115 (2013) 75–96.

[69] R.J. Marjoram, C. Guilluy, K. Burridge, Using magnets and magnetic beads to dissect signaling pathways activated by mechanical tension applied to cells, Methods 94 (2016) 19–26.

[70] L.D. Osborne, et al., TGF-β regulates LARG and GEF-H1 during EMT to affect stiffening response to force and cell invasion, Mol. Biol. Cell 25 (2014) 3528–3540, https://doi.org/10.1091/mbc.E14-05-1015.

[71] K. Luby-Phelps, D.L. Taylor, F. Lanni, Probing the structure of cytoplasm, J. Cell Biol. 102 (1986) 2015–2022.

[72] E. Moeendarbary, et al., The cytoplasm of living cells behaves as a poroelastic material, Nat. Mater. 12 (2013) 253–261.

[73] K. Luby-Phelps, Cytoarchitecture and physical properties of cytoplasm: volume, viscosity, diffusion, intracellular surface area, Int. Rev. Cytol. 192 (1999) 189–221.

[74] H. Palenzuela, et al., In vitro reconstitution of dynein force exertion in a bulk viscous medium, Curr. Biol. 30 (2020) 4534–4540.e7.

[75] F.H.C. Crick, A.F.W. Hughes, The physical properties of cytoplasm: a study by means of the magnetic particle method part I. Experimental, Exp. Cell Res. 1 (1950) 37–80.

[76] C. Garzon-Coral, H.A. Fantana, J. Howard, A force-generating machinery maintains the spindle at the cell center during mitosis, Science 352 (2016) 1124–1127.

[77] J. Xie, et al., Contribution of cytoplasm viscoelastic properties to mitotic spindle positioning, Proc. Natl. Acad. Sci. U. S. A. 119 (8) (2022), e2115593119.

[78] C.M. Hale, et al., Dysfunctional connections between the nucleus and the actin and microtubule networks in laminopathic models, Biophys. J. 95 (2008) 5462–5475.

[79] K. Hoffmann, et al., Mutations in the gene encoding the Lamin B receptor produce an altered nuclear morphology in granulocytes (Pelger-Huët anomaly), Nat. Genet. 31 (2002) 410–414.

[80] D. Zink, A.H. Fischer, J.A. Nickerson, Nuclear structure in cancer cells, Nat. Rev. Cancer 4 (2004) 677–687.

[81] T.L. Fallesen, J.C. MacOsko, G. Holzwarth, Force-velocity relationship for multiple kinesin motors pulling a magnetic bead, Eur. Biophys. J. 40 (2011) 1071–1079.

[82] Q. Xin, et al., Magnetic tweezers for the mechanical research of DNA at the single molecule level, Anal. Methods 9 (2017) 5720–5730.

[83] M.D. Szczelkun, et al., Direct observation of R-loop formation by single RNA-guided Cas9 and cascade effector complexes, Proc. Natl. Acad. Sci. U. S. A. 111 (2014) 9798–9803.

[84] Q. Feng, et al., Uptake, distribution, clearance, and toxicity of iron oxide nanoparticles with different sizes and coatings, Sci. Rep. 8 (2018) 1–13.

CHAPTER 13

Nanorobotics for investigating cell mechanics based on atomic force microscopy

Mi Li[a,b,c], Yaqi Feng[a,b,c], and Lianqing Liu[a,b,c]
[a]State Key Laboratory of Robotics, Shenyang Institute of Automation, Chinese Academy of Sciences, Shenyang, China
[b]Institutes for Robotics and Intelligent Manufacturing, Chinese Academy of Sciences, Shenyang, China
[c]University of Chinese Academy of Sciences, Beijing, China

1 Background

Probing the mechanical properties of single cells is of great significance for uncovering the underlying mechanisms guiding life activities and disease development. The cell is the fundamental structural and functional unit of life, and the behaviors and activities of cells are closely related to physiological/pathological changes in living organisms. Cells are both mechanical and biochemical systems [1], and mechanical cues play a crucial role in regulating virtually every process of cells, including migration, proliferation, differentiation, and apoptosis [2].

All cells exist within the context of a three-dimensional microenvironment (also called a niche) in which they are exposed to mechanical and physical cues [3], and a major component of the niche is the extracellular matrix (ECM), which provides the essential physical scaffolding for cells [4]. Cells sense these mechanical signals and transduce them into changes in intracellular biochemistry and gene expression in a process known as mechanotransduction [5]. The interactions between cells and ECM are reciprocal [6]. On the one hand, cells are constantly creating, breaking down, or rearranging ECM components to change the compositions and structures as well as properties of ECM. On the other hand, any changes in the ECM will, in turn, influence the behaviors of cells.

The interactions between cells and microenvironments during physiological or pathological processes often result in the formation of unique cell mechanics. In recent decades, numerous experimental results obtained on individual cells have shown that cell mechanics is an effective biomarker for indicating cell states [7] particularly for the process of disease

Robotics for Cell Manipulation and Characterization
https://doi.org/10.1016/B978-0-323-95213-2.00011-9
289

development. For example, studies have shown that cancerous cells are significantly softer and more deformable than normal cells [8,9] and invasive cancerous cells are even softer than benign cancerous cells [10,11]. For malaria, studies have shown that the progression during the parasite development stages (ring stage, trophozoite stage, schizont stage) leads to a considerable stiffening and decreased deformability of the host red blood cells compared with healthy red blood cells [12].

It is now increasingly apparent in the field of biomedicine that combining nanomechanical information with conventional biochemical assays (e.g., morphology analysis, molecular expression analysis) may provide an entirely new platform for cancer detection and monitoring [13]. In addition, studies have shown that targeting mechanical cues, by preventing or reversing the abnormal changes in tissue/cell mechanics, provides novel therapeutic avenues for cancer treatments with translational significance [14]. Hence, developing nanorobotics for single-cell manipulation and mechanical analysis will not only benefit understanding the role of mechanical forces in life processes, but also may promote the clinical diagnosis and treatment of diseases.

2 Atomic force microscopy (AFM)-based nanomanipulator

Nanorobotics based on atomic force microscopy (AFM) offers novel possibilities for probing the mechanical properties of single cells from the perspective of robotics. Nanorobotics involves developing robotic devices to perform tasks (e.g., actuation, sensing, manipulation, propulsion, signaling, information processing, intelligence, and swarm behaviors) at the nanoscale [15]. Nanorobotic systems may themselves be miniature in size (often called nanorobot), or they may be designed specifically to interact with nano-sized structures (often called nanomanipulator) [16]. Nanorobots (the overall size of which is in the nanometer scale) and nanomanipulators (of which the overall size is in the macroscale but their end effectors can perform robotic manipulations on nano-sized objects) represent two categories of nanorobotic techniques, and various nanorobots and nanomanipulators have been developed in recent years for biomedical applications [17].

So far, nano-sized robots have proved particularly suited for in vivo applications as therapeutic carriers to deliver drug molecules to specific sites within the body [18,19], and there remain severe technical challenges before the realization of untethered nanorobots applicable for medical use (e.g., soft functional materials, robot locomotion and tracking, medical functions, learning and control [20]). Nanomanipulators are suited for detecting the structures and properties of biological samples in vitro, which is meaningful for revealing the underpinnings of life activities and benefits precise diagnosis of diseases. Integrating robotics with AFM yields the AFM nanomanipulator. The invention of AFM provides an exciting multifunctional tool for single-cell mechanical studies. The unique merit of AFM over other single-cell mechanical detection techniques (e.g., optical tweezers, magnetic tweezers, micropipette aspiration, microfluidics, parallel plate, particle tracking micro-rheology (PTM), and traction force microscopy (TFM) [21,22]) is that AFM is able to simultaneously acquire the structures and mechanical properties of individual living cells with unprecedented spatiotemporal resolution (nano-meter spatial resolution and millisecond temporal resolution) in aqueous conditions [23], facilitating correlating the mechanical properties of cells with cellular structures as well as cellular functions. Therefore, the AFM nanomanipulator is particularly suited for biomedical applications.

The typical configuration of an AFM-based human-in-the-loop nanoma-nipulator system [24] is shown in Fig. 13.1A, including an AFM (the AFM probe acts as the end effector), a joystick (used for haptic feedback), the aug-mented reality user interface (used for visual feedback), and the operator (to generate control commands). The operator controls the AFM probe as the end effector to perform robotic tasks on nano-sized objects via the joystick. AFM nanomanipulator has been widely used to characterize the structures and mechanical properties of single living cells in recent years, and this has yielded novel insights into the role of mechanical cues in tumor development and clinical treatment [25–27]. The prerequisite of utilizing an AFM nanoma-nipulator to detect single cell is immobilizing cells onto a hard support, for which the immobilization methods are diverse, depending on the cell types.

Adherent mammalian cells could naturally attach to and spread on the substrate, and thus, AFM can be used to directly detect this type of cell after they grow and spread on the substrate with no extra immobilization methods required. For microbial cells, which are small and have stiff cell walls, either poly-L-lysine electrostatic adsorption or porous polymer mem-brane mechanical trapping [28] works for the studies of AFM. For suspended mammalian cells, which cannot naturally attach to the substrate and have large size and soft cell surface, immobilization methods are required for AFM manipulations. Our studies have shown that living suspended mam-malian cells could be effectively immobilized by microfabricated silicon pil-lar array coated with poly-L-lysine layer for AFM topographical imaging and mechanical detection (I–III in Fig. 13.1B) [29,30]. Furthermore, for utiliz-ing AFM nanomanipulator to measure the mechanical properties of cells,

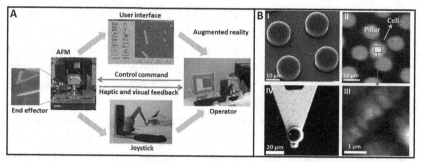

Fig. 13.1 AFM nanomanipulator for single-cell analysis. (A) Configuration of an AFM-based robotic nanomanipulation system. (B) Detecting individual cells by AFM nanorobotics. (I) SEM image of microfabricated silicon pillar array. (II) Topographical image of a living cell trapped in the pillar array obtained by AFM nanomanipulator. (III) Surface nanostructures of living cells imaged by AFM nanomanipulator. (IV) SEM image of an AFM probe with a spherical tip. *Panel A: Reprinted with permission from M. Li, L. Liu, N. Xi, Y. Wang, Biological applications of a nanomanipulator based on AFM: in situ visualization and quantification of cellular behaviors at the single-molecule level. IEEE Nanotechnol. Mag. 9 (3) (2015) 25–35. Copyright 2015 IEEE.; panel B: (I, II) are reprinted with permission from M. Li, L. Liu, N. Xi, Y. Wang, Z. Dong, O. Tabata, X. Xiao, W. Zhang, Imaging and measuring the rituximab-induced changes of mechanical properties in B-lymphoma cells using atomic force microscopy. Biochem. Biophys. Res. Commun. 404 (2) (2011) 689–694. Copyright 2010 Elsevier Inc.; (III) is reprinted with permission from M. Li, L. Liu, N. Xi, Y. Wang, Z. Dong, X. Xiao, W. Zhang, Drug-induced changes of topography and elasticity in living B lymphoma cells based on atomic force microscopy. Acta Phys. Chim. Sin. 28 (6) (2012) 1502–1508. Copyright 2012 Editorial Office of Acta Physico-Chimica Sinica; (IV) is reprinted with permission from M. Li, L. Liu, N. Xi, Y. Wang, Atomic force microscopy studies on cellular elastic and viscoelastic properties. Sci. China Life Sci. 61(1) (2018) 57–67. Copyright 2017 Science China Press and Springer-Verlag GmbH Germany.*

often an AFM probe with spherical tip is used (IV in Fig. 13.1B) [31]. The spherical tip has well-defined geometry, which facilitates theoretical modeling. A spherical tip of large size also benefits characterizing the mechanics of the whole cell.

3 Methodology of characterizing cellular mechanics by AFM nanomanipulator

AFM is able to simultaneously measure the elastic (Young's modulus) and viscoelastic properties (relaxation time) of single cells by vertically performing approach–dwell–retract cycles on cells [32]. A commercial AFM is often integrated with an optical microscope (Fig. 13.2A), which makes it easy to move AFM probe to the target cell under the guidance of optical

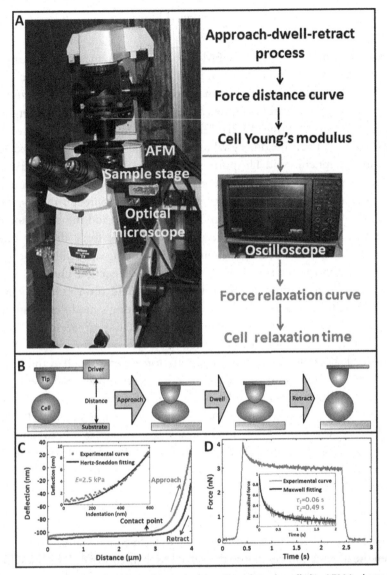

Fig. 13.2 Measuring multiple mechanical properties of single cells by AFM indentation assay. (A) Actual photograph of an AFM, which is set on an inverted optical microscope. An oscilloscope is linked to the AFM to record the relaxation curve. (B) Schematic of controlling AFM tip to vertically perform an approach-dwell-retract cycle on cells. (C) A typical force curve obtained during the approach-dwell-retract process. The indentation curve is fitted by a Hertz-Sneddon model to extract cellular Young's modulus (*inset*). (D) A typical relaxation curve obtained during the approach-dwell-retract process. The curve is fitted by a two-order Maxwell model to extract cellular relaxation times (*inset*). *Panels A, C, D are reprinted with permission from M. Li, L. Liu, X. Xiao, N. Xi, Y. Wang, Viscoelastic properties measurement of human lymphocytes by atomic force microscopy based on magnetic beads cell isolation. IEEE Trans. Nanobiosci. 15 (5) (2016) 398–411. Copyright 2016 IEEE; panel B is reprinted with permission from M. Li, N. Xi, Y. Wang, L. Liu, Atomic force microscopy in probing tumor physics for nanomedicine. IEEE Trans. Nanotechnol. 18 (2019) 83–113. Copyright 2018 IEEE.*

microscopy. After moving the AFM probe to the cell, mechanical measurements can be performed on the cell. In an approach-dwell-retract cycle, the AFM probe, which is far away from the cell, is first driven by the piezoelectric tube to approach and compress the cell until the maximum preset loading force is achieved (Fig. 13.2B) [22]. The AFM probe then dwells on the cell for a period of time, during which the vertical position of the piezoelectric tube is kept constant. The purpose of the dwelling stage is to observe the relaxation process of the cell. Subsequently, the AFM probe retracts from the cell surface to the original position. During the approach-dwell-retract cycle, the vertical displacements of the AFM probe and the deflections of the probe cantilever are obtained, which yield the force distance curve (often called force curve, for simplicity). By linking an oscilloscope (Fig. 13.2A) to the output signal of the AFM, the deflection changes of the AFM cantilever vs time are recorded, which yield the relaxation curve. The force curve is used to calculate cellular Young's modulus, while the relaxation curve is used to calculate cellular relaxation time.

Cellular Young's modulus is obtained by analyzing the force curves. There are several theoretical models for analyzing force curves, such as the Hertz-Sneddon model, Johnson-Kendall-Roberts (JKR) model, and Derjaguin-Muller-Toporov (DMT) model. The Hertz-Sneddon model does not consider the forces between the AFM-specimen contact surfaces, the JKR model considers the adhesive forces inside the contact area between AFM tip and specimen, and the DMT model considers the adhesive forces outside the contact area [33]. For practical reasons, the Hertz-Sneddon model is the most widely used [34]. Notably, the Hertz-Sneddon model is based on several assumptions of the specimen being indented, including homogeneity, isotropy, and infinite thickness. It is clear that cell does not meet these assumptions, but studies have shown that the Hertz-Sneddon model works well when the indentation depth into the cell is less than 10% of cell thickness [35].

Each force curve is composed of two parts, an approach curve and a retract curve (Fig. 13.2C). Cellular Young's modulus can be obtained by analyzing the approach curve with Hertz-Sneddon model, while the retract curve is often used for analyzing the cellular adhesive forces. After converting the approach curve into the indentation curve according to the contact point (the indentation depth is equal to the difference between cantilever deflection and probe vertical displacement), the Hertz-Sneddon model (Hertz model is suited for spherical tip and Sneddon model is suited for conical tip) is used to fit the indentation curve [26]:

$$F_{\text{spherical}} = \frac{4E\delta^{1.5}\sqrt{R}}{3(1 - v^2)} \tag{13.1}$$

$$F_{\text{conical}} = \frac{2E\delta^2 \tan\theta}{\pi(1 - v^2)} \tag{13.2}$$

where v is the Poisson ratio of the cell (cells are often considered as incompressible materials and thus $v = 0.5$), F is the loading force of AFM probe, δ is the indentation, E is the cellular Young's modulus, θ is the half-opening angle of the conical tip, and R is the radius of spherical tip.

The loading force F is obtained according to the Hooke's law:

$$F = kx \tag{13.3}$$

where k is the spring constant of the cantilever and x is the deflection of the cantilever.

Fitting the indentation curve with the Hertz-Sneddon model gives cellular Young's modulus (Fig. 13.2C). Particularly, cellular Young's modulus measured by AFM is dependent on the experimental conditions, including tip shape (spherical tip or conical tip), tip size, probe approach rate, environmental temperature, cell growth medium, cell state, cell positions being probed (central area, or peripheral area), and so on [22], and thus, experimental conditions should be maintained identical to make the results comparable.

Cellular relaxation time is obtained by analyzing the relaxation curves. Fitting the relaxation curve with Maxwell spring-dashpot model gives cellular relaxation time [26]:

$$F(t) = A_0 + A_i \sum_{i=1}^{n} e^{-t/\tau_i} \tag{13.4}$$

$$\tau_i = \frac{\eta_i}{E_i} \tag{13.5}$$

where F is the loading force of the AFM probe, A_0 is the instantaneous (purely elastic) response, A_i is the ith force amplitudes, τ_i is the ith cellular relaxation time, η_i is the ith cellular viscosity, and E_i is the ith cellular Young's modulus.

The cellular relaxation time (τ) is the ratio of cellular viscosity (η) to cellular Young's modulus (E), and thus, cellular relaxation time reflects the viscoelastic properties of cells. Studies have shown that a one-order Maxwell model often does not fit the relaxation curve well, while a two-order

Maxwell model matches the relaxation curve well [36]. Intracellular structures are highly heterogeneous, which probably causes the one-order Maxwell model to be unable to characterize the relaxation process of the cell well. Fitting the relaxation curve with a two-order Maxwell model gives two cellular relaxation times (τ_1 and τ_2) (Fig. 13.2D). The first-order cellular relaxation time τ_1 (the fast relaxation time) and the second-order cellular relaxation time τ_2 (the slow relaxation time) may reflect the rheological properties of different intracellular structures, respectively. Notably, cellular relaxation times measured by AFM are dependent on the measurement parameters, such as surface dwell time of AFM probe on cell and the approach rate of AFM probe to the cell [36], and thus, measurement parameters should be identical to make the obtained results comparable.

4 Applications of AFM nanomanipulator in detecting cell mechanics

AFM has been widely used to detect the mechanical properties of cancerous cells. In 1999, Lekka et al. [37] first used AFM to explore the mechanical differences between cancerous cells and normal cells, and the results significantly showed that that the Young's modulus of normal bladder cells is about 10 times larger than that of cancerous bladder cells. In 2008, Li et al. [38] used AFM to analyze the differences in cell mechanics between cancerous breast cells and normal breast cells, showing that the Young's modulus of nonmalignant breast cells (MCF-10A) is significantly larger (1.4–1.8 times) than that of their malignant counterparts (MCF-7). These studies were performed on adherent mammalian cancerous cell lines.

In 2012, we have used AFM to investigate the mechanical properties of suspended mammalian cancerous cell lines [11], as shown in Fig. 13.3A. Three types of hemal tumor cells with different aggressive capabilities were used, including Burkitt's lymphoma Raji cell, cutaneous lymphoma Hut cell, and chronic myeloid leukemia K562 cell. The experimental results showed that aggressive cancerous cells (Raji) were significantly softer than indolent cancerous cells (K562 and Hut), indicating that softening plays an important role in cancer progression. Interestingly, all three types of cells (Raji, Hut, K562) had a much smaller Young's modulus than that of red blood cells (RBCs). RBCs do not have cell nucleus and organelles, which produces the very soft and highly deformable features of RBCs. During tumor metastasis, primary cancerous cells need to physically penetrate and move through various obstacles, including tumor basement membrane,

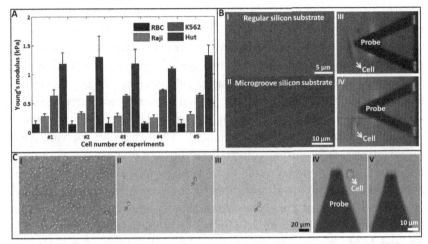

Fig. 13.3 Applications of AFM in detecting cell mechanics. (A) Histogram of Young's modulus of RBCs and three types of lymphoma cells with different aggressive capabilities measured by AFM. (B) Hierarchical micro/nanotopography-regulated cell mechanics detected by AFM. (I) Regular silicon substrate. (II) Microgroove silicon substrate. (III, IV) Optical images of controlling AFM probe to detect the cells grown on regular silicon substrates (III) or on microgroove silicon substrates (IV). (C) Detecting the mechanics of primary B lymphocytes by AFM. (I, II, III) Optical images of PBMCs (I) and isolated B lymphocytes (II, III). (IV, V) Under the guidance of optical microscopy, AFM tip was moved to the isolated individual B lymphocyte to perform mechanical measurements. (IV) Before and (V) after moving the tip to the cell. *Panel A: Reprinted with permission from M. Li, L. Liu, N. Xi, Y. Wang, Z. Dong, X. Xiao, W. Zhang, Atomic force microscopy imaging and mechanical properties measurement of red blood cells and aggressive cancer cells. Sci. China Life Sci. 55 (11) (2012) 968–973. Copyright 2012 The Author(s); panel B: Reprinted with permission from M. Li, N. Xi, L. Liu, Hierarchical micro-/nanotopography for tuning structures and mechanics of cells probed by atomic force microscopy. IEEE Trans. Nanobiosci. 20 (4) (2021) 543–553. Copyright 2021 IEEE; panel C: Reprinted with permission from M. Li, L. Liu, X. Xiao, N. Xi, Y. Wang, Viscoelastic properties measurement of human lymphocytes by atomic force microscopy based on magnetic beads cell isolation. IEEE Trans. Nanobiosci. 15(5) (2016) 398–411. Copyright 2016 IEEE.*

extracellular matrix, and blood vessel endothelial cells [39]. Softness and deformability may, therefore, facilitate the movement of cancerous cells in the process of metastasis.

AFM topographical imaging and mechanical analysis benefit understanding the mechanical cues in regulating cell-substrate interactions. Cells respond to the nanoscale surface features and the interaction of nanotopographical features with integrin receptors in the cells' focal adhesions alters how the cells adhere to surfaces and define cell fate through changes in both cell biochemistry and cell morphology [40]. Therefore, precise nanotopographical control of cellular behaviors will probably allow better understanding of signaling and cellular functions and inspire novel strategies to regulate cell activities [41].

Current studies of nanotopography-cell interactions are commonly focused on fabricating different types of nanotopographical surfaces, which are then used to grow cells [42,43], whereas, the effects of cell growth medium on nanotopography are neglected. Our studies of applying AFM to investigate cell-substrate interactions have shown that cell growth medium itself can significantly form nanogranular surfaces during cell growth [44,45], which was previously undiscovered, and suggests that cell growth medium is not a passive bystander but has important effects in establishing nanotopographical surfaces for cell growth. Based on this discovery, we used AFM to investigate the regulatory role of hierarchical micro/nanotopography in tuning cell structures and mechanics [46], as shown in Fig. 13.3B. With the use of photolithography technology, microgroove silicon substrates were fabricated. After incubating the microgroove substrates in cell growth medium, nanogranular surfaces were formed on the ridges of microgrooves, and thus, hierarchical micro/nanotopography was obtained. AFM was then used to image the morphology and measure the mechanical properties of cells grown on the hierarchical micro/nanotopography substrate or regular substrate (control group), respectively, showing that the hierarchical micro/nanotopography induced significant changes in both cell morphology (cells became longer and narrower) and mechanics (cells became softer).

We have also explored utilizing AFM manipulators to investigate the mechanical properties of primary cells. So far, studies of cellular mechanical assays by AFM have commonly been performed on cell lines grown in vitro. Cell lines grown in vitro are quite different from the in vivo cells in the human body and extensive evidence has shown that results obtained on cell lines cannot faithfully reflect the behaviors of the same cells in vivo [47]. Although a few studies have used AFM to measure the mechanical properties of primary cancerous cells and tissues [9,10], information on the mechanics of primary cells is still limited and, particularly, studies of utilizing AFM to detect the mechanical properties of primary blood cells are scarce.

We have established the methodology of applying AFM to characterize the mechanical properties of single primary B lymphocytes from the peripheral blood of lymphoma patients [32], as shown in Fig. 13.3C. Firstly, the peripheral blood mononuclear cells (PBMCs) (I in Fig. 13.3C) were isolated from the anticoagulated blood by density gradient centrifugation. Subsequently, B lymphocytes were extracted from PBMCs by CD19 antibody magnetic beads cell isolation. CD19 is a cell surface molecule that is exclusively expressed on B lymphocytes but not on other cells, and thus, magnetic beads coated with CD19 antibodies could specifically bind to B lymphocytes. After incubating PBMCs with CD19 antibody magnetic beads, PBMCs were controlled to pass through a magnetic field, and B lymphocytes were captured in the magnetic field while other cells flowed out of the magnetic field. After isolating the B lymphocytes (II, III in Fig. 13.3C) from PBMCs, fluorescence experiments were performed to examine the viability of the cells and also to confirm whether the isolated cells were really B lymphocytes. Then, the mechanical properties of single primary B lymphocytes could be measured by AFM under the guidance of optical microscopy (IV, V in Fig. 13.3C).

5 Combining AFM nanorobotics with micropipette for precise drug-induced cellular mechanical analysis

Combining AFM with micropipette allows single-cell precise drug delivery and simultaneous cell mechanics measurement. AFM has become a standard tool for characterizing the mechanics of single cells. Nevertheless, due to the fact that it is difficult for the regular AFM tip to deliver drug molecules to cells, so far, when utilizing AFM to probe the mechanical changes of cells in response to drug molecules, drug molecules have been added into the dish in which cells are grown to treat all of the cells in the dish. The disadvantage of this measurement strategy is that it is difficult to monitor the real-time mechanical changes of single cells after drug stimulation. By linking biomolecules to the surface of a specific AFM tip (such as a nanoneedle tip [48]), the biomolecules can be delivered into single cells after inserting the AFM tip into the cell. However, this method requires linking drug molecules onto the tip surface via chemical modification and is strongly dependent on the reversible chemical reactions, causing the weak controllability and limited applications of this method. Furthermore, chemical modification will inevitably reduce the fidelity of the drug molecules, and drug molecules are not native, which may further affect the interactions between drug molecules and cells. By using a microchanneled probe, drug molecules can be precisely delivered to single cells [49], but the microchanneled architecture of the probe will inevitably influence the force sensitivity of the probe.

As an alternative strategy, we have recently combined micropipette with AFM to achieve single-cell drug delivery and simultaneous cell mechanics measurement [50], as shown in Fig. 13.4. The micropipette-based single-cell microinjection system (Fig. 13.4C) was built on an inverted fluorescent microscope, using a 3D manipulator, a micropump, a syringe, a PTFE tube, and a micropipette. The micropipette was prepared from a glass capillary by using a micropipette puller. The effects of micropipette tip size on injection were examined and the experimental results showed that injection process could cause mechanical damage to the cell if the outer diameter of the micropipette tip is larger than 1 μm. The effectiveness of the established method was verified by using propidium iodide (PI) staining solution to stimulate individual cells and then measure cell mechanics, as shown in Fig. 13.5. With the use of micropipette injection, PI solution was successfully delivered to a single cell and the cell exhibited red fluorescence (Fig. 13.5A–C). Under the guidance of fluorescence (Fig. 13.5D), the AFM probe was then moved to the cell to obtain force curves (Fig. 13.5E) to measure cellular Young's modulus. With the established method, we further measured the dynamic changes of Young's modulus of single cells after the treatment of chemotherapy drug molecules, showing the potential of combining micropipette with AFM for single-cell mechanical analysis in response to ultra-trace drugs.

6 Summary

Investigating life activities at the single-cell level is of remarkable significance for uncovering the underlying mechanisms guiding life activities and diseases. The development of the AFM nanomanipulator makes it possible to perform robotic manipulations on single cells with nanometer spatial resolution, and the achievements of AFM manipulators in life sciences is significantly expanding the field of medical robotics at single-cell level. The AFM nanomanipulator is able to not only visualize the nanostructures on single living cells, but also can simultaneously acquire the multiple cellular mechanical properties, offering novel possibilities for establishing the relationship between cell structures and cell mechanics.

Applications of AFM nanomanipulator to address biological issues have yielded numerous novel insights into the role of mechanical forces in physiological and pathological processes, and particularly, utilizing AFM nanomanipulator to detect the mechanical properties of primary cells from clinical cancer patients benefits developing new methods of tumor diagnosis

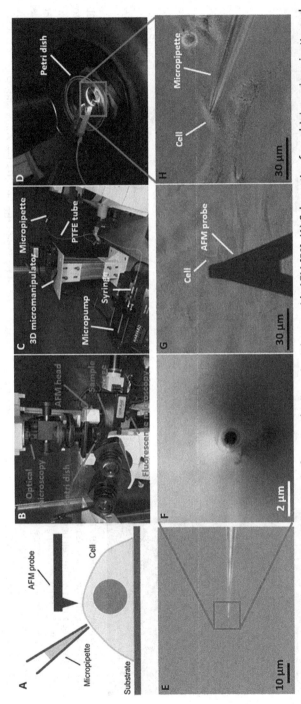

Fig. 13.4 Single-cell analysis experimental platform combining micropipette with AFM [50]. (A) Schematic of combining micropipette and AFM for single-cell drug delivery and simultaneous cell mechanics measurement. (B) Actual photograph of the AFM system. (C) Actual photograph of the micropipette system. (D) Single-cell injection by micropipette under the guidance of optical microscopy. (E, F) Optical bright field image (E) and scanning electron microscope (SEM) image (F) of a micropipette. (G) Moving AFM probe to the targeted cell under the guidance of optical microscopy. (H) Optical image showing single-cell injection by micropipette.

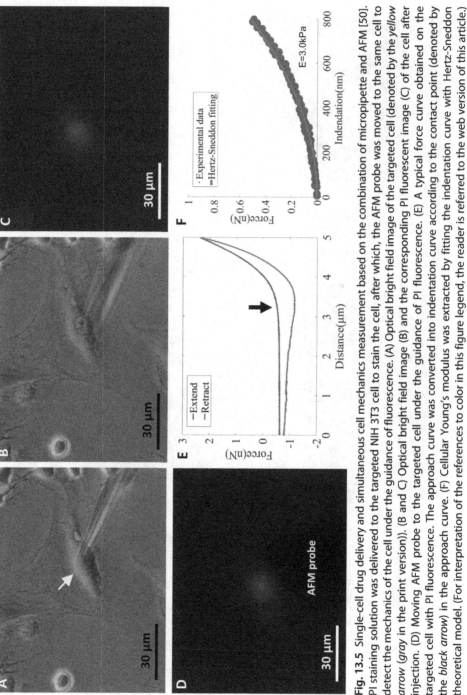

Fig. 13.5 Single-cell drug delivery and simultaneous cell mechanics measurement based on the combination of micropipette and AFM [50]. PI staining solution was delivered to the targeted NIH 3T3 cell to stain the cell, after which, the AFM probe was moved to the same cell to detect the mechanics of the cell under the guidance of fluorescence. (A) Optical bright field image of the targeted cell (denoted by the *yellow arrow* (*gray* in the print version)). (B and C) Optical bright field image (B) and the corresponding PI fluorescent image (C) of the cell after injection. (D) Moving AFM probe to the targeted cell under the guidance of PI fluorescence. (E) A typical force curve obtained on the targeted cell with PI fluorescence. The approach curve was converted into indentation curve according to the contact point (denoted by the *black arrow*) in the approach curve. (F) Cellular Young's modulus was extracted by fitting the indentation curve with Hertz-Sneddon theoretical model. (For interpretation of the references to color in this figure legend, the reader is referred to the web version of this article.)

and treatment. Combining AFM nanomanipulator with micropipette allows single-cell precise drug delivery and simultaneous cell mechanics measurement, which is meaningful for observing the real-time interactions between individual cells and ultra-trace drug molecules with high precision. In future, as more biological systems are investigated by AFM manipulator alone or combined with other complementary techniques, we will see increasing advances in the field of biomedicine.

Acknowledgments

This work was supported by the National Natural Science Foundation of China (62273330, 61922081, 61873258), the Key Research Program of Frontier Sciences CAS (ZDBS-LY-JSC043), and the LiaoNing Revitalization Talents Program (XLYC1907072).

References

[1] A. Diz-Munoz, O.D. Weiner, D.A. Fletcher, In pursuit of the mechanics that shape cell surfaces, Nat. Phys. 14 (7) (2018) 648–652.
[2] D.E. Jaalouk, J. Lammerding, Mechanotransduction gone awry, Nat. Rev. Mol. Cell Biol. 10 (1) (2009) 63–73.
[3] C.C. DuFort, M.J. Paszek, V.M. Weaver, Balancing forces: architectural control of mechanotransduction, Nat. Rev. Mol. Cell Biol. 12 (5) (2011) 308–319.
[4] C. Frantz, K.M. Stewart, V.M. Weaver, The extracellular matrix at a glance, J. Cell Sci. 123 (24) (2010) 4195–4200.
[5] N. Wang, J.D. Tytell, D.E. Ingber, Mechanotransduction at a distance: mechanically coupling the extracellular matrix with the nucleus, Nat. Rev. Mol. Cell Biol. 10 (1) (2009) 75–82.
[6] P. Lu, V.M. Weaver, Z. Werb, The extracellular matrix: a dynamic niche in cancer progression, J. Cell Biol. 196 (4) (2012) 395–406.
[7] D. Di Carlo, A mechanical biomarker of cell state in medicine, J. Lab. Autom. 17 (1) (2012) 32–42.
[8] S. Suresh, Biomechanics and biophysics of cancer cells, Acta Biomater. 3 (4) (2007) 413–438.
[9] S.E. Cross, Y.S. Jin, J. Rao, J.K. Gimzewski, Nanomechanical analysis of cells from cancer patients, Nat. Nanotechnol. 2 (12) (2007) 780–783.
[10] M. Plodinec, M. Loparic, C.A. Monnier, E.C. Obermann, R. Zanetti-Dallenbach, P. Oertle, J.T. Hyotyla, U. Aebi, M. Bentires-Alj, R.Y.H. Lim, C.A. Schoenenberger, The nanomechanical signature of breast cancer, Nat. Nanotechnol. 7 (11) (2012) 757–765.
[11] M. Li, L. Liu, N. Xi, Y. Wang, Z. Dong, X. Xiao, W. Zhang, Atomic force microscopy imaging and mechanical properties measurement of red blood cells and aggressive cancer cells, Sci. China Life Sci. 55 (11) (2012) 968–973.
[12] D.A. Fedosov, B. Caswell, S. Suresh, G.E. Karniadakis, Quantifying the biophysical characteristics of plasmodium-falciparum-parasitized red blood cells in microcirculation, Proc. Natl. Acad. Sci. U. S. A. 108 (1) (2011) 35–39.
[13] W. Yu, S. Sharma, J.K. Gimzewski, J. Rao, Nanocytology as a potential biomarker for cancer, Biomark. Med 11 (3) (2017) 213–216.

[14] M.C. Lampi, C.A. Reinhart-King, Targeting extracellular matrix stiffness to attenuate disease: from molecular mechanisms to clinical trials, Sci. Transl. Med. 10 (422) (2018), eaao0475.

[15] C. Mavroidis, A. Ferreira, Nanorobotics: past, present, and future, in: Nanorobotics, Springer-Verlag, Berlin, Germany, 2013, pp. 3–27.

[16] K. Oldham, D. Sun, Y. Sun, Focused issue on micro-/nano-robotics, Int. J. Intell. Robot. Appl. 2 (4) (2018) 381–382.

[17] M. Li, N. Xi, Y. Wang, L. Liu, Progress in nanorobotics for advancing biomedicine, IEEE Trans. Biomed. Eng. 68 (1) (2021) 130–147.

[18] S. Li, Q. Jiang, S. Liu, Y. Zhang, Y. Tian, C. Song, J. Wang, Y. Zou, G.J. Anderson, J.-Y. Han, Y. Chang, Y. Liu, C. Zhang, L. Chen, G. Zhou, G. Nie, H. Yan, B. Ding, Y. Zhao, A DNA nanorobot functions as a cancer therapeutic in response to a molecular trigger in vivo, Nat. Biotechnol. 36 (3) (2018) 258–264.

[19] P.L. Venugopalan, B. Esteban-Fernandez de Avila, M. Pal, A. Ghosh, J. Wang, Fantastic voyage of nanomotors into the cell, ACS Nano 14 (8) (2020) 9423–9439.

[20] M. Sitti, Miniature soft robots—road to the clinic, Nat. Rev. Mater. 3 (6) (2018) 74–75.

[21] E. Moeendarbary, A.R. Harris, Cell mechanics: principles, practices, and prospects, WIREs Syst. Biol. Med. 6 (5) (2014) 371–388.

[22] M. Li, D. Dang, L. Liu, N. Xi, Y. Wang, Atomic force microscopy in characterizing cell mechanics for biomedical applications: a review, IEEE Trans. Nanobiosci. 16 (6) (2017) 523–540.

[23] M. Li, N. Xi, Y. Wang, L. Liu, Advances in atomic force microscopy for single-cell analysis, Nano Res. 12 (4) (2019) 703–718.

[24] M. Li, L. Liu, N. Xi, Y. Wang, Biological applications of a nanomanipulator based on AFM: in situ visualization and quantification of cellular behaviors at the single-molecule level, IEEE Nanotechnol. Mag. 9 (3) (2015) 25–35.

[25] M. Li, L. Liu, N. Xi, Y. Wang, Applications of micro/nano automation technology in detecting cancer cells for personalized medicine, IEEE Trans. Nanotechnol. 16 (2) (2017) 217–229.

[26] M. Li, N. Xi, Y. Wang, L. Liu, Atomic force microscopy in probing tumor physics for nanomedicine, IEEE Trans. Nanotechnol. 18 (2019) 83–113.

[27] M. Li, N. Xi, Y. Wang, L. Liu, Atomic force microscopy for revealing micro/nanoscale mechanics in tumor metastasis: from single cells to microenvironmental cues, Acta Pharmacol. Sin. 42 (3) (2021) 323–339.

[28] Y.F. Dufrene, Atomic force microscopy and chemical force microscopy of microbial cells, Nat. Protoc. 3 (7) (2008) 1132–1138.

[29] M. Li, L. Liu, N. Xi, Y. Wang, Z. Dong, O. Tabata, X. Xiao, W. Zhang, Imaging and measuring the rituximab-induced changes of mechanical properties in B-lymphoma cells using atomic force microscopy, Biochem. Biophys. Res. Commun. 404 (2) (2011) 689–694.

[30] M. Li, L. Liu, N. Xi, Y. Wang, Z. Dong, X. Xiao, W. Zhang, Drug-induced changes of topography and elasticity in living B lymphoma cells based on atomic force microscopy, Acta Phys. Chim. Sin. 28 (6) (2012) 1502–1508.

[31] M. Li, L. Liu, N. Xi, Y. Wang, Atomic force microscopy studies on cellular elastic and viscoelastic properties, Sci. China Life Sci. 61 (1) (2018) 57–67.

[32] M. Li, L. Liu, X. Xiao, N. Xi, Y. Wang, Viscoelastic properties measurement of human lymphocytes by atomic force microscopy based on magnetic beads cell isolation, IEEE Trans. Nanobiosci. 15 (5) (2016) 398–411.

[33] M. Li, L. Liu, N. Xi, Y. Wang, Nanoscale monitoring of drug actions on cell membrane using atomic force microscopy, Acta Pharmacol. Sin. 36 (7) (2015) 769–782.

[34] S. Kasas, G. Longo, G. Dietler, Mechanical properties of biological specimens explored by atomic force microscopy, J. Phys. D. Appl. Phys. 46 (13) (2013), 133001.

[35] C. Rotsch, K. Jacobson, M. Radmacher, Dimensional and mechanical dynamics of active and stable edges in motile fibroblasts investigated by using atomic force microscopy, Proc. Natl. Acad. Sci. U. S. A. 96 (3) (1999) 921–926.

[36] M. Li, L. Liu, X. Xu, X. Xing, D. Dang, N. Xi, Y. Wang, Nanoscale characterization of dynamic cellular viscoelasticity by atomic force microscopy with varying measurement parameters, J. Mech. Behav. Biomed. Mater. 82 (2018) 193–201.

[37] M. Lekka, P. Laidler, D. Gil, J. Lekki, Z. Stachura, A.Z. Hrynkiewicz, Elasticity of normal and cancerous human bladder cells studied by scanning force microscopy, Eur. Biophys. J. 28 (4) (1999) 312–316.

[38] Q.S. Li, G.Y.H. Lee, C.N. Ong, C.T. Lim, AFM indentation study of breast cancer cells, Biochem. Biophys. Res. Commun. 374 (4) (2008) 609–613.

[39] D. Wirtz, K. Konstantopoulos, P.C. Searson, The physics of cancer: the role of physical interactions and mechanical forces in metastasis, Nat. Rev. Cancer 11 (7) (2011) 512–522.

[40] M.J. Dalby, N. Gadegaard, R.O.C. Oreffo, Harnessing nanotopography and integrin-matrix interactions to influence stem cell fate, Nat. Mater. 13 (6) (2014) 558–569.

[41] D.H. Kim, P.P. Provenzano, C.L. Smith, A. Levchenko, Matrix nanotopography as a regulator of cell function, J. Cell Biol. 197 (3) (2012) 351–360.

[42] W. Chen, Y. Shao, X. Li, G. Zhao, J. Fu, Nanotopographical surfaces for stem cell fate control: engineering mechanobiology from the bottom, Nano Today 9 (6) (2014) 759–784.

[43] R. Changede, H. Cai, S.J. Wind, M.P. Sheetz, Integrin nanoclusters can bridge thin matrix fibers to form cell-matrix adhesions, Nat. Mater. 18 (12) (2019) 1366–1375.

[44] M. Li, N. Xi, Y. Wang, L. Liu, Nanotopographical surfaces for regulating cellular mechanical behaviors investigated by atomic force microscopy, ACS Biomater. Sci. Eng. 5 (10) (2019) 5036–5050.

[45] M. Li, N. Xi, Y. Wang, L. Liu, In situ high-resolution AFM imaging and force probing of cell culture medium-forming nanogranular surfaces for cell growth, IEEE Trans. Nanobiosci. 19 (3) (2020) 385–393.

[46] M. Li, N. Xi, L. Liu, Hierarchical micro-/nanotopography for tuning structures and mechanics of cells probed by atomic force microscopy, IEEE Trans. Nanobiosci. 20 (4) (2021) 543–553.

[47] L.B. Oddershede, Force probing of individual molecules inside the living cell in now a reality, Nat. Chem. Biol. 8 (11) (2012) 879–886.

[48] X. Chen, A. Kis, A. Zettl, C.R. Bertozzi, A cell nanoinjector based on carbon nanotubes, Proc. Natl. Acad. Sci. U. S. A. 104 (20) (2007) 8218–8222.

[49] M. Li, L. Liu, T. Zambelli, FluidFM for single-cell biophysics, Nano Res. 15 (2) (2022) 773–786.

[50] Y. Feng, P. Yu, J. Shi, M. Li, Combining micropipette and atomic force microscopy for single-cell drug delivery and simultaneous cell mechanics measurement, Prog. Biochem. Biophys. 49 (2) (2022) 420–430.

CHAPTER 14

Robotic manipulation of zebrafish larvae for disease therapy

Songlin Zhuang[a,b], Xinghu Yu[c], and Huijun Gao[d]
[a]Yongjiang Laboratory, Ningbo, China
[b]Department of Mechanical Engineering, University of Victoria, Victoria, BC, Canada
[c]Ningbo Institute of Intelligent Equipment Technology Co. Ltd., Ningbo, China
[d]Research Institute of Intelligent Control and Systems, Harbin Institute of Technology, Harbin, China

1 Introduction

Zebrafish (*Danio rerio*) is a popular vertebrate model animal that has been increasingly utilized as an ideal experimental object in both biomedical research and clinical applications [1–3]. Its preeminent status in almost all the important fields of modern biology mainly comes from the following two advantages [4–6]:

(1) The transparency of its larva allows the live imaging of whole-organ pathology and development, and

(2) The high production and short development period enable large-scale and short-term biological research and experiments.

The other advantages, like the low cost for breeding and the high genetic similarity to human beings (82% of disease-related genes from human beings have at least one zebrafish ortholog), also attract many researchers' attention [7–9]. For instance, zebrafish can be regarded as patient avatars in cancer biology and precision cancer therapy [10], as amenable tools for large-scale drug testing by different kinds of delivery methods [11], and as biological vectors for high-throughput content screening [12].

Compared to cells or embryos, which present ball-shaped contours, clear inner structures, and few motility features, zebrafish larvae have more complex morphological characteristics, more flexible motile behaviors, and more diverse mechanical properties. The comparisons among human cells, mouse embryos, and zebrafish larvae are shown in Fig. 14.1. The zebrafish larvae present more differentiated organs, so imaging under microscopy or accessing each organ using end–effectors often requires distinctly different technical solutions and hardware supports. The unique complexity and diversity require superb manipulation skills of human operators who are

Robotics for Cell Manipulation and Characterization
https://doi.org/10.1016/B978-0-323-95213-2.00003-X

307

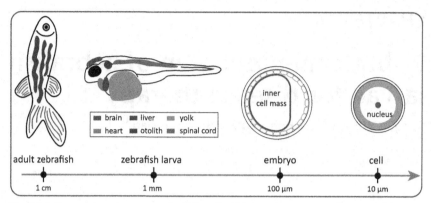

brain   liver   yolk
heart   otolith   spinal cord

adult zebrafish   zebrafish larva   embryo   cell

1 cm   1 mm   100 μm   10 μm

Fig. 14.1 Comparisons among zebrafish, zebrafish larvae, mouse embryos, and human cells.

often in training for at least 6 months before becoming a qualified operator. In particular, different experimental purposes need different hardware setups [13–18], and many personnel skills like rotating a larva to a desired orientation and holding it gently but firmly by a superfine hair or eyelash limit their widespread usage and lower the experimental efficiency. Hence, it is greatly desirable to automate or semiautomate the overall procedures of manipulating zebrafish larvae so as to shorten the whole research period.

The contradiction between the great desire for using zebrafish larvae as experimental objects and the complicated manipulation process in conducting the experiments strongly motivates the introduction of robotic techniques, which actually have already been rapidly developed in the macro world [19–21]. Similar to the commonly used cell-related robotic manipulation techniques such as cell delivery [22], large-scale immobilization [23], precise orientation control [24, 25], batch microinjection [26], etc., the robotic techniques for manipulating zebrafish larvae also highly rely on using nontraditional robotic actuators, like systems driven by different kinds of fields besides the traditional robotic arms, with the purpose of providing high manipulation stability, robustness, accuracy, efficiency, and consistency. However, due to the diversities among individual zebrafish larvae, the existing automatic cell manipulation techniques cannot be directly embedded into zebrafish-oriented systems. The modules in a system that must be updated cover the mechanical design (i.e., how to design an appropriate mechanical structure adaptively to zebrafish larvae of different sizes and shapes), the feedback channel (including both the vision-based and the force-based sensing), and the control part (e.g., the dynamics for transporting zebrafish larvae in a microfluidic channel is distinctively different from delivering a cell or embryo).

Based on the aforementioned issues and considerations, this chapter moves on to introduce robotic manipulation of zebrafish larvae for disease therapy. The rest of this chapter is organized as follows. In Section 2, we introduce the robotic transportation of zebrafish larvae initially out of order in the reservoir or multiwell plates to controlled sequences. Section 3 is devoted to summarizing the popular techniques generally adopted in the laboratory for immobilizing zebrafish larvae. In Section 4, different approaches for performing robotic orientation of zebrafish larvae are discussed and compared, and representative control schemes ranging from traditional robotic manipulation to novel field-driven methods are also summarized. Section 5 aims to introduce existing robotic injection methods that are capable of releasing a quantitative volume of foreign materials to desired positions inside the developing larvae so as to affect their development process. Then we conclude this chapter in Section 6.

2 Robotic transportation

Transportation is generally the very first step of all the related experiments using zebrafish larvae. Controllable, precise, and efficient transportation of a group zebrafish larvae highly relies on a stable and task-oriented transportation system, which plays a vital role for subsequent high-throughput manipulations.

The fundamental requirements of such a transportation system, particularly for groups of zebrafish larvae, include:

(1) Transferring the zebrafish larvae to the field of view under the microscope, and more importantly,

(2) Converting the zebrafish larvae with initial random positions/orientations in the reservoir or multiwell plates into ordered sequences.

In most experiments, zebrafish larvae that are several hours old after being hatched are utilized. Considering the fact that the larval body at this stage is tiny and delicate, transportation systems are supposed to manipulate the larva with great care to avoid unexpected damage or irreversible deformation, which can affect the subsequent experiment results unknowingly. On the other hand, the active motility of larval zebrafish also requires the system to restrict their motion. To this end, low-collateral damage, high-efficiency, and cost-effective approaches that are able to transport the larvae flexibly are needed.

Methods for transporting zebrafish embryos and larvae to desired positions and orientations under microscopy mainly rely on two kinds of actuators: microfluidic systems and robotic manipulators, depending on the different experimental purposes and hardware configurations.

As one of the most popular micromanipulation platforms for zebrafish larvae, microfluidic systems have advantages including their microscale flow-through, multiplexing capability, and flexible microstructure. Leveraging the powerful hydrodynamic force, a customized microfluidic chip can achieve a high-precision and high-resolution positioning and orientation of zebrafish larvae and meet the requirements of specific experimental tasks [27, 28]. A properly designed microfluidic chip also has the potential for enabling complex tasks such as accurate organ-specific imaging, electroporation, and microinjection, which can obviously speed up the larva-handling and zebrafish-based drug development process. Most importantly, the hydrodynamic force-based driving method, offered by microfluidic systems, is almost the same as the real living environment of zebrafish larvae in nature [29, 30].

A lot of customized microfluidic platforms designed for transport of zebrafish larvae have been developed. For instance, a two-plate droplet-based "digital" microfluidic technology was developed in [31]. Using this technique, the researchers can leverage an electrowetting-on-dielectric (EWOD)-mediated electromechanical force to realize on-chip transporting of zebrafish embryos. Despite the fact that droplet-based embryo transport is effective, digital electronics is required by the driven force to transport the embryo. Meanwhile, the electronics rely on a complex circuit design and multistep photolithography of electrode (array). A large-scale simultaneous crosswalk-free transport of embryo-incorporated droplets is also problematic. Many other applications can be found in [32–35]. In the last decade, microfluidics has also been adopted for larvae transportation. In [36], the authors reported a high-throughput vertebrate automated screening technology (VAST). This system has the ability to guide the movement of larvae, enabling it to transport from reservoirs or multiwell plates to some specific downstream pipelines. The whole process can be completed automatically. It should be noted that the whole transport process was finely tuned but cooperated among fluidic valves, programmed syringe pumping force, and restricted region of the capillary. Selecting the multiwell plates as the input ports of the whole system can guarantee a stable and controllable delivery efficiency; in contrast, zebrafish larvae in reservoirs often swim freely, so it is uncertain when a larva would be extracted from the reservoir by the syringe pump.

Robotic manipulators together with other mechanical tools can also be used to transfer zebrafish larvae. In [37], the authors used a holding pipette connected to a motorized pump to grasp the larval yolk. Once the larva is immobilized at the end of the pipette tip, the position of the larva is controllable by steering the micromanipulator. The technical challenges in the robotic manipulator-based schemes are threefold:

(1) Aspirating the zebrafish larvae from the bottom surface requires a large negative pressure of the pump, but it may cause turbulence that further affects the larval postures and, thus, results in failures.

(2) The inner pressure inside the pipette may change dramatically due to the movement of the manipulator or the grasped larva, also causing damage to the larval yolk if there is a pressure difference between the inside and outside of the pipette.

(3) The release process is not a stable process from the control subject since it is essentially an underactuated system.

The major drawbacks of robotic manipulator-based transportation are its very low throughput and the urgent requirement for a real-time visual feedback mechanism, making its use for large-scale transportation tasks impossible. However, together with the XY platform integrated in [38], it may become a promising solution for high-throughput microinjection.

The advantages and limitations of the aforementioned two mainstream transportation methods of zebrafish larvae are summarized in Fig. 14.2. The intrinsic difference between the two methods lies in whether the in situ repositioning capability is crucial for the target objectives. Robotic manipulators, integrated with the optical microscope, establish a complete closed-loop system that can be embedded with artificial intelligence algorithms for guaranteed performance, such as the success rate, the efficiency, etc. However, for large-scale transportation tasks that put throughput in the first place

transportation	microfluidic systems	robotic manipulators
Advantages	- high throughput - multiplexing capability - flexible microstructure - same to the original living environment	- organism-targeted manipulation - performance-guaranteed closed-loop system - high precision - high controllability
Limitations	- lack in situ repositioning capability - open-loop	- low throughput - urgent requirement of visual feedback - possible damage from end-effectors

Fig. 14.2 Pros and cons of two mainstream transportation methods for zebrafish larvae.

(higher than success rate), microfluidic systems are generally preferred. Thus, a future development in the field of zebrafish larvae transportation may be the integration of the two methods to fulfill different needs in distinct stages in an experiment.

3 Robotic immobilization

Immobilizing a zebrafish larva in a gentle but firm manner is essential for subsequent manipulations such as imaging or injection. The "gentle" indicates that the immobilization of the zebrafish larva cannot cause large damage or deformation that affects its future development after operations. The "firm" means the disturbance rejection capability of the immobilization method particularly when the immobilized larva struggles or is under external perturbations.

Different from cells or zebrafish embryos, which are generally motionless throughout the whole experiment, zebrafish larvae have strong motility and unique motion characteristics so that their trajectories are not easy to predict by advanced visual detection and tracking approaches, even more difficult than the trajectory extraction of adult zebrafish [39]. The high dynamics of their motions, together with their similar appearance, frequent occlusions, and highly discontinuous kinematics, prevent the micromanipulator from grasping, since they often swim away rapidly when the end-effector mounted on the micromanipulator approaches them [40]. The motility parameters for zebrafish larvae are summarized in Table 14.1. It is seen that manipulation of zebrafish larvae, especially operations inside the larval body (in situ imaging, microinjection, microsurgery, etc.) require immobilizing the larva before actions. Until now, the main ideas of methods for zebrafish larva immobilization are to lower their motility (using agarose, anesthetic) or restrict their swimming paths (using hydrogel droplets, microfluidic systems, or glass needles).

Table 14.1 Motility parameters for swimming zebrafish larvae.

Variables for turning analysis	Swimming-specific variables
Direction change	Bend location
Time point of maximal bending	Mean bend amplitude
Head-tail angle	Yaw
Bend amplitude	Swimming speed
Bend duration	Tail-beat frequency
Counterbend angle	/

The authors in [41] reported an agarose-based technique that involves mounting embryos such that the agarose could include the head and yolk sac almost entirely. Furthermore, this technique would allow for the body region elongating and developing normally. This technique can be put into practical use because posterior body elongation is an essential process in embryonic development. Without this process, the formation of a large part of the body axis would not be directed properly since multiple tissue deformations cannot act together. It is desirable to observe this whole process by long-term time-lapse imaging. To achieve this goal, the researchers leverage a mounting technique, during the larvae being transferred to microscope and acquisition, which is able to provide sufficient support in order to keep the samples from deviating from the correct orientation. Thus, the agarose method is a promising candidate. After the necessary solutions and pulled glass needle are prepared, the solution containing the low melting point agarose is heated in a microwave. After that, the agarose in the region that surrounded the posterior body should be removed. At this stage, the entire procedure is finished and the embryo is left enclosed, with its head region and around half of the yolk sac being fixed.

Another popular approach to immobilizing zebrafish larvae is using anesthesia. It has been shown in [42] that by chilling the water gradually, the general metabolic activity of fish will be reduced, such as the excretion of feces and ammonia. Moreover, the movements of zebrafish larvae will also be considerably slowed down. Hence, the amount of oxygen consumed by the larvae is reduced, and simultaneously, the dissolved oxygen in the water is increased. Nevertheless, the nerve conduction of larvae is not completely blocked, although it is greatly reduced. A typical example would be the use of MS-222. Even though MS-222 is widely used as a foreign material to induce anesthesia in fish, recently, researchers have noticed that, in many cases, it actually acts as a neuromuscular blocking agent, which would be aversive to at least a small number of species. The research work is still ongoing, so researchers should be cautious when using it as an anesthetic, and keep themselves updated about the latest findings.

The hydrogel droplet is a kind of material that changes its stiffness with the environmental temperature. This feature is leveraged in [38], where the delivery system outputs hydrogel droplets, and each droplet contains a larva. At the beginning of the experiment, all the zebrafish larvae are placed in a heated plate, which contains embryo medium mixed with 1% ultra-low gelling temperature agarose. The agarose-based hydrogel keeps the liquid phase at the normal temperature for zebrafish larvae living (about 25°C), and changes to solid phase if the temperature drops to 17°C. It is said by the authors that a very short exposure to this low temperature does not affect health of zebrafish larvae [44]. After the larva is rotated to the desired orientation, a thermoelectric module is started for cooling the temperature to 17°C so that the droplet is solidified, which further restricts any larva motions. The closed channel tubes of the microfluidic systems can naturally prevent zebrafish larvae within them from swimming away, and thus, all the larvae are immobilized. In particular, the pressure difference inside the channels drives the larvae to the outlets and the customized size of these channels further limits the larval motions for firm immobilization [45].

Many applications can also be found where different types of microfluidic systems are integrated for zebrafish larvae immobilization. The main mechanism is to leverage the pressure difference generated from the inside and the outside of the microfluidic channel outlets, so as to suppress the larval body at the outlets. Please see Refs. [46–49] for more details.

Another novel immobilization method is inspired by a practical case in nature where octopuses utilize precise pressure control to capture other marine animals. This kind of biological behavior is modeled and learned by a zebrafish larva capture system reported in [43]. The main idea therein is to control the pressure difference between the liquid environment and the inner side of the pipette tip to inhale and grasp part of the larval body. The hardware setup of the aspiration device is simple and only contains a pump and a pipette that has a tip of diameter adjusted to the holding target, such as the larval yolk, as illustrated in Fig. 14.3a. The position of the pump piston, denoted by $S_p(t)$, is the control input $u(t)$ of the capture system. The aspirated length of the larval yolk, denoted by $x(t)$, is the system output to be controlled to a desired position x_a^d from an initial length $x_a(0)$. As illustrated in Fig. 14.3b, the aspirated length, practically the distance between the pipette tip (x_{tip}) and the larval skin (x_{skin}), needs to be detected in real time to establish the entire closed-loop mechanism. An image thresholding approach can be utilized here since the image intensity of the larval yolk covered or uncovered by the pipette tip differs a lot. The overall control diagram is illustrated in Fig. 14.3c, where a proportional controller is equipped to the system with a positive proportional gain K_d to be designed.

4 Robotic orientation

Precise orientation of a zebrafish larva is an important and challenging operation and has found many important applications in organ-targeted imaging

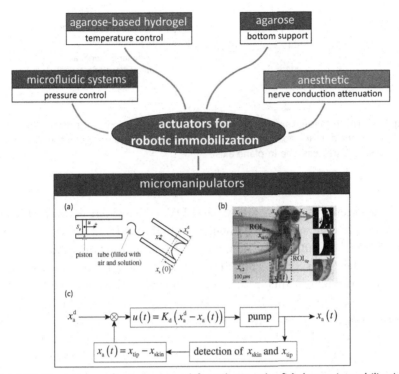

Fig. 14.3 Five classes of actuators used for robotic zebrafish larvae immobilization: Agarose, agarose-based hydrogel, microfluidic systems, anesthetic, and the micromanipulators. (a)–(c) The octopus-inspired immobilization method [43]. (a) The aspiration model. (b) The visual feedback based on the captured field of view image under the microscope. (c) The overall control diagram.

or microinjection [38, 50, 51]. Manual orientation requires human operators to use flexible end-effectors to tentatively poke the larval body, which is laborious, time-consuming, and can only perform rotation about the microscope optical axis. To achieve three-dimensional orientation, a lot of important robotic techniques and automatic control methods have been developed for rotating a larva to any desired orientation.

In practice, zebrafish larva orientation is generally performed about two axes, as illustrated in Fig. 14.4:

(1) the anterior-posterior (AP) axis (i.e., the out-of-plane axis), and

(2) the dorsal-ventral (DV) axis (i.e., the in-plane axis).

The two-axis orientation enables the accessibility of end-effectors on the micromanipulator to any organs in distinct positions inside the larval body, and the generally popular organs include the forebrain, midbrain, eyes, heart, kidney, liver, tail muscles, spinal cord, ventricles, etc., just to mention a few.

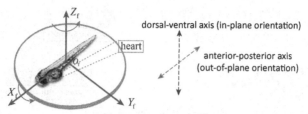

Fig. 14.4 Two rotational axes for three-dimensional orientation of a zebrafish larva. The O_f-X_f axis is the anterior-posterior axis (the out-of-plane orientation), and the O_f-Z_f axis is the dorsal-ventral axis (the in-plane axis).

The orientation about the AP and DV axes is often performed using microfluidic systems [32]. For instance, in [53], the dorsal and lateral orientation control is achieved by designing different kinds of microfluidic channels on the "Fish-Trap" chip. The chip consists of two kinds of flow channels in the horizontal plane, and each channel is connected to several channels for entrapment. In particular, the tapper of the trapping channel mimics the real morphology of larval zebrafish in the period of 4–6 days postfertilization. Thus, the larva can be automatically loaded to the first trapping channel by the bulk stream like a plug, increasing the flow resistance of the first trapping channel, and the subsequent larvae are redirected to the other channels. In [50], a high-throughput hyperdimensional optical projection tomography is established with micrometer resolution. The system is comprised of a motorized syringe pump and some fluidic components. The main part of the system is an ultrathin borosilicate glass capillary, which is held by two stepper motors. This design enables the rotation of the capillary about its main axis by rotating the motors. When a zebrafish larva is detected in the capillary, the system is capable of rotating the capillary axially for 360 degrees, such that the images from multiple perspectives can be rapidly acquired. The overall system setup is shown in Fig. 14.5A. However, the channels in the microfluidic systems are predesigned and unchanged during the manipulation process, so this kind of orientation method cannot meet the requirement of reorientation in some specific experiments. To solve the problem, some contactless field–driven methods for larva orientation are presented. For example, in [51], the authors created small magnetic spheres which were further embedded into an agarose sphere together with the zebrafish larva. By adjusting the magnetic field in the manipulation space under microscopy, the larva can be rotated to any desired postures, as shown in Fig. 14.5B. Two of the main drawbacks for magnetic manipulation are

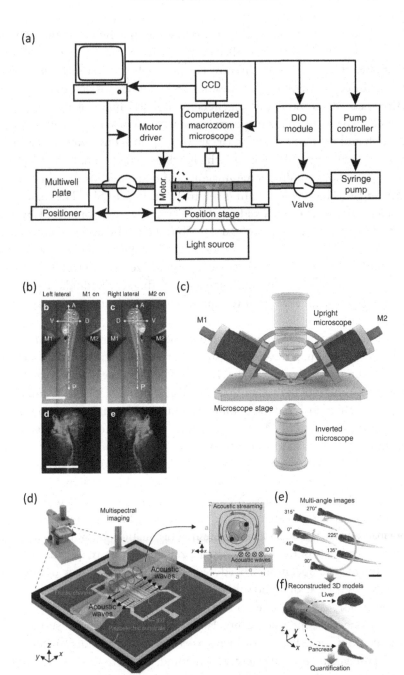

Fig. 14.5 Zebrafish larvae orientation control based on field-driven methods. (A) Two stepper motors rotate the capillary that contains the target zebrafish larva based on a microfluidic system [50]. (B, C) The magnetic field controls the magnetic bead so that the attached zebrafish larva can be rotated [51]. (D–F) The acoustic tweezers rotate a zebrafish larva axially by changing the acoustic waves [52].

(1) introducing magnetic beads that may pollute the environment, and (2) the complex hardware setups that occupy a large space that may disturb other operations. However, this novel technique has the reorientation capability of biological samples that most microfluidic systems do not have. The authors also demonstrated the generality and scalability of the proposed method by performing magnetic orientation of mouse embryos, *artemia*, live zebrafish embryos, and zebrafish larvae. In [52], an acoustofluidic rotational tweezing platform is developed, which achieves noncontact, high-speed, three-dimensional multispectral imaging and digital reconstruction of zebrafish larvae. The whole system is illustrated in Fig. 14.5D. The main idea therein is to use the polarized single streaming vortex stimulated by the acoustic waves, so the zebrafish larva within the field space is oriented axially. The authors showed by experiments that the system can rapidly (1 s per fish) and stably rotate zebrafish larvae in a wide size range. From Fig. 14.5E and F, we can see that the high accuracy of the acoustofluidic rotational tweezing platform enables three-dimensional modeling of larval inner organs.

Another kind of important orientation method, inspired by human operators' manipulation, is developed using only pipettes. Under such a setting, the two-dimensional orientation is achieved by using the pipette tip to poke different positions of the zebrafish larva, as illustrated in Fig. 14.6A and B. In [37], the authors analyzed the poking dynamics and accordingly established a switched linear system. The "switching" occurs when the pipette tip touches different positions or retreats from the larval body, and the overshoot phenomenon may happen when the liquid solution has low stiffness, undesirably leading to system divergence. To solve this problem, the authors designed the controllers of each switched mode in a unified framework, so that the switching dynamics can be eliminated, which can replace human operators and rotate the larva to any desired orientation. In [56], this method is updated to uncross the two overlapped zebrafish larvae frequently appearing in the manual operation procedures. The pipette can aspirate or release a quantitative volume of liquid solutions to perturb the overlapped fish, while simultaneously detecting the postures of each fish. The overall closed-loop control scheme stops until every larval fish is separated for subsequent manipulations. For the out-of-plane orientation, the authors in [55] proposed a three-dimensional orientation method, which requires an additional introduction of a rotational degree of freedom at the end of an XYZ micromanipulator. Leveraging the control techniques developed in [37], the larval tail can be aligned to the pipette tip, which is further used to aspirate part of

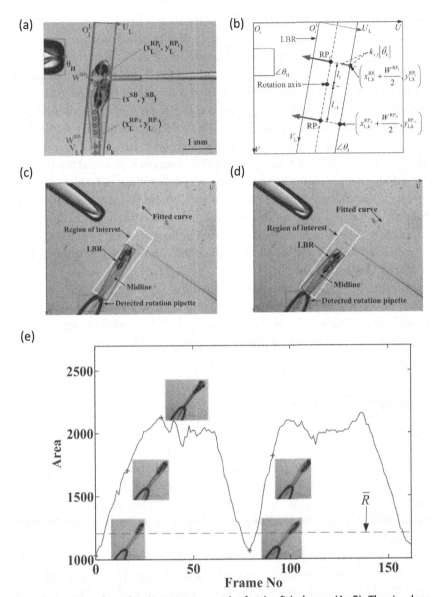

Fig. 14.6 Pipette-based orientation control of zebrafish larva. (A, B) The in-plane orientation. The two rotation points (RP) imply the contact points poked by the pipette tip [54]. (C, D) The out-of-plane orientation. The pipette rotates after aspirating the larva into the inside of the pipette, so the larva can be rotated to any desired orientation [55]. (E) The area of larval texture changes during 360 degrees orientation.

the larval tail to the inside of the pipette gently but firmly, as shown in Fig. 14.6C and D. After the tail is captured, the modified micromanipulator rotates the integrated motor such that the larval tail is also rotated about the larval body axis, as illustrated in Fig. 14.6E. However, one technical problem in this setup lies in that the imprecise assembly of the rotational motor may cause misalignment between the pipette and the motor axis, undesirably moving the captured larva out of the field of view during the orientation process. Thus, an adaptive calibration method is needed before the first use of this micromanipulator [57]. Moreover, to avoid the damage caused by inappropriate aspiration, an adaptive robust control method is integrated to the tail capture system in [58] where the aspiration model is roughly estimated as a model-known nonlinear system with unknown parameters. These parameters are updated in real time in the whole control process such that the adaptive robust controller can aspirate the larval body as a desired velocity.

Compared to field-driven orientation approaches, the robotic manipulator-based methods require very few hardware updates relative to manual orientation control, which may solve the problem proposed in [2] where the authors stated that over-customized platforms may limit the spread of existing automatic solutions.

5 Robotic injection

Microinjection is one of the most-widely utilized robotic methods for delivering a controlled tiny amount of foreign materials (e.g., DNA, RNAi, sperm, protein, and drug compounds) to desired locations inside the developing zebrafish larvae. By adjusting the injection time and site, researchers are capable of intervening in the whole development process of zebrafish larvae and accordingly designing the research from highly flexible status snapshots.

The automatic methods for performing microinjection of distinct injection sites in zebrafish larvae can be classified into two categories: the microfluidic-based and the manipulator-based approaches. The very first microfluidic-based microinjection system is reported in [38], and the main part of the system the authors relied on is a customized highly efficient larva-targeted delivery system. The input module of the delivery system is a water reservoir or multiwell plate that contains many zebrafish larvae swimming freely, and the output module is a robot-held droplet generator. An appropriate-sized glass needle is used to connect the two parts. The system starts the visual detection algorithm for judging whether a larva is within the needle while simultaneously controlling the syringe pump to continuously extract the water together with fish or debris from the input module. Once a larva is detected under the optical microscope, the system automatically distributes zebrafish larvae into hydrogel droplet arrays. The automatic orientation control is also integrated into the system: The lateral orientation is induced by mild anesthesia and the dorsal orientation is enabled by vibrational stimulation. After that, the larvae can be restricted within each droplet by lowering the substrate temperature for solidification. An autonomous batch microinjection system for zebrafish larvae is presented in [60, 61], and the main improvements lie in the use of an agarose supporting base that can rapidly immobilize larvae and align them to desired orientations. A deep learning-based machine vision framework is established for detecting different injection sites in each zebrafish larva. Experiments demonstrated that the proposed robotic system achieves a higher success rate, better stability, and better performance relative to the existing microinjection methods.

The microfluidic-based microinjection schemes focuses on designing various delivery and orientation channels such that the zebrafish larvae can be manipulated in a high-throughput manner. However, there is very little discussion of the concrete injection procedures of each larva in this kind of injection scheme; most of these works test the performance of their system only targeting larval yolk or brain, which are relatively simple organs to be injected. On the other hand, as discussed in [3], although the aforementioned systems highly improve the automation degree of the overall microinjection process, the high level of customization, including the complex mechanical structures and operating procedures, strongly limits the widespread usage of the reported systems. To address this problem, many researchers are increasingly focusing on how to automate the microinjection process based on as few hardware updates as possible relative to the standard manual microinjection system [37, 43, 54, 55, 58, 59, 62].

A standard manual microinjection system is comprised of three parts [14]: the vision part, the motion part, and the pressure part, as illustrated in Fig. 14.7A. The vision part includes an optical microscope and a digital camera, which is used to grab images of the field of view (FOV) under the microscope consecutively and send to the host computers in real time. The motion part generally involves two micromanipulators placed on the left and the right sides of the microscope, with a needle mounted on each micromanipulator. Generally, one needle is used to aspirate a small part of the larval body and thus provide support force, and the other needle is to penetrate the larval

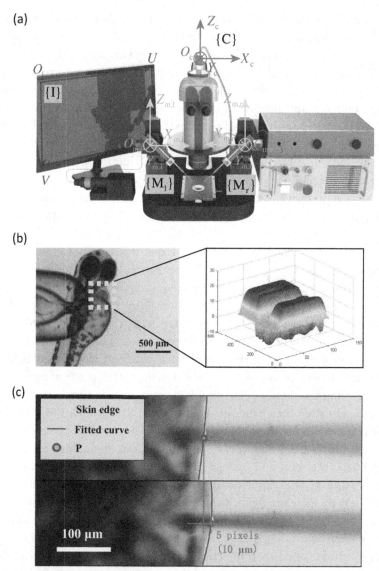

Fig. 14.7 Zebrafish larva microinjection. (A) A standard manual microinjection system for zebrafish larvae. A total of five coordinate frames are defined to establish the relations among motion modules and the microscope camera. (B) Larval heart detection after immobilization [59]. (C) A curve-evolution-based contact detection approach that precisely detects the contour changes before and after the pipette contacts the larval skin.

skin for reagent releasing and tissue extracting. The pressure module covers two pumps, respectively, being connected to the ends of the two needles, so that the aspiration or the injection volume can be precisely controlled. To automate the manual microinjection process, a total of five coordinate frames are generally established, including the camera coordinate frame $\{C\}$, image coordinate frame $\{I\}$, left and right micromanipulator coordinate frames $\{M_l\}$, $\{M_r\}$, and zebrafish body coordinate frame $\{Z\}$. The transformation matrices T_l between $\{I\}$ and $\{M_l\}$, and T_r between $\{I\}$ and $\{M_r\}$ need to be calibrated a priori. Frame $\{Z\}$ is assigned to each zebrafish placed on the petri dish to evaluate independent states.

Based on the human operators' manipulation process, the authors in [59] develop an automatic system for zebrafish larva heart microinjection. Compared to other organs microinjection, such as larval yolk injection [60] or brain injection [38], the larval heart continuously beats, which generates large motions of the surrounding larval skin. This feature also invalidates the well-established contact detection methods [38, 63, 64] that achieve better performance than human operators in other organs. To solve such a practical issue, the authors in [62] modify the motion history image (MHI) method to alleviate the noise in the visual feedback channel. In [59], a curve-evolution-based contact detection approach is presented. The variation of intensity among consecutively sampled image frames is calculated and filtered. Thereafter, the minimum description length criterion is introduced to segment the images using a global energy function, which finally converges from the initial curve to the larval skin edge in the neighborhood of the larval heart. Thus, the original contact detection problem is turned into an edge displacement detection problem, and can be sensed via a predefined threshold. Experimental results demonstrated that this method can effectively attenuate larval skin motions induced by heart beating (Fig. 14.7B and C).

Essentially, a complete microinjection process covers a series of manipulation procedures, including the aforementioned transportation, immobilization, and orientation. Particularly, the oriented larva must be immobilized firmly for penetration. Currently, to the authors' knowledge, only one publication discusses some details of penetration automation [38]. The proposed image template matching algorithm works well for the concerned brain injection, but becomes invalid when performing microinjection of organs with large motion information [59], such as the larval heart and the spinal cord, etc. The other studies in the literature conduct the penetration in an open-loop framework that may induce failures due to the huge diversities among injection sites and individual larval fish.

6 Conclusion

The emerging robotic manipulation techniques are already playing important roles in every step of procedures manipulating zebrafish larvae for disease therapy. Large-scale zebrafish larvae can be transported from initial random positions to an ordered sequence, and subsequently immobilized in a predesigned and controllable manner by using microfluidic systems. Any desired three-dimensional orientation of zebrafish larvae can be achieved by using field-driven methods like microfluidic systems, magnetic systems, or acoustic systems, or by combining traditional robotic micromanipulators, as in the macro world. On the other hand, the microinjection of zebrafish larvae has been automated or semiautomated to an advanced level by integrating high-efficiency delivery systems and advanced vision-based closed-loop control schemes. However, these still rely heavily on human operators' judgement on when and whether the larval skin is contacted or penetrated, and the experience for making such a decision also differs a lot among different organs. Thus, there still exists a high demand for developing powerful robotic manipulation techniques to facilitate the biomanipulation of zebrafish larvae. It is also believed that the robotic manipulation of zebrafish larvae will grow more and more rapidly with focuses on full automation and advanced intelligence.

Acknowledgments

This work was supported in part by the Joint Funds of the National Natural Science Foundation of China (U20A20188) and the Major Scientific and Technological Special Project of Heilongjiang Province under Grant 2021ZX05A01.

References

[1] L.I. Zon, R.T. Peterson, In vivo drug discovery in the zebrafish, Nat. Rev. Drug Discov. 4 (2005) 35–44.
[2] N.J. Silva, L.C. Dorman, I.D. Vainchtein, N.C. Horneck, A.V. Molofsky, In situ and transcriptomic identification of microglia in synapse-rich regions of the developing zebrafish brain, Nat. Commun. 12 (2021) 5916.
[3] E.E. Patton, L.I. Zon, D.M. Langenau, Zebrafish disease models in drug discovery: from preclinical modelling to clinical trials, Nat. Rev. Drug Discov. 20 (2021) 611–628.
[4] C. Laggner, D. Kokel, V. Setola, Chemical informatics and target identification in a zebrafish phenotypic screen, Nat. Chem. Biol. 8 (2012) 144–146.
[5] S. Richter, U. Schulze, P. Tomancak, Small molecule screen in embryonic zebrafish using modular variations to target segmentation, Nat. Commun. 8 (2017) 1901.
[6] E. Marquez-Legorreta, L. Constantin, M. Piber, Brain-wide visual habituation networks in wild type and *fmr1* zebrafish, Nat. Commun. 13 (2022) 895.

[7] G. Golling, A. Amsterdam, Z. Sun, Insertional mutagenesis in zebrafish rapidly identifies genes essential for early vertebrate development, Nat. Genet. 31 (2002) 135–140.

[8] E.A. Mosser, C.N. Chiu, T.K. Tamai, Identification of pathways that regulate circadian rhythms using a larval zebrafish small molecule screen, Sci. Rep. 9 (2019) 12405.

[9] A. Metzner, J.D. Griffiths, A.J. Streets, A high throughput zebrafish chemical screen reveals ALK5 and non-canonical androgen signalling as modulators of the $pkd2^{-/-}$ phenotype, Sci. Rep. 10 (2020) 72.

[10] M. Fazio, J. Ablain, Y. Chuan, D.M. Langenau, L.I. Zon, Zebrafish patient avatars in cancer biology and precision cancer therapy, Nat. Rev. Cancer 20 (2020) 263–273.

[11] C. MacRae, R. Peterson, Zebrafish as tools for drug discovery, Nat. Rev. Drug Discov. 14 (2015) 721–731.

[12] M. Ghannad-Rezaie, M. Eimon, P.M. Wu, Engineering brain activity patterns by neuromodulator polytherapy for treatment of disorders, Nat. Commun. 10 (2019) 2620.

[13] J.H. Gutzman, H. Sive, Zebrafish brain ventricle injection, J. Vis. Exp. (2009) e1218.

[14] C.C. Cosentino, B.L. Roman, I.A. Drummond, N.A. Hukriede, Intravenous microinjections of zebrafish larvae to study acute kidney injury, J. Vis. Exp. (2010) e2079.

[15] J.L. Cocchiaro, J.F. Rawls, Microgavage of zebrafish larvae, J. Vis. Exp. (2013) e4434.

[16] E.A. Harvie, A. Huttenlocher, Non-invasive imaging of the innate immune response in a zebrafish larval model of *Streptococcus iniae* infection, J. Vis. Exp. (2015) e52788.

[17] F. Ellett, D. Irimia, Microstructured devices for optimized microinjection and imaging of zebrafish larvae, J. Vis. Exp. (2017) e56498.

[18] S. Thrikawala, E.E. Rosowski, Infection of zebrafish larvae with aspergillus spores for analysis of host-pathogen interactions, J. Vis. Exp. (2020) e61165.

[19] Y. Sun, B.J. Nelson, D.P. Potasek, E. Enikov, A bulk microfabricated multi-axis capacitive cellular force sensor using transverse comb drives, J. Micromech. Microeng. 12 (2002) 832–840.

[20] Y. Sun, D. Piyabongkarn, A. Sezen, B.J. Nelson, R. Rajamani, A high-aspect-ratio tow-axis electrostatic microactuator with extended travel range, Sens. Actuators A 102 (2002) 49–60.

[21] Y. Sun, B.J. Nelson, Biological cell injection using an autonomous microrobotic system, Int. J. Rob. Res. 21 (2002) 861–868.

[22] L. Wang, A. Ernst, D. An, A bioinspired scaffold for rapid oxygenation of cell encapsulation systems, Nat. Commun. 12 (2021) 5846.

[23] G.F. Bickerstaff, Immobilization of Enzymes and Cells, Springer, 1997, pp. 1–11.

[24] X. Liu, Z. Lu, Y. Sun, Orientation control of biological cells under inverted microscopy, IEEE/ASME Trans. Mechatron. 16 (5) (2010) 918–924.

[25] M. Xie, A. Shakoor, C. Li, D. Sun, Robust orientation control of multi-DOF cell based on uncertainty and disturbance estimation, Int. J. Robust Nonlinear Control 29 (14) (2019) 4859–4871.

[26] H. Huang, D. Sun, J.K. Mills, S.H. Cheng, Robotic cell injection system with position and force control: toward automatic batch biomanipulation, IEEE Trans. Robot. 25 (3) (2009) 727–737.

[27] D. Choudhury, Fish and chips: a microfluidic perfusion platform for monitoring zebrafish development, Lab Chip 12 (2012) 892–900.

[28] E.M. Wielhouwer, Zebrafish embryo development in a microfluidic flow-through system, Lab Chip 11 (2011) 1815–1824.

[29] A. Khalili, P. Rezai, Microfluidic devices for embryonic and larval zebrafish studies, Brief. Funct. Genom. 18 (6) (2019) 419–432.

[30] Y. Lee, H.W. Seo, K.J. Lee, J.-W. Jang, S. Kim, A microfluidic system for stable and continuous EEG monitoring from multiple larval zebrafish, Sensors 20 (2020) 5903.

[31] S. Son, R.L. Garrell, Transport of live yeast and zebrafish embryo on a droplet ("digital") microfluidic platform, Lab Chip 9 (16) (2009) 2398–2401.

[32] M.F. Yanik, C.B. Rohde, C. Pardo-Martin, Technologies for micromanipulating, imaging, and phenotyping small invertebrates and vertebrates, Annu. Rev. Biomed. Eng. 13 (2011) 185–217.

[33] J.N. Wittbrodt, U. Liebel, J. Gehrig, Generation of orientation tools for automated zebrafish screening assays using desktop 3D printing, BMC Biotechnol. 14 (1) (2014) 1–6.

[34] F. Yang, C. Gao, G. Zhang, Z. Chen, Fish-on-a-chip: microfluidics for zebrafish research, Lab Chip 16 (2016) 1106–1125.

[35] R. Pulak, Tools for automating the imaging of zebrafish larvae, Methods 96 (2016) 118–126.

[36] C. Pardo-Martin, T.-Y. Chang, B.K. Koo, High-throughput in vivo vertebrate screening, Nat. Methods 7 (8) (2010) 634–636.

[37] S. Zhuang, W. Lin, H. Gao, X. Shang, L. Li, Visual servoed zebrafish larva heart microinjection system, IEEE Trans. Ind. Electron. 64 (2017) 3727–3736.

[38] T.-Y. Chang, Organ-targeted high-throughput in vivo biologics screen identifies materials for RNA delivery, Integr. Biol. 6 (2014) 926–934.

[39] A. Perez-Escudero, J. Vicente-Page, R.C. Hinz, S. Arganda, G.G. de Polavieja, idtracker: tracking individuals in a group by automatic identification of unmarked animals, Nat. Methods 11 (2014) 743–748.

[40] C. Wang, M. Tong, L. Zhao, X. Yu, S. Zhuang, H. Gao, DanioSense: automated high-throughput quantification of zebrafish larvae group movement, IEEE Trans. Autom. Sci. Eng. 19 (2) (2022) 1058–1069.

[41] E. Hirsinger, B. Steventon, A versatile mounting method for long term imaging of zebrafish development, J. Vis. Exp. (119) (2017) e55210.

[42] M. Matthews, Z. Varga, Anesthesia and euthanasia in zebrafish, ILAR J. 53 (2) (2012) 192–204.

[43] C. Qian, M. Tong, X. Yu, S. Zhuang, H. Gao, Octopus-inspired microgripper for deformation-controlled biological sample manipulation, IEEE Trans. Neural Netw. Learn. Syst. (2021), https://doi.org/10.1109/TNNLS.2021.3070631.

[44] Y. Long, G. Song, J. Yan, X. He, Q. Li, Z. Cui, Transcriptomic characterization of cold acclimation in larval zebrafish, BMC Genom. 14 (2013) 612.

[45] X. Lin, High-throughput brain activity mapping and machine learning as a foundation for systems neuropharmacology, Nat. Commun. (2018) 5142.

[46] N.C. Dubois, SIRPA is a specific cell-surface marker for isolating cardiomyocytes derived from human pluripotent stem cells, Nat. Biotechnol. 29 (2011) 1011–1018.

[47] J.A. Phillips, Y. Xu, Z. Xia, Z.H. Fan, W. Tan, Enrichment of cancer cells using aptamers immobilized on a microfluidic channel, Anal. Chem. 81 (3) (2009) 1033–1039.

[48] C. Collymore, A. Tolwani, C. Lieggi, S. Rasmussen, Efficacy and safety of 5 anesthetics in adult zebrafish (*Danio rerio*), J. Am. Assoc. Lab Anim. Sci. 53 (2) (2014) 198–203.

[49] J. Nordgreen, F.M. Tahamtani, A.M. Janczak, T.E. Horsberg, Behavioural effects of the commonly used fish anaesthetic tricaine methanesulfonate (MS-222) on zebrafish (*Danio rerio*) and its relevance for the acetic acid pain test, PLoS One 9 (3) (2014) e92116.

[50] C. Pardo-Martin, A. Allalou, J. Medina, et al., High-throughput hyperdimensional vertebrate phenotyping, Nat. Commun. 4 (2013) 1467.

[51] F. Berndt, G. Shah, R.M. Power, J. Brugués, J. Huisken, Dynamic and non-contact 3D sample rotation for microscopy, Nat. Commun. 9 (1) (2018) 1–7.

[52] C. Chen, Y. Gu, J. Philippe, Acoustofluidic rotational tweezing enables high-speed contactless morphological phenotyping of zebrafish larvae, Nat. Commun. 12 (1) (2021) 1–13.

[53] X. Lin, S. Wang, X. Yu, High-throughput mapping of brain-wide activity in awake and drug-responsive vertebrates, Lab Chip 15 (3) (2015) 680–689.

[54] S. Zhuang, Visual detection and two-dimensional rotation control in zebrafish larva heart microinjection, IEEE/ASME Trans. Mechatron. 22 (2017) 2003–2012.

[55] S. Zhuang, Visual servoed three-dimensional rotation control in zebrafish larva heart microinjection system, IEEE. Trans. Biomed. Eng. 65 (2018) 64–73.

[56] G. Zhang, X. Yu, G. Huang, D. Lei, M. Tong, An improved automated zebrafish larva high-throughput imaging system, Comput. Biol. Med. 136 (2021) 104702.

[57] S. Zhuang, W. Lin, G. Zhang, X. Shang, H. Gao, Micromanipulator control for grasping and orienting zebrafish larva, in: 2017 IEEE 56th Annual Conference on Decision and Control, 2017, pp. 1628–1633.

[58] G. Zhang, Zebrafish larva orientation and smooth aspiration control for microinjection, IEEE. Trans. Biomed. Eng. 68 (1) (2021) 47–55.

[59] G. Zhang, Visual-based contact detection for automated zebrafish larva heart microinjection, IEEE Trans. Autom. Sci. Eng. 18 (4) (2021) 1803–1813.

[60] Z. Chi, Q. Xu, N. Ai, W. Ge, Design and development of an automatic microinjection system for high-throughput injection of zebrafish larvae, IEEE Trans. Autom. Sci. Eng. (2021), https://doi.org/10.1109/TASE.2021.3119405.

[61] Z. Chi, Q. Xu, N. Ai, W. Ge, Design and implementation of an automatic batch microinjection system for zebrafish larvae, IEEE Robot. Autom. Lett. 7 (2) (2022) 1848–1855.

[62] C. Qian, M. Tong, X. Yu, S. Zhuang, CNN-based visual processing approach for biological sample microinjection systems, Neurocomputing 459 (2021) 70–80.

[63] W. Wang, X. Liu, Y. Sun, Contact detection in microrobotic manipulation, Int. J. Robot. Res. 26 (2007) 821–828.

[64] J. Liu, Robotic adherent cell injection for characterizing cell-cell communication, IEEE. Trans. Biomed. Eng. 62 (2015) 119–125.

CHAPTER 15

Acoustic field techniques for cell characterization in health monitoring

Prajwal Agrawal*, Zhiyuan Zhang*, Zahra Ghorbanikharaji, Zhan Shi, and Daniel Ahmed
Acoustic Robotics Systems Laboratory, Institute of Mechanical Systems, Department of Mechanical and Process Engineering, ETH Zurich, Rushlikon, Switzerland

1 Introduction

Health monitoring (HM) is necessary in order to detect diseases at an early stage and help reduce clinical costs and human suffering. In recent years, HM systems have become a critical component of biological applications. Introducing the timely diagnosis and treatment of diseases provides patients with alternatives to traditional clinical treatments. However, HM applications are dependent on infrastructure that enables them to receive, save, and process information from a variety of sources such as medical results, clinical records, and sensors. Some of the diseases in which HM can be of benefit for patients include hypertension, ischemic heart disease, cardiac arrhythmias, cancer (colon, lung, prostate, breast), and asthma. Parameters that might be measured include glucose levels in diabetic patients [1], electrocardiographic signals of cardiac patients [2], body temperature, other biological elements such as enzymes, cells, and antibodies, and many more besides. Furthermore, the cell characterization necessary for HM is also fundamentally important knowledge for medicine, cell biology, microbiology, and related fields.

In pathophysiologic processes, cellular mechanics are of primary importance; mechanical forces influence numerous biological processes and, therefore, affect not only a cell's capability to respond to exogenous stimuli and signals, but also significantly influence differentiation decisions during development. Historically, the classical techniques of spectroscopy or fluorescence have been used extensively to detect single molecules (e.g., their dynamics, distribution, and position), but detailed information regarding

---

* These authors contributed equally.

Robotics for Cell Manipulation and Characterization
https://doi.org/10.1016/B978-0-323-95213-2.00007-7

329

mechanical properties, the functional nature of biomolecules, and molecular interactions among biological systems is not readily available. Recent decades have seen tremendous advances in manipulating and characterizing particles and cells, both passively and actively. In passive methodologies, the structural geometry and hydrodynamic forces directly affect the behavior of particles and cells without external force. Currently available passive methods for manipulating cells and particles include immunoaffinity-based separation [3], deterministic lateral displacement [4], microfiltration [5], and inertial-based microfluidics [6]. These methods are based on lab-on-a-chip technology and microfluidic devices [7,8] and support the manipulation of particles/cells by leveraging their surface functionalization, deformability, concentration, density, shape, and size, as well as viscosity and fluid velocity. Although passive techniques are label-free, robust, and independent of external sources, they are considered to have some major drawbacks, namely a lack of controllability and the complexity involved in fabricating some geometries.

In contrast, active techniques need an external force and work by exploiting differences in the magnetic, electrical, optical, and acoustic properties of dop particles or cells [9–13]. As cells generally lack a diamagnetic or paramagnetic characteristic, the cell surface would have to be labeled with immunomagnetic particles to allow magnetism-based manipulation, which can be performed with an electromagnet or permanent magnet. Meanwhile, electrical approaches are capable of interacting with cells/particles in an electronic field. The electrical signal generating devices can be used in a wide range of applications by adjusting medium properties, electrode configuration, and applied voltage. Optical methods rely on laser beams to manipulate cells/particles; the photons are converted into energy when they collide with a surface, thereby, yielding a drive that controls the behavior of the particles.

Lastly, acoustofluidics is a highly popular technique that is employed to manipulate cells and particles in microfluidic platforms. In this technique, an acoustic wave interacts with fluids, cells, particles, etc., which, in the case of manipulation, determines the path cells or particles take. In the case of direct interaction between the acoustic wave and the fluid, the fluid must be housed in a substrate capable of transmitting the acoustic wave. Additionally, piezoelectric materials are also widely used in transducers to convert electrical voltage into mechanical stress in the form of an acoustic wave, which is then transferred to the fluid [14]. Acoustic techniques that have been efficiently employed for cell characterization and HM include acoustic traveling

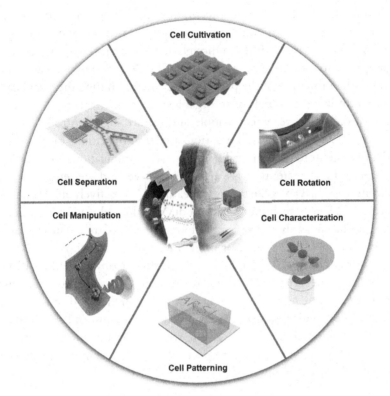

Fig. 15.1 Overview of the utility of acoustic techniques for cell characterization and manipulation.

waves (ATW), acoustic standing waves (ASW), acoustic streaming (AS), focused ultrasound (FU), holographic acoustic tweezers (HAT), surface acoustic waves (SAW), and ultrasound imaging (UI); Fig. 15.1 illustrates the uses to which such techniques have been put.

Regardless of modality, all active techniques simplify the process of manipulation by offering higher controllability and separation yields. Magnetic techniques are reliable and efficient, but are not label-free [15]. Electric techniques are label-free, highly selective, and contactless, which is practical, but they require a low-conductivity buffer, and their accuracy and efficiency decrease with increasing flow rate [16]. Optical techniques have the advantages of high efficiency and accuracy, being label-free, employing virtual electrodes, and having low-cost and a simple procedure, but are also associated with disadvantages such as low throughput, requiring opaque substrates, and requiring low-conductivity solutions [11]. Ultimately, although magnetic, electrical, and optical techniques have been successful, they all face limitations when it comes to acquiring information about all regions of a living model. Additionally, although such techniques have been successfully applied to rotate objects at microscale [17–20], most of them are dependent on the properties of the particular specimen, limiting their application to different living materials or to organisms of different sizes.

Acoustic fields are biocompatible and feature high controllability, and have also seen successful use in various practical operations on the microscale. In contrast to other modalities for active manipulation, acoustic waves propagate at a fast rate; they are also noninvasive, and the setup needed to generate acoustic waves can be easily created and is relatively affordable. Furthermore, acoustic techniques can produce biological responses under the same conditions as physiology over a longer period of time without causing any harm to the cells, tissues, or even an organism as a whole.

In this chapter, we briefly outline the various acoustic methods that can be used to characterize cells in HM. We further discuss other relevant applications of acoustic techniques, such as for cell separation and concentration, cell patterning, cultivation and intercellular interactions, and cell sonoporation, transfection, and rotation. Lastly, we consider the current advantages and limitations and future the prospects of acoustic techniques for HM and for cell characterization.

2 Acoustic techniques

Acoustic techniques are attractive modalities for in vivo medical applications because they penetrate deep into the tissue, are not affected by the opaque nature of an object, are relatively noninvasive, and can generate a broad range of forces. At present, for analyzing and manipulating cells, seven effective acoustic techniques are utilized, namely ATW, ASW, AS, FU, HAT, SAW, and UI, as described in the following subsections.

2.1 Acoustic traveling wave

In acoustic traveling wave (ATW) technique, usually, a piezoelectric transducer (PZT) (Fig. 15.2A) or an interdigital transducer (IDT) possessing a piezoelectric substrate (Fig. 15.2B) is utilized to generate ATW, which entails the conversion of electrical energy to acoustic energy when excited by alternating current signals [21]. In addition, to stimulate the ATW with the maximum acoustic potential, the resonance frequency is usually necessary (Fig. 15.2C). Moreover, based upon the gradient of the Gor'kov

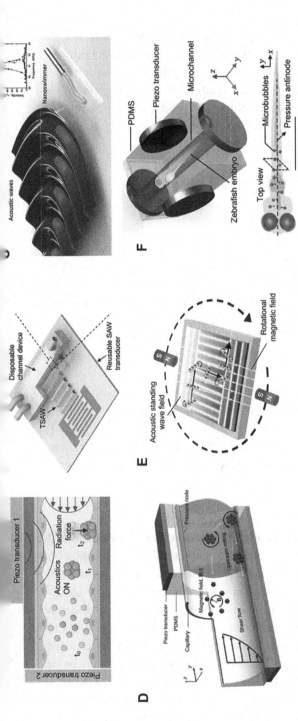

Fig. 15.2 Techniques utilizing acoustic traveling (A, B, and C) and standing (D, E, and F) waves. (A) Self-assembly and manipulation of microrobots using two piezoelectric transducers in combination. (B) A system for acoustic cell sorting. A disposable channel device is hand-placed on a SAW transducer, which is comprised of a piezoelectric substrate patterned with interdigital transducers. (C) Propulsion of artificial swimmers by acoustically-activated flagella. Inset shows the respective frequency responses of Au/Ni/PPy (nanoswimmer), Au/Ni, and Au particles. (D) Self-assembly of superparamagnetic particles injected into a capillary into a spinning microswarm due to dipole–dipole interaction with a rotating magnetic field. (E) Rolling propulsion of microchains along two-dimensional dynamic acoustic virtual walls. (F) Acoustic manipulation of microparticles inside a zebrafish embryo housed in a transparent PDMS chamber bracketed by four piezo transducers. *(Panel (A) Reprinted with permission from A.D.C. Fonseca, T. Kohler, D. Ahmed, Navigation of Ultrasound-Controlled Swarmbots under Physiological Flow Conditions, 2022. Copyright 2022, the authors. Panel (B) Reprinted with permission from Z. Ma, D.J. Collins, Y. Ai, Anal. Chem. 88 (2016) 5316–5323. Copyright 2016, American Chemical Society. Panel (C) Reprinted with permission from D. Ahmed, T. Baasch, B. Jang, S. Pane, J. Dual, B.J. Nelson, Nano Lett. 16 (2016a) 4968–4974. Copyright 2016, American Chemical Society. Panel (D) Reprinted with permission from D. Ahmed, A. Sukhov, D. Hauri, D, Rodrigue, G. Maranta, J. Harting, B.J. Nelson, Nat. Mach. Intell. 3 (2021) 116–124. Copyright 2021, Springer Nature Limited. Panel (E) Reprinted with permission from Z. Zhang, A. Sukhov, J. Harting, P. Malgaretti, D. Ahmed, Rolling Microswarms along Acoustic Virtual Wall, 2022. Copyright 2022, the authors. Panel (F) Reprinted with permission from V.M. Jooss, J.S. Bolten, J. Huwyler, D. Ahmed, Sci. Adv. 8 (2022) eabm2785. Copyright 2022, AAAS.)*

potential [22]), in an acoustic field, suspended microparticles or cells are subjected the acoustic radiation forces (ARF) that stem from both the primary and secondary radiation forces [23]. Basically, the primary radiation forces are produced by the scattering of background acoustic standing waves on the particle, while secondary radiation forces are produced by the scattered waves between adjacent particles. Thus, the time-averaged ARF acting on a single isolated incompressible particle may be expressed as:

$$\mathbf{F}_{AR} = -\frac{4\pi}{3}a^3\nabla\left[f_1\frac{1}{2\rho_0 c_0^2}\langle p_{in}^2\rangle - f_2\frac{3}{4}\rho_0\langle v_{in}^2\rangle\right], \qquad (15.1)$$

$$f_1 = 1 - \tilde{k}, \text{with } \tilde{k} = \frac{k_p}{k_0}, \qquad (15.2)$$

$$f_2 = \frac{2(\tilde{\rho}-1)}{2\tilde{\rho}+1}, \text{with } \tilde{\rho} = \frac{\rho_p}{\rho_0}, \qquad (15.3)$$

where a, p_{in}, and v_{in} are the radius of the particle, incident acoustic pressure, and incident acoustic velocity, respectively; f_1 and f_2 are dimensionless scattering coefficients; ρ_0 and ρ_p denote the density of the liquid and particle, respectively; and κ_0 and κ_p denote the compressibility of the liquid and the particle, respectively.

By utilizing the associated radiation force, which increases with particle volume, ATW has been utilized for self-assembling microswarms [24], microrobot upstream propulsion [25], and high-resolution imaging [26]. As a relatively safe, noninvasive, and relatively inexpensive procedure, high frequency acoustic fields at moderate levels of pressure offer great potential for clinical diagnostics and therapeutics at the cellular level.

2.2 Acoustic standing wave

Typically, as shown in Fig. 15.2D, for the creation of a one-dimensional (1D) acoustic standing wave (ASW) field a single transducer is utilized with an opposing reflector, or a pair of transducers are used [27]. In an ASW field, the nodes of acoustic pressure are found at definite points (where the acoustic pressure is zero) whereas the corresponding maximum pressure position is termed as the antinode. The number of pressure nodes and antinodes in the ASW can be tailored by adjusting the frequency. In addition, as a result of the ARF, particles in the field will be trapped at either pressure nodes or antinodes, which is dependent on the density and compressibility of the particle and the surrounding medium, for example in a water medium microparticles such as polystyrene beads (15 µm) are trapped at the nodal position, whereas, microbubbles are trapped at the antinodal position [28]. Likewise, as shown in Fig. 15.2E, in a two-dimensional (2D) standing wave field, acoustic pressure is used to produce dots/islands [29–31]. Furthermore, within the three-dimensional (3D) standing wave field, by creating an array of different configurations, the acoustic pressure may be generated in the form of sheets, lines, or dots/islands [32]. Moreover, by modulating acoustic parameters, such as phase shifts and amplitude modulations [33], as shown in Fig. 15.2F, ASW can yield highly efficient and versatile patterns for cell generation and transport [34].

2.3 Acoustic streaming

When an acoustic wave propagates in a fluid medium, two types of acoustic streaming (AS) are produced. This is due to the fact that spatial attenuation of waves in free space produces an acoustic pressure gradient in the direction of propagation [35–37]. As the gradient increases over time, it produces a net force in the same direction, considering the nonlinear effects, and following a net flow. A similar streaming is referred to as a "quartz wind" and it is usually related to high Reynolds numbers. The second mechanism, as shown in Fig. 15.3A–C, is the result of friction between vibrating interfaces, such as fluid–solid interfaces (wave guides [29,38], pillars [39,40], and sharp edges [41–43]), fluid–fluid interfaces (standing wave fields [44]), and liquid–gas interfaces (liquid drops in gas and gas bubbles in liquid [45–49]). In particular, this type of streaming is associated with a lower Reynolds number flow confined within a thin viscous boundary layer (Stokes layer) $\delta \approx \sqrt[2]{\nu/\omega}$, where ω is the angular frequency and ν is the kinematic viscosity [50]. According to Stokes drag, the force emanating from acoustic microstreaming can be calculated in acoustofluidic systems as:

$$F_{AS} = 6\pi\eta au, \tag{15.4}$$

where η is the dynamic viscosity of the liquid and u is the particle velocity.

According to Eqs. (15.1) and (15.4), the ARF scales with particle volume, while the streaming force scales linearly with its radius. Therefore, for a small particle below the transition size, the streaming force dominates its motion, causing it to follow the streaming. In the case of a particle that exceeds the transition size, the ARF will dominate. On the basis of this principle, AS has also been widely used for the separation and purification of fluidic mixtures [51–53], trapping and rotating micro-/nano-objects in three dimensions [51,54,55], and design of microrobots with solid sharp edge or integrated bubbles [51,56,57].

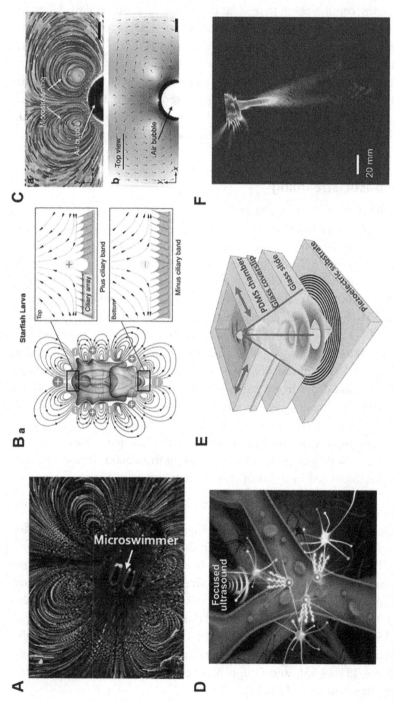

Fig. 15.3 See figure legend on opposite page

(Continued)

2.4 Focused ultrasound

In order to achieve selective manipulation of a single cell, focused ultrasound (FU), also known as acoustic tweezers, is a promising technique. Usually, the existence of an acoustic vortex is observed in a small area, as shown in Fig. 15.3D, where the acoustic pressure is lower (or higher) than the surrounding boundary; in this circumstance, a single cell can be trapped in the vortex. At present, two types of FU devices have been developed, passive FU and active FU. Passive FU devices have fixed acoustic phases, such as a gigahertz high-power transducer that can induce high-speed acoustic flow beams [58], a spiral transducer (Fig. 15.3E) [59–61] or a flat transducer integrating a concave lens [60,61] to contract acoustic waves, a transducer array in which transducers have different spatial locations [64], or a custom Fresnel lenses that can form a special phase pattern [65].

Unlike passive FU devices, active FU devices produce a dynamic acoustic field by programming the excitation sequence [66,67], making the shapes, dimensions, and locations of vortex adjustable. For example, a

---

Fig. 15.3, cont'd   Techniques utilizing acoustic streaming (A, B, and C) and focused ultrasound (D, E, and F). (A) Generation of microstreaming in water through acoustic oscillation of microswimmer bubbles. (B) Generation of a complex flow profile of counter-rotating vortices by a starfish larva through a series of + and − ciliary bands arranged on its body protuberances. (C) Acoustic streaming induced by microbubble oscillation, visualized by optical microscopy and computational simulation. (D) Sono-optogenetic neural modulation via ultrasound-triggered light emission from nanoparticles circulating in the blood. (E) Schematic of the Archimedes-Fermat acoustical tweezers, which can be used for particle capture. (F) Projection of a single-element ultrasound beam for imaging within a skull cavity, visualized in axial profile. *(Panel (A) Reprinted with permission from D. Ahmed, M. Lu, A. Nourhani, P.E. Lammert, Z. Stratton, H.S. Muddana, V.H. Crespi, T.J. Huang, Sci. Rep. 5 (2015) 9744. Copyright 2015, Springer Nature Limited. Panel (B) Reprinted with permission from C. Dillinger, N. Nama, D. Ahmed, Nat. Commun. 12 (2021) 6455. Copyright 2022, Springer Nature Limited. Panel (C) Reprinted with permission from D. Ahmed, A. Ozcelik, N. Bojanala, N. Nama, A. Upadhyay, Y. Chen, W. Hanna-Rose, T.J. Huang, Nat. Commun. 7 (2016b) 11085. Copyright 2016, Springer Nature Limited. Panel (D) Reprinted with permission from X. Wu, X. Zhu, P. Chong, J. Liu, L.N. Andre, K.S. Ong, K. Brinson, A.I. Mahdi, J. Li, L.E. Fenno, H. Wang, G. Hong, Proc. Natl. Acad. Sci. 116 (2019b) 26332–26342. Copyright 2019, NAS. Panel (E) Reprinted with permission from M. Baudoin, J.-C. Gerbedoen, A. Riaud, O.B. Matar, N. Smagin, J.-L. Thomas, Science Advances 5 (2019) eaav1967. Copyright 2019, AAAS. Panel (F) Reprinted with permission from K. Yu, X. Niu, B. He, Adv. Funct. Mater. 30 (2020) 1908999. Copyright 2020, John Wiley and Sons Inc.)*

transducer ring can be used to generate Bessel-shaped acoustic vortices, while a transducer array can be used to produce helical acoustic fields [68]. The passive and active methods of FU can be used in complex situations. With the use of high-intensity focused ultrasound (HIFU), as shown in Fig. 15.3F, one can generate a wide range of forces, enabling precise manipulation, in situ assemblies, and contactless levitation [69,70]. HIFU can also produce a large amount of heat energy in a short period, which has been used in noninvasive tumor thermal ablation [71,72]. Due to the inertial cavitation effect of high-power HIFU, it is possible to use it to safely deliver a biological agent to the brain [73,74], and HIFU is being evaluated as a potential treatment for deep-seated conditions. These include the dissolution of blood clots [75], the relief of vasospasm [76], the enhancement of drug delivery [77], imaging guidance [78], and neuroregulation [79].

2.5 Holographic acoustic tweezers

To create multiple simultaneous acoustic vortices and to realize 3D vortex modulations, the holographic acoustic tweezers (HAT) technique has been developed (Fig. 15.4A). Using its powerful manipulation capabilities and phase programming driving, the HAT can generate new acoustic structures, like tweezer-like twin traps and bottle-shaped traps (Fig. 15.4B), for non-touching assembly in the air [80], location and orientation of multiple objects, and to create virtual objects that can be addressed with similar levels of feedback and confirmation as those given by the real thing [81–83]. In conventional acoustic devices, trapped particles were usually surrounded by acoustic elements [84–86], whereas, HAT has been shown to perform 3D capture, translation, and rotation using a single-sided array; the reconfigurable capabilities of HAT provide an enticing vision for building holographic micromanipulation platforms. Compared with optical tweezers [87,88], HAT has unique characteristics, such as higher capture power, the ability to manipulate multiscale objects and capture across media, and so on. In the future, HAT will be used for new display and biomedical applications, as well as offering an appealing boost for artificial intelligence through acoustic techniques [89]. For instance, a deep neural network (DNN) was used to generate spatially variable sound fields in arbitrarily shaped microchannels [90], Wright et al. built acoustic deep physical neural networks (PNN) [91] and Lu et al. developed a human-microrobot interface based on local acoustic streaming to transport targeted cancer cells [92].

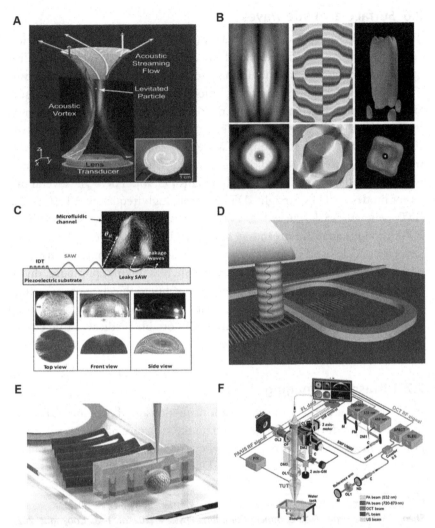

Fig. 15.4 Techniques utilizing holographic acoustic tweezers (A and B), surface acoustic waves (C and D), and acoustic imaging (E and F). (A) Particle levitation via creation of a focused acoustic vortex for in-plane particle trapping. (B) Simulation of focused acoustics generated, which can trap the cells. (C) Experimental and numerical modeling for a droplet positioned at the center with the SAW propagation direction. (D) Surface-acoustic wave-photonic devices for demonstrate and detect the surface-acoustic waves in silicon on insulator. (E) Device for particle manipulation inside a channel, consisting of an acoustic wave-generating piezo-transducer, glued onto a glass slide adjacent to a polydimethylsiloxane (PDMS) channel. (F) Schematic of a seamlessly integrated quadruple fusion imaging system for quadruple ultrasound, photoacoustic, optical coherence, and fluorescence fusion imaging.

(Continued)

2.6 Surface acoustic waves

The concept of surface acoustic waves (SAW) was developed and widely adopted to avoid the limitations of bulk acoustic waves, such as low frequency, rapid decay, and difficult miniaturization. On the surface, as shown in Fig. 15.4C, SAW intensity decays more slowly, whereas, on the subsurface, the amplitude fell rapidly with depth [50]. A SAW device is comprised of IDTs on the surface of a piezoelectric substrate, such as lithium niobate (LiNbO3) [93,94]. In addition, SAW devices can easily be integrated with microfluidic systems (4d), allowing versatile lab-on-a-chip manipulation with a combination of flexible IDT designs [95,96]. There are several different modes of IDTs: straight IDTs generate high frequency ATW, paired IDTs generate ASW, curved IDTs can focus acoustic waves, slanted-finger IDTs produce acoustic streaming, and chirp IDTs (variable finger spacing) generate a wide band of frequencies. SAWs have demonstrated remarkable performance on operations of microfluids such as mixing, pumping, dividing, and nebulizing, and in manipulations of microobjects such as sorting, trapping, concentrating, merging, patterning, aligning, and focusing [97–99]. SAW permits the development of portable medical testing equipment as well as drug delivery devices [100].

2.7 Ultrasound imaging

Based on the nature of acoustic radiation, an ultrasound imaging (UI) technique has been developed to demonstrate the structure of objects and components. In the event that high-frequency ($\sim$GHz) sound waves encounter obstacles, echoes will be generated. Due to the fact that obstacles with

---

Fig. 15.4, cont'd   *(Panel (A) Reprinted with permission from J. Li, A. Crivoi, X. Peng, L. Shen, Y. Pu, Z. Fan, S.A. Cummer. Commun. Phys. 4 (2021a) 1–8. Copyright 2021, Springer Nature Limited. Panel (B) Reprinted with permission from A. Marzo, S.A. Seah, B.W. Drinkwater, D.R. Sahoo, B. Long, S. Subramanian, Nat. Commun. 6 (2015) 8661. Copyright 2015, Springer Nature Limited. Panel (C) Reprinted with permission from M. Alghane, B.X. Chen, Y.Q. Fu, Y. Li, J.K. Luo, A.J. Walton, J. Micromech. Microeng. 21 (2010) 015005. Copyright 2010, IOP Publishing. Panel (D) Reprinted with permission from D. Munk, M. Katzman, M. Hen, M. Priel, M. Feldberg, T. Sharabani, S. Levy, A. Bergman, A. Zadok, Nat. Commun. 10 (2019) 4214. Copyright 2019, Springer Nature Limited. Panel (E) Reprinted with permission from N.F. Läubli, N. Shamsudhin, H. Vogler, G. Munglani, U. Grossniklaus, D. Ahmed, B.J. Nelson, Small Methods 3 (2019) 1800527. Copyright 2019, John Wiley and Sons, Inc. Panel (F) Reprinted with permission from J. Park, B. Park, T.Y. Kim, S. Jung, W.J. Choi, J. Ahn, D.H. Yoon, J. Kim, S. Jeon, D. Lee, U. Yong, J. Jang, W.J. Kim, H.K. Kim, U. Jeong, H.H. Kim, C. Kim, Proc. Natl. Acad. Sci. 118 (2021) e1920879118. Copyright 2021, NAS.)*

different shapes, structures, and properties generate different echoes, collecting and displaying them enables the reconstruction and analysis of the tested objects [101–103]. Currently, several UI techniques, as shown in Fig. 15.4E–F, have been implemented to diagnose and treat human diseases. These include A-type ultrasounds, B-type ultrasounds, and Doppler ultrasounds [104] (Acoustic Imaging, [105]). UI or an acoustic microscope can provide clear images of various organ sections without causing damage. Nevertheless, UI-based observations and monitoring of cell-level and subcellular processes are rarely reported, due to the limitations imposed by excitation frequency and acoustic attenuation.

3 Cell analysis

Cell analysis is the basis of modern biological and medical research, providing significant insights and guidance for clinical diagnosis and treatment. Acoustic techniques have promoted the innovation and development of cell analysis with the advantages of portable setup, maintaining cell activity, allowing high-throughput operations, and allowing multicell interactions. To date, multifunctional acoustic analysis has played an important role in the research of cell separation, cell patterning, cell transport, cell mechanical characterization and so on.

3.1 Cell separation and concentration

Many innovative medical techniques are based on the separation and concentration of cells from complex mixtures [106]. The presence of abnormal amounts of each component in the blood indicates a wide range of diseases, for example, excess lipid particles suggests obstruction of blood flow [107]. Hence, to prevent this, we need an effective separation technique that provides valuable information for early detection. This will help in understanding the differences in forces experienced by cells suspended in a liquid. Thus, acoustic methods, particularly ATW and ASW (including bulk and SAW devices), can separate particles and cells of different sizes, as well as particles with different physical or mechanical properties (Fig. 15.5A), such as the separation of lipid droplets, platelets, *Escherichia coli* and blood cells, and live and dead cells [13,108].

When compared to traditional separation methods, acoustic separation is active, label-free, biocompatible, contactless, and high throughput (Fig. 15.5B). Moreover, being a highly scalable process, acoustic separation can be used to separate biological particles ranging from 10 nm in diameter to hundreds of microns in diameter, including exosomes (30–150 nm in

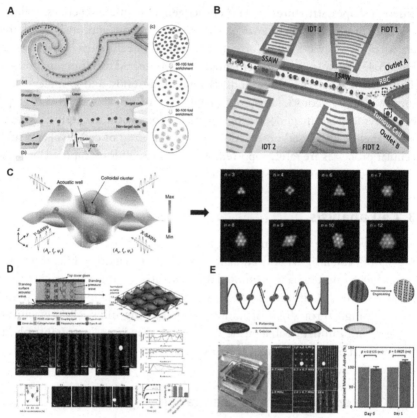

Fig. 15.5 Application of acoustic manipulation to cell separation and concentration (A and B), cell patterning (C), and cell cultivation (D and E). (A) Hybrid sorting device combining inertial focusing and acoustic manipulation for rare cell isolation. (B) Schematic illustration of a multistage device and its use for tumor cell isolation. (C) Illustration of harmonic acoustic wells for tuning of cell spacing [100]. Copyright 2020, Springer Nature Limited. (D) Acoustophoretic system for tissue fabrication. (E) Schematic of an acoustic cell patterning device. *(Panel (A) Reprinted with permission from Y. Zhou, Z. Ma, Y. Ai, RSC Adv. 9 (2019) 31186–31195. Copyright 2019, The Royal Society of Chemistry. PAnel (B) Reprinted with permission from K. Wang, W. Zhou, Z. Lin, F. Cai, F. Li, J. Wu, L. Meng, L. Niu, H. Zheng, Sensors Actuators B Chem. 258 (2018) 1174–1183. Copyright 2020, Springer Nature Limited. Panel (D) Reprinted with permission from B. Kang, J. Shin, H.-J. Park, C. Rhyou, D. Kang, S.-J. Lee, Y. Yoon, S.-W. Cho, H. Lee, Nat. Commun. 9 (2018) 5402. Copyright 2018, Springer Nature Limited. Panel (E) Reprinted with permission from J.P.K. Armstrong, J.L. Puetzer, A. Serio, A.G. Guex, M. Kapnisi, A. Breant, Y. Zong, V. Assal, S.C. Skaalure, O. King, T. Murty, C. Meinert, A.C. Franklin, P.G. Bassindale, M.K. Nichols, C.M. Terracciano, D.W. Hutmacher, B.W. Drinkwater, T.J. Klein, A.W. Perriman, M.M. Stevens, Adv. Mater. 30 (2018) 1802649. Copyright 2018, John Wiley and Sons, Inc.)*

diameter) and CTCs (8–20 μm in diameter) [109,110]. Furthermore, since cells pass through the node in sequence, the node can also be utilized for cell counting [111] and cell bio-printing [112] as well. Due to its smaller size and lower cost, acoustic separation and concentration are increasingly used in biomedical research, especially in clinical diagnosis and therapy, and are capable of developing into point-of-care, multifunctional, and immediate diagnostic platforms for cells (Fig. 15.5C).

3.2 Cell patterning, cultivating, and intercellular interaction

The development and functionalization of cell spheroids and organoid microtissues have become an increasingly vital area of research in biomedicine. This might open up new potential avenues for cell treatment. Benefitting from the ordered pressure nodes and antinodes, static ASW patterns have been utilized to induce stem cell differentiation, cell migration, tissue development, neurite outgrowth, and intercellular communication (Fig. 15.5D) [113–117]. The ultrasonic patterns capture a high proportion of cells and the patterned cell clusters possess a highly dense structure and a uniform size and shape (Fig. 15.5E). When combined with techniques such as dielectrophoresis, magnetic manipulation, and bioprinting, ultrasonic patterning enables the creation of controlled cell cultures and engineered tissues.

3.3 Cell sonoporation, transfection, and rotation

Sonoporation, transfection, and rotation of cells can have profound medical implications in the delivery of drugs intracellularly and in genetic editing and therapy. Due to its lower toxicity and immunogenicity than viral vectors, sonoporation has been receiving increasing attention as a highly efficient nonviral delivery technology [118–121]. Through the use of acoustic methods, foreign substances can be introduced into a cell through transient and repairable pores in the cell membrane (Fig. 15.6A). It is the mechanical force of TAW and FU that leads to sonoporation through the stable cavitation-induced collapse of the bubble [122,123]. In order to improve the efficiency of sonoporation, geometric structures with sharp edges have been reported so that both cellular trapping and sonoporation can occur simultaneously [124]. In principle, oscillating microbubble arrays with the same amplitude and resonant frequency can be used to ensure high-throughput and uniform sonoporation (Fig. 15.6B). By using sonoporation, drug delivery and gene therapy can be enhanced without creating any damage, using small doses, and can be carried out in vivo. Additionally, as cells

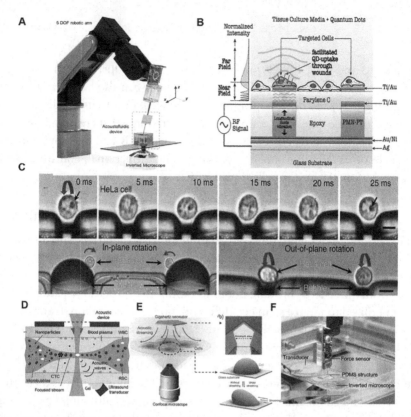

Fig. 15.6 Application of acoustic techniques to cell sonoporation and transfection (A), cell trapping (B), cell rotation (C), and cell characterization (D, E, and F). (A) Experimental setup for robot assisted acoustofluidic end effector for cell trapping. (B) Lateral-resolution ultrasonic microtransducer arrays for assessment of single-cell sonoporation. (C) Experimental images showing rotation of cells driven by oscillating bubbles. (D) Schematic showing cell-attached functionalized microparticles in vivo used for cell characterization. (E) Platform for controllable cell deformation using acoustics. (F) Device used for mechanical characterization of specimens, consisting of a PDMS structure and a piezoelectric transducer, along with the force sensor. *(Panel (A) Reprinted with permission from J. Durrer, P. Agrawal, et al., A robot-assisted acoustofluidic end effector, Nat. Commun. 13 (2022) 6370, https://doi.org/10.1038/s41467-022-34167-y. Copyright 2022, Springer Nature Limited. Panel (B) Reprinted with permission from M. Thein, A. Cheng, P. Khanna, C. Zhang, E.-J. Park, D. Ahmed, C.J. Goodrich, F. Asphahani, F. Wu, N.B. Smith, C. Dong, X. Jiang, M. Zhang, J. Xu, Site-specific sonoporation of human melanoma cells at the cellular level using high lateral-resolution ultrasonic micro-transducer arrays, Biosens. Bioelectron. 27 (1) (2011) 25–33, https://doi.org/10.1016/j.bios.2011.05.026. Copyright © 2011 Elsevier B.V. Panel (C) Reprinted with permission from D. Ahmed, A. Ozcelik, N. Bojanala, N. Nama, A. Upadhyay, Y. Chen, W. Hanna-Rose, T.J. Huang, Nat. Commun. 7 (2016b) 11085. Copyright 2016, Springer Nature Limited. Panel (D) Reprinted with permission from E.I. Galanzha, M.G. Viegas, T.I. Malinsky, A.V. Melerzanov, M.A. Juratli, M. Sarimollaoglu, D.A. Nedosekin, V.P. Zharov, Sci. Rep. 6 (2016) 21531. Copyright 2016, Springer Nature Limited. Panel (E) Reprinted with permission from X. Guo, M. Sun, Y. Yang, H. Xu, J. Liu, S. He, Y. Wang, L. Xu, W. Pang, X. Duan, Adv. Sci. 8 (2021) 2002489. Copyright 2020, John Wiley and Sons, Inc. PAnel (F) Reprinted with permission from N.F. Läubli, J.T. Burri, J. Marquard, H. Vogler, G. Mosca, N. Vertti-Quintero, N. Shamsudhin, A. deMello, U. Grossniklaus, D. Ahmed, B.J. Nelson, Nat. Commun. 12 (2021a) 2583. Copyright 2021, Springer Nature Limited.)*

are trapped, they also rotate. Moreover, by adjusting the oscillation mode of the bubble, both in-plane and out-of-plane rotation of cells can be achieved [46,125], which makes it possible to perform 3D imaging and mechanical force characterization of cells (Fig. 15.6C).

3.4 Cell mechanical characterization

Cellular mechanical properties (hardness and deformability) are one measure of cellular health and are crucial for the understanding of many cellular processes, including cell motility, growth, and adaptation [126]. Recent studies have used acoustic force spectroscopy (AFS) techniques to continuously and noninvasively measure the stiffness of cells [127,128]. Using controlled acoustic forces, AFS attracts cell-attached functionalized microparticles (Fig. 15.6D). Therefore, the mechanical properties of the cells can be described in terms of how they respond to motion.

Among its many applications, AFS has been utilized to study the effects of temperature and drug treatment on the mechanical properties of cellular membranes [129,130], to improve research into the mechanics of adherent cells [131,132], and to test how the mechanical properties of cells differ under the effects of diseases such as malaria and cancer [133,134]. As a result of this method, the mechanical dynamics of multiple individual cells can be tracked simultaneously in a fast and precise manner, which cannot be observed at the population level. In addition, AS has been used for cell shape control and cell deformation analysis (Fig. 15.6E).

In some instances, exposing cells to shear stress created by AS can promote the typical phenotype of mature endothelial cells, allow repetitive testing of cell deformability, and be used to selectively manipulate the movement of individual cells [135,136]. By altering the processing time, intensity, and waveform of the flow, AS is able to modulate cell stiffness (Fig. 15.6F) and control behavior at a microscopic scale [137–139]. Cell mechanical characterization using acoustic field technology features unique advantages such as high flexibility, high precision, and high throughput. These characteristics indicate its potential use as a reliable tool for cell characterization, early health monitoring, and drug development [140,141].

4 Current challenges and future direction

The analysis above suggests that acoustic fields can be utilized for various applications, such as cell characterization/manipulation, HM, lab-on-a-chip, and so on. Generally, designs that are based on solid properties do not allow more visual effects to be produced by reorienting samples due to a limitation in the generation of in-plane vortices [142]. This challenge can be overcome by using microstreaming, which is generated by oscillating micro-bubbles (i.e., using bubbles by various acoustic methods such as ASWs or AS). By altering the frequency of excitation, microbubble microstreaming patterns can be converted from out-of-plane to in-plane, allowing for 3D rotation of organisms and cells [46,47,55,143,144], mixing, and pumping (i.e., SAWs technique). As compared to other acoustic techniques, micro-bubbles can create greater disruptions in fluids while requiring less input power. In addition, a microchannel system that allows for out-of-plane streaming greatly facilitates the manipulation of the specimen.

The principal disadvantage of using microbubbles is their lack of stability over time. In fact, a quick cleaning, drying, and refilling of the system can make it a completely nonproblematic process [145]. Furthermore, the com-bination of microbubbles with ultrasound can achieve a reversible manipu-lation (i.e., controlled cross-, up-, and down-motion) of these bubbles within a cerebral blood vessel or an intersegmental vessel, so that micropar-ticles that are loaded with drugs can be used effectively for targeted drug administration. Even though the exact node position may pose a challenge to the precise movement of trapped microbubbles with the desired fre-quency without visual feedback, more sophisticated control of the applied frequency with a tool with autocalibration may make it simpler to plan the path of microbubbles inside the body [28].

Moreover, most of the acoustofluidic platforms, such as AS and ASWs, rely on a closed microchannel to incorporate acoustic excitation into a hydrodynamic flow, so direct access is difficult for some animals. Con-versely, the inability to manipulate and access various parts of the body pre-vents a full direct interaction between experimental setups and exterior tools, resulting in inaccurate measurements, since data is only collected from a single region [146–151]. Therefore, the analysis of various parts of the body must be done via experimental setups in order to overcome this limitation.

Using medical transducers in ATW and SAW techniques has a variety of applications, such as fully characterizing pressure fields. While high testing flexibility for various beamforming sequences and imaging has been accom-plished using only computational effort, addressing some systematic errors related to beamforming parameters (e.g., apodization, focus depth, etc.) and calculating the pressure field for high-power models without hydro-phone damage is challenging and requires more experimental/numerical efforts [127].

The majority of acoustic techniques enable noninvasive, continuous monitoring of a single cell over several generations, which becomes too subtle at the population level, owing to cellular heterogeneity. To minimize potential drawbacks of the cortex, it would be helpful to determine how cortex thickness, modules, and tension affect the dynamic characteristics of cell stiffness during mitosis [141]. Furthermore, they can be helpful for health monitoring by reducing costs, increasing the effectiveness of treatment, and improving conditions for long-term health monitoring [152].

In some methods based on ATWs and AS, acoustic microrobots can oscillate under low amplitudes and generate a controlled flow via a different method, such as source and sink adjustment, in conjunction with nonlinear acoustics. These methods have the potential to efficiently circulate, trap, or transfer microparticles in a manner that mimics the mechanism for feeding, thereby, overcoming the dependency on external control. Since ultrasound attenuation is very small in a biological medium, these microrobots may be useful for administering and manipulating drugs locally in the body, especially in the stomach. In view of this, studying the performance of microrobots in non-Newtonian liquids such as shear-thinning gels, blood, and examining more complex microchannels with 3D designs for a more realistic visual representation of their performance in the body is of high importance [24,41].

In conclusion, as discussed in the section on "Cell Patterning, Cultivation, and Intercellular Interaction," combining acoustic fields with other techniques, especially the magnetic field, has recently attracted attention for the purpose of manipulating, clustering, and rolling microparticles or microchains. On the whole, rolling along a physical wall can have some challenges that can be reduced by rolling along an acoustic-virtual wall, providing an effective solution to driving over rugged terrain. Even though the method permits optional stimulation of objects in the microscale, adept handling, efficient extension into the animal or human body, and effective actuation, it does not predict microchain thickness and length, nor some factors related to rolling motions from this combined field, such as magnetic intensity, acoustic frequency, and acoustic voltage. Therefore, further studies should be conducted to examine the coupling effects of this combined field [33].

References

[1] A. Tirkey, A. Jesudoss, A non-invasive health monitoring system for diabetic patients, in: 2020 international conference on communication and signal processing (ICCSP). Presented at the 2020 International Conference on Communication and Signal Processing (ICCSP), 2020, pp. 1065–1067.

[2] H.T. Yew, M.F. Ng, S.Z. Ping, S.K. Chung, A. Chekima, J.A. Dargham, IoT based real-time remote patient monitoring system, in: 2020 16th IEEE international colloquium on Signal Processing & its Applications (CSPA). Presented at the 2020 16th IEEE international colloquium on Signal Processing & its Applications (CSPA), 2020, pp. 176–179.

[3] S. Kayo, J. Bahnemann, M. Klauser, R. Pörtner, A.-P. Zeng, Lab Chip 13 (2013) 4467–4475.

[4] P. Jusková, L. Matthys, J.-L. Viovy, L. Malaquin, Chem. Commun. 56 (2020) 5190–5193.

[5] T.-H. Kim, M. Lim, J. Park, J.M. Oh, H. Kim, H. Jeong, S.J. Lee, H.C. Park, S. Jung, B.C. Kim, K. Lee, M.-H. Kim, D.Y. Park, G.H. Kim, Y.-K. Cho, Anal. Chem. 89 (2017) 1155–1162.

[6] A. Sarkar, H.W. Hou, A.E. Mahan, J. Han, G. Alter, Sci. Rep. 6 (2016) 23589.

[7] W. Tang, S. Zhu, D. Jiang, L. Zhu, J. Yang, N. Xiang, Lab Chip 20 (2020) 3485–3502.

[8] S. Zhang, Y. Wang, P. Onck, J. den Toonder, Microfluid. Nanofluid. 24 (2020) 24.

[9] Q. Wang, K.F. Chan, K. Schweizer, X. Du, D. Jin, S.C.H. Yu, B.J. Nelson, L. Zhang, Sci. Advances 7 (2021) eabe5914.

[10] F. Olm, A. Urbansky, J.H. Dykes, T. Laurell, S. Scheding, Sci. Rep. 9 (2019) 8777.

[11] X. Wang, S. Chen, M. Kong, Z. Wang, K.D. Costa, R.A. Li, D. Sun, Lab Chip 11 (2011) 3656–3662.

[12] S. Yaman, M. Anil-Inevi, E. Ozcivici, H.C. Tekin, Front. Bioeng. Biotechnol. 6 (2018).

[13] Y. Zhang, S. Wang, J. Chen, F. Yang, G. Li, BioChip J. 14 (2020) 185–194.

[14] M. Wu, A. Ozcelik, J. Rufo, Z. Wang, R. Fang, T. Jun Huang, Microsyst. Nanoeng. 5 (2019) 1–18.

[15] C.W.S. Iv, J.L. Wang, K.A. Ohiri, E.D. Essoyan, B.B. Yellen, A.J. Armstrong, G.P. López, Lab Chip 16 (2016) 3833–3844.

[16] C. Wu, R. Chen, Y. Liu, Z. Yu, Y. Jiang, X. Cheng, Lab Chip 17 (2017) 4008–4014.

[17] J. Hong, J.B. Edel, A.J. deMello, Drug Discov. Today 14 (2009) 134–146.

[18] C.W.S. Iv, C.D. Reyes, G.P. López, Lab Chip 15 (2015) 1230–1249.

[19] L. Mi, L. Huang, J. Li, G. Xu, Q. Wu, W. Wang, Lab Chip 16 (2016) 4507–4511.

[20] J.R. Rettig, A. Folch, Anal. Chem. 77 (2005) 5628–5634.

[21] A. Ozcelik, J. Rufo, F. Guo, Y. Gu, P. Li, J. Lata, T.J. Huang, Nat. Methods 15 (2018) 1021–1028.

[22] Hydrodynamics, in: Selected Papers of Lev P Gor'kov, World Scientific, 2011, pp. 307–317.

[23] L.V. King, Proc. R. Soc. Lond. A Math. Phys. Sci. 147 (1934) 212–240.

[24] A.D.C. Fonseca, T. Kohler, D. Ahmed, Navigation of Ultrasound-Controlled Swarmbots under Physiological Flow Conditions, 2022.

[25] High shear rate propulsion of acoustic microrobots in complex biological fluids [WWW Document], n.d. URL https://www.science.org/doi/10.1126/sciadv.abm5126 (accessed 6.13.22).

[26] G. Jin, H. Bachman, T.D. Naquin, J. Rufo, S. Hou, Z. Tian, C. Zhao, T.J. Huang, ACS Nano 14 (2020) 8624–8633.

[27] D. Ahmed, A. Sukhov, D. Hauri, D. Rodrigue, G. Maranta, J. Harting, B.J. Nelson, Nat. Mach. Intell. 3 (2021) 116–124.

[28] V.M. Jooss, J.S. Bolten, J. Huwyler, D. Ahmed, Sci. Adv. 8 (2022) eabm2785.

[29] Acoustic tweezer with complex boundary-free trapping and transport channel controlled by shadow waveguides [WWW Document], n.d. https://www.science.org/doi/10.1126/sciadv.abi5502 (accessed 6.13.22).

[30] On-chip manipulation of single microparticles, cells, and organisms using surface acoustic waves | PNAS [WWW Document], n.d. https://www.pnas.org/doi/10. 1073/pnas.1209288109 (accessed 6.13.22).

[31] J. Shi, X. Mao, D. Ahmed, A. Colletti, T. Jun Huang, Lab Chip 8 (2008) 221–223.

[32] Y. Yang, T. Ma, S. Li, Q. Zhang, J. Huang, Y. Liu, J. Zhuang, Y. Li, X. Du, L. Niu, Y. Xiao, C. Wang, F. Cai, H. Zheng, Research 2021, 2021.

[33] Z. Zhang, A. Sukhov, J. Harting, P. Malgaretti, D. Ahme, Rolling microswarms along acoustic virtual wall, Nat. Commun. 13 (2022) 7347, https://doi.org/10.1038/ s41467-022-35078-8.

[34] Fabrication of Micropatterned Dipeptide Hydrogels by Acoustic Trapping of Stimulus-Responsive Coacervate Droplets - Nichols - 2018 - Small - Wiley Online Library [WWW Document], n.d. https://onlinelibrary.wiley.com/doi/10.1002/ smll.201800739 (accessed 6.13.22).

[35] C.P. Lee, T.G. Wang, J. Acoust. Soc. Am. 88 (1990) 2367–2375.

[36] M. Wiklund, R. Green, M. Ohlin, Lab Chip 12 (2012) 2438–2451.

[37] J. Durrer, P. Agrawal, et al., A robot-assisted acoustofluidic end effector, Nat. Commun. 13 (2022) 6370, https://doi.org/10.1038/s41467-022-34167-y.

[38] H. Zhang, Z. Tang, Z. Wang, S. Pan, Z. Han, C. Sun, M. Zhang, X. Duan, W. Pang, Phys. Rev. Appl. 9 (2018), 064011.

[39] A. Ghanbari, V. Nock, S. Johari, R. Blaikie, X. Chen, W. Wang, J. Micromech. Microeng. 22 (2012), 095009.

[40] G. Zhang, W. Cui, W. Pang, S. Liu, S. Ning, X. Li, M. Reed, Adv. Mater. Interfaces 8 (2021) 2101334.

[41] C. Dillinger, N. Nama, D. Ahmed, Nat. Commun. 12 (2021) 6455.

[42] A.A. Doinikov, M.S. Gerlt, A. Pavlic, J. Dual, Microfluid. Nanofluid. 24 (2020) 32.

[43] M. Kaynak, P. Dirix, M.S. Sakar, Adv. Sci. 7 (2020) 2001120.

[44] F. Guo, Z. Mao, Y. Chen, Z. Xie, J.P. Lata, P. Li, L. Ren, J. Liu, J. Yang, M. Dao, S. Suresh, T.J. Huang, Proc. Natl. Acad. Sci. 113 (2016) 1522–1527.

[45] D. Ahmed, M. Lu, A. Nourhani, P.E. Lammert, Z. Stratton, H.S. Muddana, V.H. Crespi, T.J. Huang, Sci. Rep. 5 (2015) 9744.

[46] D. Ahmed, A. Ozcelik, N. Bojanala, N. Nama, A. Upadhyay, Y. Chen, W. Hanna-Rose, T.J. Huang, Nat. Commun. 7 (2016) 11085.

[47] N.F. Läubli, J.T. Burri, J. Marquard, H. Vogler, G. Mosca, N. Vertti-Quintero, N. Shamsudhin, A. deMello, U. Grossniklaus, D. Ahmed, B.J. Nelson, Nat. Commun. 12 (2021) 2583.

[48] L. Meng, X. Liu, Y. Wang, W. Zhang, W. Zhou, F. Cai, F. Li, J. Wu, L. Xu, L. Niu, H. Zheng, Adv. Sci. 6 (2019) 1900557.

[49] Q. Tang, F. Liang, L. Huang, P. Zhao, W. Wang, Biomed. Microdevices 22 (2020) 13.

[50] T. Laurell, A. Lenshof, Microscale Acoustofluidics, Royal Society of Chemistry, 2014.

[51] D. Ahmed, T. Baasch, B. Jang, S. Pane, J. Dual, B.J. Nelson, Nano Lett. 16 (2016) 4968–4974.

[52] D. Ahmed, X. Mao, B.K. Juluri, T.J. Huang, Microfluid. Nanofluid. 7 (2009) 727.

[53] D. Ahmed, X. Mao, J. Shi, B.K. Juluri, T.J. Huang, Lab Chip 9 (2009) 2738–2741.

[54] C. Chen, Y. Gu, J. Philippe, P. Zhang, H. Bachman, J. Zhang, J. Mai, J. Rufo, J.F. Rawls, E.E. Davis, N. Katsanis, T.J. Huang, Nat. Commun. 12 (2021) 1118.

[55] N.F. Läubli, N. Shamsudhin, H. Vogler, G. Munglani, U. Grossniklaus, D. Ahmed, B.J. Nelson, Small Methods 3 (2019) 1800527.

[56] M. Kaynak, A. Ozcelik, A. Nourhani, P.E. Lammert, V.H. Crespi, T.J. Huang, Lab Chip 17 (2017) 395–400.

[57] L. Ren, N. Nama, J.M. McNeill, F. Soto, Z. Yan, W. Liu, W. Wang, J. Wang, T.E. Mallouk, Sci. Advances 5 (2019) eaax3084.
[58] X. Guo, Z. Ma, R. Goyal, M. Jeong, W. Pang, P. Fischer, X. Duan, T. Qiu, Acoustofluidic tweezers for the 3D manipulation of microparticles, in: 2020 IEEE international conference on robotics and automation (ICRA). Presented at the 2020 IEEE International Conference on Robotics and Automation (ICRA), 2020, pp. 11392–11397.
[59] M. Baudoin, J.-C. Gerbedoen, A. Riaud, O.B. Matar, N. Smagin, J.-L. Thomas, Sci. Adv. 5 (2019) eaav1967.
[60] M. Baudoin, J.-L. Thomas, R.A. Sahely, J.-C. Gerbedoen, Z. Gong, A. Sivery, O.B. Matar, N. Smagin, P. Favreau, A. Vlandas, Nat. Commun. 11 (2020) 4244.
[61] H. Tang, Z. Chen, N. Tang, S. Li, Y. Shen, Y. Peng, X. Zhu, J. Zang, Adv. Funct. Mater. 28 (2018) 1801127.
[62] Y. Meng, K. Hynynen, N. Lipsman, Nat. Rev. Neurol. 17 (2021) 7–22.
[63] M.S. Ozdas, A.S. Shah, P.M. Johnson, N. Patel, M. Marks, T.B. Yasar, U. Stalder, L. Bigler, W. von der Behrens, S.R. Sirsi, M.F. Yanik, Nat. Commun. 11 (2020) 4929.
[64] A. Marzo, A. Barnes, B.W. Drinkwater, Rev. Sci. Instrum. 88 (2017), 085105.
[65] K. Melde, A.G. Mark, T. Qiu, P. Fischer, Nature 537 (2016) 518–522.
[66] W.-C. Lo, C.-H. Fan, Y.-J. Ho, C.-W. Lin, C.-K. Yeh, Proc. Natl. Acad. Sci. 118 (2021), e2023188118.
[67] A. Marzo, S.A. Seah, B.W. Drinkwater, D.R. Sahoo, B. Long, S. Subramanian, Nat. Commun. 6 (2015) 8661.
[68] Z. Hong, J. Zhang, B.W. Drinkwater, Phys. Rev. Lett. 114 (2015), 214301.
[69] M.A.B. Andrade, A.L. Bernassau, J.C. Adamowski, Appl. Phys. Lett. 109 (2016), 044101.
[70] M.A.B. Andrade, N. Pérez, J.C. Adamowski, Appl. Phys. Lett. 106 (2015), 014101.
[71] N.T. Sanghvi, J. Acoust. Soc. Am. 134 (2013) 4089, https://doi.org/10.1121/1.4830936.
[72] L. Xiaoping, Z. Leizhen, Int. J. Hyperth. 29 (2013) 678–682.
[73] G. Samiotaki, E.E. Konofagou, IEEE Trans. Ultrason. Ferroelectr. Freq. Control 60 (2013) 2257–2265.
[74] N. Vykhodtseva, N. McDannold, K. Hynynen, The resurgence of therapeutic ultrasound: a 21st century phenomenon, Ultrasonics 48 (2008) 279–296.
[75] S.J. Monteith, N.F. Kassell, O. Goren, S. Harnof, Neurosurg. Focus. 34 (2013) E14.
[76] R.H. Bonow, J.R. Silber, D.R. Enzmann, N.J. Beauchamp, R.G. Ellenbogen, P.D. Mourad, J. Ther. Ultrasound 4 (2016) 6.
[77] D.L. Miller, N.B. Smith, M.R. Bailey, G.J. Czarnota, K. Hynynen, I.R.S. Makin, Bioeffects Committee of the American Institute of Ultrasound in Medicine, J. Ultrasound Med. 31 (2012) 623–634.
[78] Y.-F. Zhou, World J. Clin. Oncol. 2 (2011) 8–27.
[79] J. Kubanek, Neurosurg. Focus. 44 (2018) E14.
[80] A. Marzo, B.W. Drinkwater, Proc. Natl. Acad. Sci. 116 (2019) 84–89.
[81] A volumetric display for visual, tactile and audio presentation using acoustic trapping | Nature [WWW Document], n.d. URL https://www.nature.com/articles/s41586-019-1739-5 (accessed 6.13.22).
[82] S. Ando, H. Shinoda, An acoustic tactile sensing element with five dimensional sensitivity, in: Proceedings of the international solid-state sensors and actuators conference - TRANSDUCERS '95. Presented at the proceedings of the international solid-state sensors and actuators conference - TRANSDUCERS '95, 1995, pp. 644–647.
[83] I. Rakkolainen, A. Sand, R. Raisamo, A survey of mid-air ultrasonic tactile feedback, in: 2019 IEEE international symposium on multimedia (ISM). Presented at the 2019 IEEE international symposium on multimedia (ISM), 2019, pp. 94–944.

[84] Y. Ochiai, T. Hoshi, J. Rekimoto, ACM Trans. Graph. 33 (2014) 85:1–85:13.
[85] S.A. Seah, B.W. Drinkwater, T. Carter, R. Malkin, S. Subramanian, IEEE Trans. Ultrason. Ferroelectr. Freq. Control 61 (2014) 1233–1236.
[86] W.J. Xie, C.D. Cao, Y.J. Lü, Z.Y. Hong, B. Wei, Appl. Phys. Lett. 89 (2006), 214102.
[87] J.E. Curtis, B.A. Koss, D.G. Grier, Opt. Commun. 207 (2002) 169–175.
[88] D. McGloin, Philos. Trans. Royal Soc. A 364 (2006) 3521–3537.
[89] M.J. Bianco, P. Gerstoft, J. Traer, E. Ozanich, M.A. Roch, S. Gannot, C.-A. Deledalle, J. Acoust. Soc. Am. 146 (2019) 3590–3628.
[90] S.J. Raymond, D.J. Collins, R. O'Rorke, M. Tayebi, Y. Ai, J. Williams, Sci. Rep. 10 (2020) 8745.
[91] L.G. Wright, T. Onodera, M.M. Stein, T. Wang, D.T. Schachter, Z. Hu, P.L. McMahon, Nature 601 (2022) 549–555.
[92] X. Lu, K. Zhao, W. Liu, D. Yang, H. Shen, H. Peng, X. Guo, J. Li, J. Wang, ACS Nano 13 (2019) 11443–11452.
[93] D.R. Morgan, Ultrasonics 11 (1973) 121–131.
[94] Z. Tian, S. Yang, P.-H. Huang, Z. Wang, P. Zhang, Y. Gu, H. Bachman, C. Chen, M. Wu, Y. Xie, T.J. Huang, Sci. Advances 5 (2019) eaau6062.
[95] T. Dung Luong, N. Trung Nguyen, Micro Nanosyst. 2 (2010) 217–225.
[96] D.B. Go, M.Z. Atashbar, Z. Ramshani, H.-C. Chang, Anal. Methods 9 (2017) 4112–4134.
[97] K. Wang, W. Zhou, Z. Lin, F. Cai, F. Li, J. Wu, L. Meng, L. Niu, H. Zheng, Sensors Actuators B Chem. 258 (2018) 1174–1183.
[98] G. Destgeer, H.J. Sung, Lab Chip 15 (2015) 2722–2738.
[99] X. Ding, P. Li, S.-C.S. Lin, Z.S. Stratton, N. Nama, F. Guo, D. Slotcavage, X. Mao, J. Shi, F. Costanzo, T.J. Huang, Lab Chip 13 (2013) 3626–3649.
[100] S. Ramesan, A.R. Rezk, L.Y. Yeo, Lab Chip 18 (2018) 3272–3284.
[101] Acoustic Imaging - an overview | ScienceDirect Topics [WWW Document], n.d. https://www.sciencedirect.com/topics/earth-and-planetary-sciences/acoustic-imaging (accessed 6.13.22).
[102] W. Qiu, Y. Yu, F.K. Tsang, L. Sun, IEEE Trans. Ultrason. Ferroelectr. Freq. Control 59 (2012) 1558–1567.
[103] L. Zhang, Z. Tian, H. Bachman, P. Zhang, T.J. Huang, ACS Nano 14 (2020) 3159–3169.
[104] F. Wang, P. Jin, Y. Feng, J. Fu, P. Wang, X. Liu, Y. Zhang, Y. Ma, Y. Yang, A. Yang, X. Feng, Sci. Advances 7 (2021) eabi9283.
[105] S. Pane, V. Iacovacci, E. Sinibaldi, A. Menciassi, Appl. Phys. Lett. 118 (2021), 014102.
[106] S. Yang, Z. Tian, Z. Wang, J. Rufo, P. Li, J. Mai, J. Xia, H. Bachman, P.-H. Huang, M. Wu, C. Chen, L.P. Lee, T.J. Huang, Nat. Mater. 21 (2022) 540–546.
[107] K. Vermeulen, D.R. Van Bockstaele, Z.N. Berneman, Cell Prolif. 36 (2003) 131–149.
[108] Y. Zhou, Z. Ma, Y. Ai, RSC Adv. 9 (2019) 31186–31195.
[109] A. Kulasinghe, H. Wu, C. Punyadeera, M.E. Warkiani, Micromachines 9 (2018) 397.
[110] P. Li, Z. Mao, Z. Peng, L. Zhou, Y. Chen, P.-H. Huang, C.I. Truica, J.J. Drabick, W.S. El-Deiry, M. Dao, S. Suresh, T.J. Huang, Proc. Natl. Acad. Sci. 112 (2015) 4970–4975.
[111] M.S. Gerlt, D. Haidas, A. Ratschat, P. Suter, P.S. Dittrich, J. Dual, Biomicrofluidics 14 (2020), 064112.
[112] I. Leibacher, J. Schoendube, J. Dual, R. Zengerle, P. Koltay, Biomicrofluidics 9 (2015), 024109.
[113] J.P.K. Armstrong, S.A. Maynard, I.J. Pence, A.C. Franklin, B.W. Drinkwater, M.M. Stevens, Lab Chip 19 (2019) 562–573.

[114] J.P.K. Armstrong, J.L. Puetzer, A. Serio, A.G. Guex, M. Kapnisi, A. Breant, Y. Zong, V. Assal, S.C. Skaalure, O. King, T. Murty, C. Meinert, A.C. Franklin, P.G. Bassindale, M.K. Nichols, C.M. Terracciano, D.W. Hutmacher, B.W. Drinkwater, T.J. Klein, A.W. Perriman, M.M. Stevens, Adv. Mater. 30 (2018) 1802649.

[115] D.J. Collins, B. Morahan, J. Garcia-Bustos, C. Doerig, M. Plebanski, A. Neild, Nat. Commun. 6 (2015) 8686.

[116] F. Guo, P. Li, J.B. French, Z. Mao, H. Zhao, S. Li, N. Nama, J.R. Fick, S.J. Benkovic, T.J. Huang, Proc. Natl. Acad. Sci. 112 (2015) 43–48.

[117] K. Koo, A. Lenshof, L.T. Huong, T. Laurell, Micromachines 12 (2021) 3.

[118] X. Du, J. Wang, Q. Zhou, L. Zhang, S. Wang, Z. Zhang, C. Yao, Drug Deliv. 25 (2018) 1516–1525.

[119] H.-D. Liang, J. Tang, M. Halliwell, Proc. Inst. Mech. Eng. H 224 (2010) 343–361.

[120] U.F.O. Themes, Sonoporation: Applications for Cancer Therapy. Radiology Key, 2017.

[121] J. Wu, J.P. Ross, J.-F. Chiu, J. Acoust. Soc. Am. 111 (2002) 1460–1464.

[122] J. Collis, R. Manasseh, P. Liovic, P. Tho, A. Ooi, K. Petkovic-Duran, Y. Zhu, Ultrasonics 2009 (50) (2010) 273–279. Selected Papers from ICU.

[123] Z. Fan, H. Liu, M. Mayer, C.X. Deng, Proc. Natl. Acad. Sci. 109 (2012) 16486–16491.

[124] B. Song, W. Zhang, X. Bai, L. Feng, D. Zhang, F. Arai, A novel portable cell sonoporation device based on open-source acoustofluidics, in: 2020 IEEE/RSJ international conference on intelligent robots and systems (IROS). Presented at the 2020 IEEE/RSJ International Conference on Intelligent Robots and Systems (IROS), 2020, pp. 2786–2791.

[125] N. Läubli, N. Shamsudhin, D. Ahmed, B.J. Nelson, Procedia CIRP, in: 3rd CIRP Conference on BioManufacturing, Vol. 65, 2017, pp. 93–98.

[126] J.H.-C. Wang, B.P. Thampatty, Biomech. Model. Mechanobiol. 5 (2006) 1–16.

[127] V. Romanov, G. Silvani, H. Zhu, C.D. Cox, B. Martinac, An Acoustic Platform for Single-Cell, High-Throughput Measurements of the Viscoelastic Properties of Cells, 2020.

[128] R. Sorkin, G. Bergamaschi, D. Kamsma, G. Brand, E. Dekel, Y. Ofir-Birin, A. Rudik, M. Gironella, F. Ritort, N. Regev-Rudzki, W.H. Roos, G.J.L. Wuite, MBoC 29 (2018) 2005–2011.

[129] N.S. Gov, S.A. Safran, Biophys. J. 88 (2005) 1859–1874.

[130] R. Rodríguez-García, I. López-Montero, M. Mell, G. Egea, N.S. Gov, F. Monroy, Biophys. J. 108 (2015) 2794–2806.

[131] D. Denning, W.H. Roos, Cell Adhes. Migr. 10 (2016) 540–553.

[132] A.J. Engler, S. Sen, H.L. Sweeney, D.E. Discher, Cell 126 (2006) 677–689.

[133] S. Suresh, J. Mater. Res. 21 (2006) 1871–1877.

[134] S. Suresh, J. Spatz, J.P. Mills, A. Micoulet, M. Dao, C.T. Lim, M. Beil, T. Seufferlein, Acta Biomater. 1 (2005) 15–30.

[135] S.E. Cross, Y.-S. Jin, J. Rao, J.K. Gimzewski, Nat. Nanotech. 2 (2007) 780–783.

[136] X. Guo, M. Sun, Y. Yang, H. Xu, J. Liu, S. He, Y. Wang, L. Xu, W. Pang, X. Duan, Adv. Sci. 8 (2021) 2002489.

[137] J.Y. Hwang, C.W. Yoon, H.G. Lim, J.M. Park, S. Yoon, J. Lee, K.K. Shung, Ultrasonics 63 (2015) 94–101.

[138] K.H. Lam, Y. Li, Y. Li, H.G. Lim, Q. Zhou, K.K. Shung, Sci. Rep. 6 (2016) 37554.

[139] H.G. Lim, H.-C. Liu, C.W. Yoon, H. Jung, M.G. Kim, C. Yoon, H.H. Kim, K.K. Shung, Microsyst. Nanoeng. 6 (2020) 1–12.

[140] J.Y. Hwang, C. Lee, K.H. Lam, H.H. Kim, J. Lee, K.K. Shung, IEEE Trans. Ultrason. Ferroelectr. Freq. Control 61 (2014) 399–406.

[141] J.H. Kang, T.P. Miettinen, L. Chen, S. Olcum, G. Katsikis, P.S. Doyle, S.R. Manalis, Nat. Methods 16 (2019) 263–269.

[142] A. Ozcelik, N. Nama, P.-H. Huang, M. Kaynak, M.R. McReynolds, W. Hanna-Rose, T.J. Huang, Small 12 (2016) 5120–5125.

[143] Y. Li, X. Liu, Q. Huang, T. Arai, Appl. Phys. Lett. 118 (2021), 063701.

[144] P. Marmottant, S. Hilgenfeldt, Nature 423 (2003) 153–156.

[145] N.F. Läubli, M.S. Gerlt, A. Wüthrich, R.T.M. Lewis, N. Shamsudhin, U. Kutay, D. Ahmed, J. Dual, B.J. Nelson, Anal. Chem. 93 (2021) 9760–9770.

[146] M. Backholm, O. Bäumchen, Nat. Protoc. 14 (2019) 594–615.

[147] M. Elmi, V.M. Pawar, M. Shaw, D. Wong, H. Zhan, M.A. Srinivasan, Sci. Rep. 7 (2017) 12329.

[148] C.L. Essmann, D. Martinez-Martinez, R. Pryor, K.-Y. Leung, K.B. Krishnan, P.P. Lui, N.D.E. Greene, A.E.X. Brown, V.M. Pawar, M.A. Srinivasan, F. Cabreiro, Nat. Commun. 11 (2020) 1043.

[149] B.C. Petzold, S.-J. Park, E.A. Mazzochette, M.B. Goodman, B.L. Pruitt, Integr. Biol. 5 (2013) 853–864.

[150] A. Sanzeni, S. Katta, B. Petzold, B.L. Pruitt, M.B. Goodman, M. Vergassola, elife 8 (2019) e43226.

[151] P. Song, X. Dong, X. Liu, Biomicrofluidics 10 (2016), 011912.

[152] A. Nguyen, Remote personal health monitoring with radio waves, in: Health Monitoring of Structural and Biological Systems 2008. Presented at the Health Monitoring of Structural and Biological Systems 2008, SPIE, 2008, pp. 675–679.

CHAPTER 16

Separation and characterization of cells using electrical field

Yupan Wu[a,b,c] and Yingqi Meng[d]

[a]School of Microelectronics, Northwestern Polytechnical University, Xi'an, PR China
[b]Research & Development Institute of Northwestern Polytechnical University in Shenzhen, PR China
[c]Yangtze River Delta Research Institute of NPU, Taicang, PR China
[d]Jiading District Central Hospital Affiliated Shanghai University of Medicine and Health Sciences, Shanghai, PR China

1 Introduction

In microfluidic systems, the manipulation of fluids and particles is of vital importance in the biotechnology, chemistry, and biomedical areas [1–4]. Manipulations involve rotating, sorting, mixing [5,6], and trapping, which can be achieved by various approaches, such as, acoustic waves, optical method, magnetic method, and electric method [7]. Cell sorting is indispensable in a variety of applications ranging from cancer diagnostics to cell-based therapies [8]. Characterization of bioparticles within microfluidic chips offers a quantitative and analytical way to interrogate the physicochemical properties of bioparticles. The application of electric fields to manipulate and characterize the particles or cells in suspension using microelectrode structure is a well-established technique. The electric method does not rely on biomarkers or other labels and is an efficient, noninvasive approach to manipulate, sort, and characterize various cells based on the properties of those cells and the suspending medium, as well as the electrode geometry.

To analyze the behavior of targeted cells, the separation of targeted cells is usually the first step, followed by a second stage for cell characterization. In this chapter, we describe the theoretical background of AC electrokinetics, whereby electric fields are used to sort and characterize bioparticles. We show the developments of AC electrokinetics-based cell sorting, and characterization based on cell electrophysiological properties. The main focus of this chapter is devoted to presenting the operating principles and diverse applications of AC electrokinetics techniques. Examples of recent research progresses in cell sorting and characterization using AC electrokinetics are

Robotics for Cell Manipulation and Characterization
https://doi.org/10.1016/B978-0-323-95213-2.00017-X
355

offered, as is a discussion of dielectrophoresis, which has been employed to manipulate and characterize cancer cells, T cells, red blood cells, bacteria, algae, fungi, viruses, and DNA to analyze their electrophysiological properties [9–13]. Lastly, current opportunities and challenges are discussed in diverse biomedical fields.

2 Theoretical background

The phenomenon of dielectrophoresis (DEP) is caused by the interaction of the induced dipole moment with the external electric field. Depending on the difference between the polarizability of the particle and suspending medium, an induced dipole moment will form across the particle, causing a net particle DEP motion. A neutral particle in a uniform electric field, showing the direction of dipole moment, is shown in Fig. 16.1A. To generate DEP force, the electric field must be nonuniform, as shown in Fig. 16.1B. When the particle polarizability is larger than the medium, the particle behaves as a conductor and will move to strong electric field areas due to positive DEP force. The field vectors bend around the particles, which behave as an insulator when the particle polarizability is less than the suspended medium, causing negative DEP force. The particle moves toward the weak electric field region due to negative dielectrophoresis (nDEP) force [14].

The time averaged DEP force acting on the particle can be written as follows [15–18]:

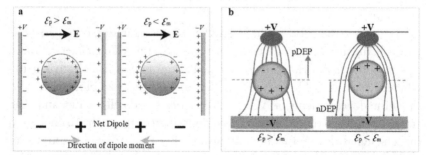

Fig. 16.1 (A) A neutral particle in a uniform electric field showing the direction of dipole moment. (B) A neutral particle in a nonuniform electric field: positive DEP (left) and negative DEP (right).

$$\langle F_D \rangle = \pi r^3 \varepsilon_m \operatorname{Re}[K(w)] \nabla \left(\tilde{E} \cdot \tilde{E}^* \right)$$
$$- 2\pi r^3 \varepsilon_m \operatorname{Im}[K(w)] \left(\nabla \times \operatorname{Re}(\tilde{E}) \times \operatorname{Im}(\tilde{E}) \right)$$
$$= \frac{1}{4} v_p \operatorname{Re}[\tilde{\alpha}] \nabla \left(\tilde{E} \cdot \tilde{E}^* \right) - \frac{1}{2} v_p \operatorname{Im}[\tilde{\alpha}] \left(\nabla \times \operatorname{Re}(\tilde{E}) \times \operatorname{Im}(\tilde{E}) \right). \quad (16.1)$$

The DEP force vector depends on the nonuniformity degree of the electric field and the magnitude of the Clausius-Mossotti factor:

$$K(w) = \left(\varepsilon_p^* - \varepsilon_m^* \right) / \left(\varepsilon_p^* + 2\varepsilon_m^* \right) \text{ with } \varepsilon^* = \varepsilon - j(\sigma/\omega). \quad (16.2)$$

Sometimes, it can also be written in terms of a complex effective polarizability $\tilde{\alpha}$ and the volume of the particle v_p:

$$\tilde{\alpha} = 3\varepsilon_m \left(\varepsilon_p^* - \varepsilon_m^* \right) / \left(\varepsilon_p^* + 2\varepsilon_m^* \right) = 3\varepsilon_m K(w), \quad (16.3)$$

where ε_p^* and ε_m^* are the frequency-dependent complex permittivities of the particles and its suspending medium, ε is the permittivity, σ is the conductivity, r is the particle radius, and $\operatorname{Im}(K(w))$ and $\operatorname{Re}(K(w))$ denote the imaginary and real parts, respectively, of the factor $K(w)$.

The first term in Eq. (16.1) is the conventional DEP (cDEP) force. Fig. 16.2 displays a plot of the real and imaginary parts of the Clausius-Mossotti factor for

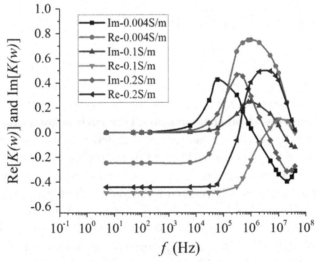

Fig. 16.2 Plot showing a spectrum of the real and imaginary components of the Clausius-Mossotti factor for a yeast cell in an electrolyte for different suspending medium conductivities.

a yeast cell in solutions with different conductivities. The yeast cell will be attracted to or repelled from the high electric filed region, depending on whether $\text{Re}(K(w))$ is positive or negative. The former and the latter are called positive DEP (pDEP) and negative DEP (nDEP) force, respectively. The second term in Eq. (16.1) denotes the traveling wave DEP (twDEP) force. Depending on the polarity of the factor $\text{Im}(K(w))$, the yeast cells will be directed toward the regions where the phases of the field component are smaller $(\text{Im}(K(w)) < 0)$ or larger $(\text{Im}(K(w)) > 0)$. By equating Eq. (16.1) and the drag force, $F_D = 6\pi\eta r v$, the motion of particles can be deduced.

The interaction between the electric field and the dipole moment will generate a torque on the particle when a dipole sits in a field. A finite time or phase delay between the electric field application and the generation of the dipole moment exists. The particle will rotate once the field vector rotates. The time-average electrorotational torque can also be calculated as follows, when applying a rotating electric field to the driving electrodes:

$$\langle T \rangle = -4\pi r^3 \text{Im}[K(w)]\tilde{E}^2 \tag{16.4}$$

Note that the torque depends on the $\text{Im}[K(w)]$ and the positive or negative torque implies that the particle can rotate in the same direction as the field rotation or in the opposite direction.

3 Cell sorting using the electric field

In biological applications, cell sorting is essential for subsequent cell analysis and disease diagnosis. Based on AC electrokinetics, cells can be manipulated and characterized well without damaging the biological activity of the sample. Due to its higher specificity in dielectric properties among cell types and facile integration with other components, DEP has been used widely for cell sorting. DEP-based systems can also be classified as insulator-based dielectrophoresis (iDEP) and electrode-based dielectrophoresis (eDEP).

Using iDEP, particles can be manipulated by exploiting devices with insulating structures. iDEP has been employed for the separation of biological cells. For example, Blanca et al. [19] concentrated and sorted a mixture of bacteria and yeast cells by insulator-based DEP in a microchannel containing an array of cylindrical insulating structures, as shown in Fig. 16.3B. Li et al. [20] demonstrated the iDEP separation of white blood cells and breast cancer cells by an insulating hurdle fabricated within a microchannel. The cell trajectories are deflected by DEP force near the hurdle, resulting in

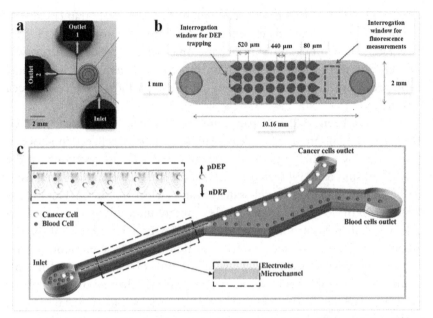

Fig. 16.3 (A) Image of the double spiral microchannel designed for sorting cells and superimposed images for continuous dep sorting. (B) Schematic representation of microchannel showing reservoirs and electrodes. (C) Illustration of the sorting platform. *(Panel (A) Reproduced with permission from J. Zhu, T.R. Tzeng, X. Xuan, Continuous dielectrophoretic separation of particles in a spiral microchannel, Electrophoresis 31 (2010) 1382–1388, Copyright 2010 John Wiley and Sons. Panel (B) Reproduced with permission from H. Moncada-Hernandez, B.H. Lapizco-Encinas, Simultaneous concentration and separation of microorganisms: insulator-based dielectrophoretic approach, Anal. Bioanal. Chem. 396 (2010) 1805–1816. Copyright 2010 Springer Nature. Panel (C) Reproduced with permission from A. Alazzam, B. Mathew, F. Alhammadi, Novel microfluidic device for the continuous separation of cancer cells using dielectrophoresis, J. Sep. Sci. 40 (2017) 1193–1200. Copyright 2017 John Wiley and Sons.)*

sorting large and small cells into different streams. This chip is easy to fabricate and does not require imbedded microelectrodes inside the channel. Separation of cells with different sizes can be completed by adjusting the applied signal without needing a new chip structure and altered dimensions. Xuan et al.[21] developed a novel particle sorting method by DEP force in a DC electrokinetic flow through a planar double spiral microchannel, as shown in Fig. 16.3A. A mixture of 5 and 10 μm particles are separated continuously without needing in-channel microelectrodes or insulators, as well as additional pressure pumping. In iDEP devices, joule heating and electrothermal flow [22] have been exploited to selectively trap circulating tumor cells and significantly enhance the separation performance in a higher flow rate. The elimination of in-channel microelectrodes or insulators will simplify the fabrication and operation process. Furthermore, the channel depth does not influence the sorting efficiency, which enhances the cell throughput.

DEP has been used to separate mixtures of cells on electrode arrays. Becker et al. [23] sorted cancer cells from normal blood cells on a microelectrode array by balancing the hydrodynamic and DEP forces acting on the cells. Petrosino et al. [9] sorted *Escherichia coli* and *Saphylococcus auretus* from blood by combing membraneless dialysis and dielectrohporesis. This microfluidic system combines membraneless microfluidic dialysis and DEP to achieve label-free isolation and concentration of bacteria from whole blood. Alazzam et al. [24] reported a novel microfluidic chip for continuously sorting cancer cells from a heterogeneous mixture of cancerous and normal blood cells using DEP (Fig. 16.3C). Separation is achieved by using planar electrodes, which protrude into the microchannel at one side. To obtain a high-efficiency DEP separation, gravitational sedimentation-based sheathless prefocusing is introduced prior to separation [25]. This method used an oblique interdigitated electrode array situated at the bottom of the channel for cell manipulation by DEP force, as displayed in Fig. 16.4A.

DEP field-flow fractionation (DEP-FFF) [26] was exploited to separate a nucleated cell fraction from cell debris and the bulk of the erythrocyte population by balancing the DEP force and the hydrodynamic lift force. Cheng et al. [27] continuously and effectively separated bacteria (*S. aureus*) and blood cells (RBC) using traveling wave DEP forces at a high throughput rate, and noted that this method repelled cells away from the chip surface by choosing a field frequency based on a nDEP force (as shown in Fig. 16.4B), which prevents cell sticking and lysing near the electrode edges or surface. Our group [28] previously separated polystyrene (PS) microbeads and cells on their own for very small and nonflowing sample volumes by combining the conventional dielectrophoresis (cDEP) forces and traveling wave DEP (twDEP) forces.

Another active approach based on optically induced DEP can be used to separate cells. Wu et al. [29] developed a microfluidic system for purification of circulating tumor cells by integrating optically induced DEP, which is capable of effectively and easily sorting a specific cell species from a cell sample. The cells can also be sorted based on the electroosmosis flow [30], Chen et al. [31] exploited an induced charge electroosmotic vortex-based method

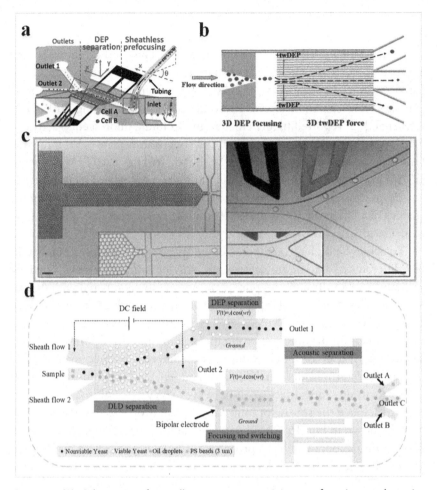

Fig. 16.4 (A) Schematic of a cell separator containing prefocusing and sorting components. (B) Top view of electrode structure, in which particles experience twDEP forces, which transport them into relative outlets. (C) Generation and separation droplets by triggering droplet fluorescence. (D) Schematic illustration of an integrated microfluidic system designed to achieve multitarget separation. *(Panel (A) Reproduced with permission from T. Luo, L. Fan, Y. Zeng, Y. Liu, S. Chen, Q. Tan, R.H. W. Lam, D. Sun, A simplified sheathless cell separation approach using combined gravitational-sedimentation-based prefocusing and dielectrophoretic separation, Lab Chip (2018). Copyright 2018 Royal Society of Chemistry. Panel (B) Reproduced with permission from I.F. Cheng, V.E. Froude, Y. Zhu, H.C. Chang, H.C. Chang, A continuous high-throughput bioparticle sorter based on 3D traveling-wave dielectrophoresis, Lab Chip 9 (2009) 3193–3201. Copyright 2009 Royal Society of Chemistry. Panel (C) Reproduced with permission from J.C. Baret, O.J. Miller, V. Taly, M. Ryckelynck, A. El-Harrak, L. Frenz, C. Rick, M.L. Samuels, J.B. Hutchison, J.J. Agresti, D.R. Link, D.A. Weitz, A. Griffiths, Fluorescence-activated droplet sorting (FADS): efficient microfluidic cell sorting based on enzymatic activity, Lab Chip 9 (2009) 1850–1858. Copyright 2009 Royal Society of Chemistry. Panel (D) Reprinted with permission from Y.Wu, R. Chattaraj, Y. Ren, H. Jiang, D. Lee, Label-free multitarget separation of particles and cells under flow using acoustic, electrophoretic, and hydrodynamic forces, Anal. Chem. 93 (2021) 7635–7646. Copyright [2021] American Chemical Society.)*

for separating targeted particles from a mixture. This method shows several advantages, such as contact free separation and ease of integration. Compared to direct separation of the cells in a solution, many researchers have exploited the separation of cells encapsulated in droplets using DEP. Griffiths et al. [32] reported highly efficient microfluidic fluorescence-activated droplet sorting (FADS) by integrating microtiter plate screening and fluorescence-activated cell sorting, as shown in Fig. 16.4C. Weitz et al. [33,34] developed an ultrahigh-throughput screening device by using aqueous drops dispersed in oil as picoliter-volume reaction vessels, allowing the analysis of a library of 108 in about 10 h, using a total reagent volume of less than 150 μL.

To achieve a high throughput and multiple parameter separation, combining different separation modalities can provide a number of advantages, including higher throughput, purity, and specificity, compared with standalone modules. Soh et al. [35] exploited an integrated DEP-magnetic activated cell sorting (iDMAS) by combining dielectrophoresis and magnetophoresis in a single microfluidic chip for highly efficient multitarget separation. Wu et al. [36] developed four different particle manipulation modules—deterministic lateral displacement (DLD), DEP, bipolar electrode (BPE) focusing, and surface acoustic waves (SAW)—within a single platform to achieve the sorting of solid particles, oil droplets, and live and dead cells, as shown in Fig. 16.4D. The integrated sorting technology provided a potential and biocompatible approach for sample preparation in many applications since it is capable of noninvasively separating various cells depending on electrical properties (surface charge, conductivity, permittivity, etc.) and mechanical properties (size, density, and compressibility) without requiring any labels, opening up the potential for on-chip analysis and diagnostics. Furthermore, the integration of three sorting mechanisms could potentially be used in liquid biopsies for cancer diagnosis by enabling isolation and characterization of circulating tumor cells (CTCs) and extracellular vesicles (EVs). Thus, the sorting technique by DEP force is envisioned to be integrated, with different functional components, into microfluidic devices for a variety of applications. There are many more examples of DEP- or electroosmosis (EO)-based cell separation in addition to those mentioned above.

4 Cell characterization

Cell properties, including size, stiffness, and impedance, are key parameters in analyzing the cellular processes. The interpretation of the electric properties has been recognized as a viable method for medical diagnosis and treatment influence evaluation [37]. The mechanical phenotype of a cell is an inherent and valuable indicator of its function and state. The changes in the mechanical properties [38] of a cell can be linked to cell cycle progression, stem cell differentiation, and cancer metastasis [39]. Methods developed for detecting these properties of cells include atomic force microscopy, micropipe aspiration, optical tweezers, etc. However, these techniques generally suffer from low throughput with high cost. In this context, microfluidic systems have been exploited to probe these properties of cells. In contrast with various noninvasive methods for sorting and identification of cells at the micro- and nanoscale levels, dielectrophoresis is recognized as an approach with unique advantages and great potential in manipulating and characterizing bioparticles such as bacteria, viruses, DNA, and cells. We aim to offer an overview of actual DEP approaches for cell characterization based on dielectric properties, which may contribute to therapeutic treatment evaluation and be used to determine whether cells are resistant or sensitive to a drug.

To realize rapid evaluation of the electric properties of bioparticles, electrical impedance spectroscopy is a method for characterizing various types of cell based on the electrophysiological properties of cells in the frequency domain. Jang et al. [40] investigated the dielectric properties of four kinds of cell (HeLa, A549, MCF-7, and MDA-MB-231) based on impedance measurements after trapping single cells by DEP force. In addition, AC electrokinetic methods, including dielectrophoreis and electrorotation, have been reported widely [41]. In the 1980s, DEP devices were developed using planar electrodes within the sample chamber. The main factors affecting the dielectric properties of a cell or particle are the membrane capacitance, the surface charge, the conductivity of the membrane, and cytoplasm. The DEP technique can be used to distinguish different types of cells, to detect changes in cell cytoplasmic properties, and to sort viable and nonviable yeast cells [42]. By exploiting the behavior of cells across a broad frequency range, the dielectric properties of cells can be analyzed based on the determination of a frequency where the DEP force is zero. This approach has been used to investigate cellular responses to different toxicants, membrane changes during apoptosis, and cancer cell transformation.

Information about the membrane and cytoplasm can be obtained by analyzing the DEP behavior of cells over a broad frequency range. The changes in the cytoplasm during apoptosis can be detected using the dielectric properties. Hughes et al. [43] developed a method for extraction of dielectric properties of multiple populations from the dielectrophoretic collection spectrum. Stoicheva et al. [44] measured dielectrophoretic coefficients of chloroplasts and investigated the effect of surface charge at different frequencies in a device where the AC field was generated between two concentric electrodes. DEP has been employed in a variety of cancer cell studies and is often used to characterize cell subpopulations and quantify cell death. A DEP analysis method has been reported to accurately detect apoptosis progression and toxicity using Hela and Jurkat cells [45]. Salmanzadeh et al. studied the dielectrophoretic responses of progressive stages of mouse ovarian surface epithelial (MOSE) cells, as well as mouse fibroblast and macrophage cell lines. The DEP technique does not need any cell staining, has lower complexity and operating costs, and has verified good sensitivity to physiological changes in cells at different stages of apoptosis.

The DEP crossover frequency can be achieved by analyzing the cell position when changing the applied AC frequency. The frequency at which the direction alters (or cell is in a stationary state without any net movement) is the DEP crossover frequency. This response for a typical yeast cell for different values of the conductivity of the suspending medium is shown in Fig. 16.2. Gascoyne et al. [46] reported a DEP crossover frequency technique for analyzing cellular responses to toxicants in a microchamber; changes in the membrane capacitance and conductance are detectable using this approach. Sawai et al. [47] discriminated between viable and nonviable hepatic carcinoma cells after treatment with cytotoxic agents, demonstrating that the exposure to cytotoxic agents generated a significant shift from nDEP to pDEP methods in cancer screening.

The DEP-based method [47] can also be used to differentiate hepatocellular carcinoma (HCC) cells from the normal liver cells, and to discriminate nonviable and viable HCC cells after treatment with plant-based cytotoxic agent konjac glucomannan. This DEP-based approach is a potential complementary method to be applied alongside other biochemical and molecular methods in cancer research and screening. Furthermore, Li et al. [48] investigated four typical types of cells (Raji cells, MCF-7 cells, HEK 293 cells, and K562 cells) based on the crossover frequency spectra. This technique allows the noninvasive determination of cell membrane capacitance and conductance by optically induced DEP. Most cells own two different crossover frequencies and these second crossover frequencies are difficult to obtain because of operation limitations including the maximum frequency of the signal generator, and the medium conductivity.

An extension of crossover methods is to explore a larger frequency-dependent spectrum. Cell parameters including membrane conductance, membrane capacitance, and cytoplasmic conductivity can be obtained by fitting the cell model using the measured data across a full range of frequencies. The DEP spectra of MCF7 breast cancer cells and polymorphonuclear and mononuclear leukocytes can be obtained by using reciprocal V-shaped planar electrodes at different frequencies, as shown in Fig. 16.5A, providing a generic platform for identifying the DEP responses of different cells [49]. However, such assays usually take a long time to collect sufficient data to obtain valid interpretations from a single DEP spectrum, limiting the application in areas from drug discovery to diagnostics.

Kirby et al. [50] reported an automated experimental method by measuring the effects of ethambutol treatment on the dielectrophoretic response of *M. smegmatis*. This method, based on DEP measurements, shows the potential for detecting changes in mycobacterial membrane properties associated with genetic mutation and chemical treatments. Full-frequency spectra can be also be obtained by measuring cell collection or concentration at each frequency along the spectrum. Labeed et al. [51] realized real-time electrophysiology by developing a programmable and multichannel system using a dot microelectrode array. Particle motion within the dot area can be quantified in terms of the shifts in absorbed light intensities. To better realize DEP as a common method, Hughes et al. [52] developed a novel approach by using three dimensional electrodes drilled from laminates of copper and further reported a commercial DEP cytometry platform, the 3DEP, as shown in Fig. 16.5B. This device allows simultaneous analysis of thousands of cells, obtaining data at high speed.

The electrotation technique [53] is used widely for cell characterization, based on the imaginary part of the Clausius-Mossotti factor. The frequency at which rotation direction alters in the ROT assays is the crossover frequency. Becker et al. [23] conducted electrorotation measurements and demonstrated significant differences between the dielectric properties of MDA231 cells, erythrocytes, and T lymphocytes. The DEP mean crossover frequencies obtained in the experiments were 15, 58, and 95 kHz for these three types of cells, respectively, facilitating the DEP separation by selecting frequencies appropriately. Gasperis et al. [54] reported a system for cell

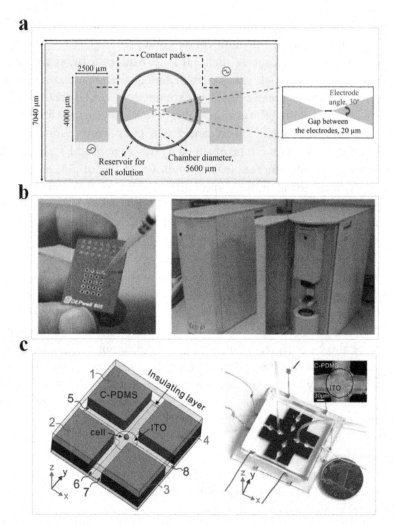

Fig. 16.5 (A) Schematic illustration of a DEP spectrum device, with an enlarged view of the electrode tips. (B) The 3DEP system comprises two components. (C) The device for 3D cell electrorotation. *(Panel (A) Reproduced with permission from Z. Caglayan, Y. Demircan Yalcin, H. Kulah, Examination of the dielectrophoretic spectra of MCF7 breast cancer cells and leukocytes, Electrophoresis 41 (2020) 345–352. Copyright 2020 John Wiley and Sons. Panel (B) Reproduced with permission from K.F. Hoettges, E.A. Henslee, R.M. Torcal Serrano, R.I. Jabr, R.G. Abdallat, A.D. Beale, A. Waheed, P. Camelliti, C.H. Fry, D.R. van der Veen, F.H. Labeed, M.P. Hughes, Ten-second electrophysiology: evaluation of the 3DEP platform for high-speed, high-accuracy cell analysis, Sci. Rep. 9, 2019, 19153. Hughes et al. 2019. Panel (C) Reproduced with permission from L. Huang, F. Liang, Y. Feng, A microfluidic chip for single-cell 3D rotation enabling self-adaptive spatial localization, J. Appl. Phys. 126 (2019) 234702. Copyright 2019 AIP Publishing.)*

motion estimation for real-time ROT measurements and ROT spectral analysis using a polynomial electrode configuration. This work enables faster and far less laborious determination of cell properties.

The ROT method has also been exploited to conduct single cell assays by integrating cell imaging in microfluidic devices [55]. Shaked et al. [55] combined DEP and tomographic interferometry for capturing the 3D refractive index map of single cells in a noninvasive manner. This integrated system can be used for cell sorting, and for multiple single cells' rotation at a time, for label-free 3D imaging. Huang et al. [56] reported a microfluidic chip for single cell 3D rotation by using four sidewall electrodes and four planar electrodes via dielectrophoresis and reconstructed the 3D morphology of single cells (Fig. 16.5C). Wang et al. [57] combined 3D electrodes and planar electrodes to realize 3D rotation on the vertical plane for measuring cells' properties and studied several cell lines, including Hela, B lymphocyte, C3H10, and HepaRG, based on ROT spectra. This work quantified the membrane capacitance and cytoplasm conductivity for the four cells and measured the difference of geometric parameters based on cell 3D electrorotation and imaging.

Swami et al. [58] demonstrated that variation in cytoplasmic conductivity estimated using cell electrorotational analysis can serve as a sensitive and rapid method for detecting the live *C. difficile* subpopulation after vancomycin treatment. However, it is difficult to simultaneously rotate multiple single cells in a single chip. Keim et al. [59] realized multiple trapping and rotation of single cells by independently addressed single quadrupole cages in a chamber. The ROT method [60] can also be used for monitoring the permeabilization of cells based on their dielectric properties by combing DEP and electrorotation in situ experiments. They analyzed the electrophysiological properties of human leukemic T-cell lymphoblast and murine melanoma cells before and after treatment by pulsed electric field. Thus, the electrophysiological fingerprint of the cells would be useful for analyzing cells under different conditions.

However, due to the lack of automation, ROT usually requires a long operation time without viability assessment. Morphology analysis is also limited for cells with concaved surfaces, which makes it unsuitable for particular cell types. To improve the analysis of ROT characterization, many authors combined automation, an on-chip analysis module, and estimation of cytoplasmic properties. Keim et al. [59] demonstrated multiple trapping and rotation of single cells by independently addressed single quadrupole cages. Some of these cellular parameters, including the membrane capacitance and cytoplasmic conductivity, are summarized in Table 16.1.

Table 16.1 Summary of recent DEP- or ROT-measured dielectric properties of cells.

Cell type	Radius (μm)	Cytoplasm conductivity (S/m)	Cytoplasm permittivity	Membrane conductivity (S/m)	Membrane capacitance (mF/m²)	Membrane permittivity
Hela [61]	8.13	0.56			14.6	
Hela [59]	8.74	0.84			17.5	
MCF-7 [61]	8.53	0.53	60		10.7	
K562 [51]	9.96				5.6–8.2	5.8–7.9
K562 [62]	8	0.23	40	1.8×10^{-6}		8
B16F1 [63]	10	0.5	4.97	10^{-7}		80.2
B16F10 [60]		0.1	44.7	0.49×10^{-5}		8.2
Jurkat E 6.1 [60]		0.19	40	0.4×10^{-4}		6.3
HEK293 [59]	6.75	0.5	78		9.81	
C3H10 [57]	14	0.31			14.73	
Blymphocyte [57]	8.2	0.55			10.14	
HepaRG [57]	12	0.26			15.83	

The discussion of ROT characterization can also be found in several recent reviews [64,65].

5 Conclusion and future perspectives

The AC electrokinetic-based microfluidic platform is advantageous, since it is contact-free, noninvasive, easy to use, low cost, and label-free. DEP-based cell characterization needs to be conducted in a low-conductivity solution, which may influence the cell viability, generate unpredictable changes in properties because of the media changes, and require more time and steps to wash cells from the culture media. To address these limitations, the analysis time of cells in low conductivity media can be reduced by rapid characterization. Furthermore, these methods are capable of easily integrating with other components to form a multifunctional platform on an integrated chip. To exploit integrated on-chip devices with multiple functions has been a recent focus of various applications by combining different downstream steps including separation, trapping, or enrichment. In order to push the DEP method toward more end-user applications, the chip should be easy to use, and the fabrication process needs to be simple and cost effective.

Hopefully readers of this chapter will see more applications for DEP-based characterization and separation. It is noted that a huge amount of information regarding DEP characterization exists in the previously published literature and new studies rely heavily on the published data. Due to numerous efforts by researchers, the AC electrokinetic-based microfluidic platform has come a long way in the biomedical field recently. By addressing the abovementioned challenges in the future, this approach will be sure to move to real-world applications. Further advances in this field will most likely come from integrated teams including engineers, biologists, chemists, and clinicians.

Acknowledgments

This work is supported by the National Natural Science Foundation of China (No.62104195), Basic Research Programs of Taicang, 2020 (Grant No. TC2020JC02), and Guang Dong Basic and Applied Basic Research Foundation (Grant No. 2020A1515110747), Key Research and Development Program of Shaanxi (ProgramNo.2022SF-111), the Fundamental Research Funds for the Central Universities (D5000210625, G2021KY05101), Industrial Development and Foster Project of Yangtze River Delta Research Institute of NPU, Taicang (CY20210204).

References

[1] Y. Wu, B. Hu, X. Ma, H. Zhang, W. Li, Y. Wang, S. Wang, Generation of droplets with adjustable chemical concentrations based on fixed potential induced-charge electro-osmosis, Lab Chip 22 (2022) 403–412.

[2] E.K. Sackmann, A.L. Fulton, D.J. Beebe, The present and future role of microfluidics in biomedical research, Nature 507 (2014) 181–189.

[3] G.M. Whitesides, The origins and the future of microfluidics, Nature 442 (2006) 368–373.

[4] T. Thorsen, S.J. Maerkl, S.R. Quake, Microfluidic large-scale integration, Science 298 (2002) 580–584.

[5] Y. Wu, Y. Ren, Y. Tao, L. Hou, Q. Hu, H. Jiang, A novel micromixer based on the alternating current-flow field effect transistor, Lab Chip 17 (2016) 186–197.

[6] Y. Wu, Y. Ren, H. Jiang, Enhanced model-based design of a high-throughput three dimensional micromixer driven by alternating-current electrothermal flow, Electrophoresis 38 (2017) 258–269.

[7] A. Ramos, Electrokinetics and Electrohydrodynamics in Microsystems, Vol. 530, Springer Science & Business Media, 2011.

[8] S.A. Faraghat, K.F. Hoettges, M.K. Steinbach, D.R. van der Veen, W.J. Brackenbury, E.A. Henslee, F.H. Labeed, M.P. Hughes, High-throughput, low-loss, low-cost, and label-free cell separation using electrophysiology-activated cell enrichment, Proc. Natl. Acad. Sci. U. S. A. 114 (2017) 4591–4596.

[9] L. D'Amico, N.J. Ajami, J.A. Adachi, P.R. Gascoyne, J.F. Petrosino, Isolation and concentration of bacteria from blood using microfluidic membraneless dialysis and dielectrophoresis, Lab Chip 17 (2017) 1340–1348.

[10] W. Waheed, A. Alazzam, A.N. Al-Khateeb, E. Abu-Nada, Multiple particle manipulation under dielectrophoresis effect: modeling and experiments, Langmuir 36 (2020) 3016–3028.

[11] A. Barik, Y. Zhang, R. Grassi, B.P. Nadappuram, J.B. Edel, T. Low, S.J. Koester, S.H. Oh, Graphene-edge dielectrophoretic tweezers for trapping of biomolecules, Nat. Commun. 8 (1867) 2017.

[12] N. Kumar, W. Wang, J.C. Ortiz-Marquez, M. Catalano, M. Gray, N. Biglari, K. Hikari, X. Ling, J. Gao, T. van Opijnen, K.S. Burch, Dielectrophoresis assisted rapid, selective and single cell detection of antibiotic resistant bacteria with G-FETs, Biosens. Bioelectron. 156 (2020), 112123.

[13] A. Mustafa, E. Pedone, L. Marucci, D. Moschou, M.D. Lorenzo, A flow-through microfluidic chip for continuous dielectrophoretic separation of viable and non-viable human T-cells, Electrophoresis (2021).

[14] Y. Wu, Y. Ren, Y. Tao, L. Hou, H. Jiang, High-throughput separation, trapping, and manipulation of single cells and particles by combined dielectrophoresis at a bipolar electrode array, Anal. Chem. 90 (2018) 11461–11469.

[15] Y. Zhao, U.C. Yi, S.K. Cho, Microparticle concentration and separation by traveling-wave dielectrophoresis (twDEP) for digital microfluidics, J. Microelectromech. Syst. 16 (2007) 1472–1481.

[16] T.B. Jones, Basic theory of dielectrophoresis and electrorotation, IEEE Eng. Med. Biol. Mag. 22 (2003) 33–42.

[17] X.-B. Wang, Y. Huang, F. Becker, P. Gascoyne, A unified theory of dielectrophoresis and travelling wave dielectrophoresis, J. Phys. D. Appl. Phys. 27 (1994) 1571.

[18] Y. Wu, Y. Ren, Y. Tao, L. Hou, H. Jiang, Large-scale single particle and cell trapping based on rotating electric field induced-charge electroosmosis, Anal. Chem. 88 (2016) 11791–11798.

[19] H. Moncada-Hernandez, B.H. Lapizco-Encinas, Simultaneous concentration and separation of microorganisms: insulator-based dielectrophoretic approach, Anal. Bioanal. Chem. 396 (2010) 1805–1816.

[20] Y. Kang, D. Li, S.A. Kalams, J.E. Eid, DC-dielectrophoretic separation of biological cells by size, Biomed. Microdevices 10 (2008) 243–249.

[21] J. Zhu, T.R. Tzeng, X. Xuan, Continuous dielectrophoretic separation of particles in a spiral microchannel, Electrophoresis 31 (2010) 1382–1388.

[22] A. Aghilinejad, M. Aghaamoo, X. Chen, J. Xu, Effects of electrothermal vortices on insulator-based dielectrophoresis for circulating tumor cell separation, Electrophoresis 39 (2018) 869–877.

[23] F.F. Becker, X.B. Wang, Y. Huang, R. Pethig, J. Vykoukal, P.R. Gascoyne, Separation of human breast cancer cells from blood by differential dielectric affinity, Proc. Natl. Acad. Sci. U. S. A. 92 (1995) 860–864.

[24] A. Alazzam, B. Mathew, F. Alhammadi, Novel microfluidic device for the continuous separation of cancer cells using dielectrophoresis, J. Sep. Sci. 40 (2017) 1193–1200.

[25] T. Luo, L. Fan, Y. Zeng, Y. Liu, S. Chen, Q. Tan, R.H.W. Lam, D. Sun, A simplified sheathless cell separation approach using combined gravitational-sedimentation-based prefocusing and dielectrophoretic separation, Lab Chip (2018).

[26] J. Vykoukal, D.M. Vykoukal, S. Freyberg, E.U. Alt, P.R. Gascoyne, Enrichment of putative stem cells from adipose tissue using dielectrophoretic field-flow fractionation, Lab Chip 8 (2008) 1386–1393.

[27] I.F. Cheng, V.E. Froude, Y. Zhu, H.C. Chang, H.C. Chang, A continuous high-throughput bioparticle sorter based on 3D traveling-wave dielectrophoresis, Lab Chip 9 (2009) 3193–3201.

[28] Y.P. Wu, Y.K. Ren, Y. Tao, H.Y. Jiang, Fluid pumping and cells separation by DC-biased traveling wave electroosmosis and dielectrophoresis, Microfluid. Nanofluid. 21 (2017).

[29] T.K. Chiu, W.P. Chou, S.B. Huang, H.M. Wang, Y.C. Lin, C.H. Hsieh, M.H. Wu, Application of optically-induced-dielectrophoresis in microfluidic system for purification of circulating tumour cells for gene expression analysis—cancer cell line model, Sci. Rep. 6 (2016) 32851.

[30] S. Hosic, S.K. Murthy, A.N. Koppes, Microfluidic sample preparation for single cell analysis, Anal. Chem. 88 (2016) 354–380.

[31] X. Chen, Y. Ren, L. Hou, X. Feng, T. Jiang, H. Jiang, Induced charge electro-osmotic particle separation, Nanoscale 11 (2019) 6410–6421.

[32] J.C. Baret, O.J. Miller, V. Taly, M. Ryckelynck, A. El-Harrak, L. Frenz, C. Rick, M.L. Samuels, J.B. Hutchison, J.J. Agresti, D.R. Link, D.A. Weitz, A.D. Griffiths, Fluorescence-activated droplet sorting (FADS): efficient microfluidic cell sorting based on enzymatic activity, Lab Chip 9 (2009) 1850–1858.

[33] J.J. Agresti, E. Antipov, A.R. Abate, K. Ahn, A.C. Rowat, J.C. Baret, M. Marquez, A.-M. Klibanov, A.D. Griffiths, D.A. Weitz, Ultrahigh-throughput screening in drop-based microfluidics for directed evolution, Proc. Natl. Acad. Sci. U. S. A. 107 (2010) 4004–4009.

[34] L. Mazutis, J. Gilbert, W.L. Ung, D.A. Weitz, A.D. Griffiths, J.A. Heyman, Single-cell analysis and sorting using droplet-based microfluidics, Nat. Protoc. 8 (2013) 870–891.

[35] U. Kim, H.T. Soh, Simultaneous sorting of multiple bacterial targets using integrated dielectrophoretic-magnetic activated cell sorter, Lab Chip 9 (2009) 2313–2318.

[36] Y. Wu, R. Chattaraj, Y. Ren, H. Jiang, D. Lee, Label-free multitarget separation of particles and cells under flow using acoustic, electrophoretic, and hydrodynamic forces, Anal. Chem. 93 (2021) 7635–7646.

[37] I. Turcan, M.A. Olariu, Dielectrophoretic manipulation of cancer cells and their electrical characterization, ACS Comb. Sci. 22 (2020) 554–578.

[38] M. Urbanska, H.E. Munoz, J. Shaw Bagnall, O. Otto, S.R. Manalis, D. Di Carlo, J. Guck, A comparison of microfluidic methods for high-throughput cell deformability measurements, Nat. Methods 17 (2020) 587–593.

[39] K. Jiang, L. Liang, C.T. Lim, Engineering confining microenvironment for studying cancer metastasis, iScience 24 (2021), 102098.

[40] J.-L. Hong, K.-C. Lan, L.-S. Jang, Electrical characteristics analysis of various cancer cells using a microfluidic device based on single-cell impedance measurement, Sens. Actuators B Chem. 173 (2012) 927–934.

[41] R. Hoque, H. Mostafid, M.P. Hughes, Rapid, low-cost dielectrophoretic diagnosis of bladder cancer in a clinical setting, IEEE J. Transl. Eng. Health Med. 8 (2020) 4300405.

[42] N. Lewpiriyawong, K. Kandaswamy, C. Yang, V. Ivanov, R. Stocker, Microfluidic characterization and continuous separation of cells and particles using conducting poly(dimethyl siloxane) electrode induced alternating current-dielectrophoresis, Anal. Chem. 83 (2011) 9579–9585.

[43] L.M. Broche, F.H. Labeed, M.P. Hughes, Extraction of dielectric properties of multiple populations from dielectrophoretic collection spectrum data, Phys. Med. Biol. 50 (2005) 2267–2274.

[44] N. Stoicheva, D.S. Dimitrov, A. Ivanov, Chloroplast dielectrophoresis, Eur. Biophys. J. 14 (1987).

[45] E.A. Henslee, R.M. Torcal Serrano, F.H. Labeed, R.I. Jabr, C.H. Fry, M.P. Hughes, K.F. Hoettges, Accurate quantification of apoptosis progression and toxicity using a dielectrophoretic approach, Analyst 141 (2016) 6408–6415.

[46] K. Ratanachoo, P.R.C. Gascoyne, M. Ruchirawat, Detection of cellular responses to toxicants by dielectrophoresis, Biochim. Biophys. Acta - Biomembr. 1564 (2002) 449–458.

[47] S. Sawai, N. Ahmad Shukri, M.S. Mohktar, W.K. Zaman, W. S., Dielectrophoresis-based discrimination of hepatic carcinoma cells following treatment with cytotoxic agents, engineering science and technology, an, Int. J. 25 (2022), 100990.

[48] W. Liang, Y. Zhao, L. Liu, Y. Wang, W.J. Li, G.B. Lee, Determination of cell membrane capacitance and conductance via optically induced electrokinetics, Biophys. J. 113 (2017) 1531–1539.

[49] Z. Caglayan, Y. Demircan Yalcin, H. Kulah, Examination of the dielectrophoretic spectra of MCF7 breast cancer cells and leukocytes, Electrophoresis 41 (2020) 345–352.

[50] B.G. Hawkins, C. Huang, S. Arasanipalai, B.J. Kirby, Automated dielectrophoretic characterization of mycobacterium smegmatis, Anal. Chem. 83 (2011) 3507–3515.

[51] H.O. Fatoyinbo, N.A. Kadri, D.H. Gould, K.F. Hoettges, F.H. Labeed, Real-time cell electrophysiology using a multi-channel dielectrophoretic-dot microelectrode array, Electrophoresis 32 (2011) 2541–2549.

[52] K.F. Hoettges, E.A. Henslee, R.M. Torcal Serrano, R.I. Jabr, R.G. Abdallat, A.D. Beale, A. Waheed, P. Camelliti, C.H. Fry, D.R. van der Veen, F.H. Labeed, M.P. Hughes, Ten-second electrophysiology: evaluation of the 3DEP platform for high-speed, high-accuracy cell analysis, Sci. Rep. 9 (2019) 19153.

[53] W.M. Arnold, U. Zimmermann, Electro-rotation: development of a technique for dielectric measurements on individual cells and particles, J. Electrost. 21 (1988) 151–191.

[54] G.D. Gasperis, X. Wang, J. Yang, F.F. Becker, P.R.C. Gascoyne, Automated electro-rotation: dielectric characterization of living cells by real-time motion estimation, Meas. Sci. Technol. 9 (1998) 518–529.

[55] M. Habaza, M. Kirschbaum, C. Guernth-Marschner, G. Dardikman, I. Barnea, R. Korenstein, C. Duschl, N.T. Shaked, Rapid 3D refractive-index imaging of live cells in suspension without labeling using dielectrophoretic cell rotation, Adv. Sci. 4 (2017) 1600205.

[56] L. Huang, F. Liang, Y. Feng, A microfluidic chip for single-cell 3D rotation enabling self-adaptive spatial localization, J. Appl. Phys. 126 (2019), 234702.

[57] L. Huang, P. Zhao, W. Wang, 3D cell electrorotation and imaging for measuring mul-
      tiple cellular biophysical properties, Lab Chip (2018).
[58] A. Rohani, J.H. Moore, Y.H. Su, V. Stagnaro, C. Warren, N.S. Swami, Single-cell
      electro-phenotyping for rapid assessment of Clostridium difficile heterogeneity under
      vancomycin treatment at sub-MIC (minimum inhibitory concentration) levels, Sens.
      Actuators B Chem. 276 (2018) 472–480.
[59] K. Keim, M.Z. Rashed, S.C. Kilchenmann, A. Delattre, A.F. Goncalves, P. Ery, C.
      Guiducci, On-chip technology for single-cell arraying, electrorotation-based analysis
      and selective release, Electrophoresis 40 (2019) 1830–1838.
[60] C.I. Trainito, O. Francais, B. Le Pioufle, Monitoring the permeabilization of a single
      cell in a microfluidic device, through the estimation of its dielectric properties based on
      combined dielectrophoresis and electrorotation in situ experiments, Electrophoresis
      36 (2015) 1115–1122.
[61] Z. Zhang, T. Zheng, R. Zhu, Characterization of single-cell biophysical properties and
      cell type classification using dielectrophoresis model reduction method, Sens. Actuators
      B Chem. 304 (2020), 127326.
[62] Y. Demircan, A. Koyuncuoglu, M. Erdem, E. Ozgur, U. Gunduz, H. Kulah, Label-free
      detection of multidrug resistance in K562 cells through isolated 3D-electrode dielectro-
      phoresis, Electrophoresis 36 (2015) 1149–1157.
[63] J. Oblak, D. Krizaj, S. Amon, A. Macek-Lebar, D. Miklavcic, Feasibility study for cell
      electroporation detection and separation by means of dielectrophoresis, Bioelectro-
      chemistry 71 (2007) 164–171.
[64] W. Liang, X. Yang, J. Wang, Y. Wang, W. Yang, L. Liu, Determination of dielectric
      properties of cells using AC electrokinetic-based microfluidic platform: a review of
      recent advances, Micromachines 11 (2020).
[65] E.A. Henslee, Review: dielectrophoresis in cell characterization, Electrophoresis
      41 (2020) 1915–1930.

Index

Note: Page numbers followed by *f* indicate figures, *t* indicate tables and *b* indicate boxes.

Printed in the United States
by Baker & Taylor Publisher Services